T0073178

Classical and Dynamical Markov and Lagrange Spectra

Dynamical, Fractal and Arithmetic Aspects

Davi Lima

Universidade Federal de Alagoas, Brazil

Carlos Matheus

CNRS, France

Carlos G Moreira

IMPA, Brazil

Sergio Romaña

Universidade Federal do Rio de Janeiro, Brazil

NEW JERSEY · LONDON · SINGAPORE · BEIJING · SHANGHAI · HONG KONG · TAIPEI · CHENNAI · TOKYO

Published by

World Scientific Publishing Co. Pte. Ltd.
5 Toh Tuck Link, Singapore 596224
USA office: 27 Warren Street, Suite 401-402, Hackensack, NJ 07601
UK office: 57 Shelton Street, Covent Garden, London WC2H 9HE

Library of Congress Cataloging-in-Publication Data
Names: Lima, Davi, author.
Title: Classical and dynamical Markov and Lagrange spectra : dynamical, fractal and
    arithmetic aspects / Davi Lima, Universidade Federal de Alagoas, Brazil,
    Carlos Matheus, CNRS, France, Carlos G. Moreira, IMPA, Brazil,
    Sergio Romaña, Universidade Federal do Rio de Janeiro, Brazil.
Description: First. | New Jersey : World Scientific, [2021] | Includes bibliographical references.
Identifiers: LCCN 2020040401 | ISBN 9789811225284 (hardcover) |
    ISBN 9789811225291 (ebook for institutions) | ISBN 9789811225307 (ebook for individuals)
Subjects: LCSH: Markov spectrum. | Lagrange spectrum.
Classification: LCC QA242 .L67 2021 | DDC 512.7/2--dc23
LC record available at https://lccn.loc.gov/2020040401

British Library Cataloguing-in-Publication Data
A catalogue record for this book is available from the British Library.

For any available supplementary material, please visit
https://www.worldscientific.com/worldscibooks/10.1142/11965#t=suppl

Dedication Page

We dedicate this book to Christian Mauduit and
Jean-Christophe Yoccoz, in memoriam.
We also dedicate this book to Rosa, Ivanildo, Manoel, Aline,
Marie-Ines, Carminha, Carlos, Josy and Cecilia.

Preface

This book on classical and dynamical Markov and Lagrange spectra was motivated by the renewed interest on these subjects related by the recent research activity by several authors, including the authors of this book. The classical Markov and Lagrange spectra are natural objects related to Diophantine approximations, and their study combines naturally several areas of Mathematics, as Number Theory, Combinatorics, Differential Geometry, Dynamical Systems and Fractal Geometry. They have natural dynamical generalizations in several contexts, some of them discussed in this book.

We start the book with a relatively quick review of results on the classical Markov and Lagrange spectra, many of them already discussed in the excellent book [CF (1989)] by Cusick and Flahive. We perform a more detailed discussion of some recent results on the classical spectra, including those by Gayfulin on attainable numbers and left endpoints of gaps in the Lagrange spectrum, by Matheus and Moreira on fractal properties of the set difference $M \setminus L$ between the classical Markov and Lagrange spectra, by Delecroix, Matheus and Moreira on numerical approximations of these spectra and Moreira's theorem on the fractal structure of Markov and Lagrange spectra between their discrete beginning and the so-called Hall's ray.

In the part of this book dedicated to dynamical Markov and Lagrange spectra, we introduce general notions of Markov and Lagrange spectra and present recent results related to them, as the theorem by Cerqueira, Matheus and Moreira on continuity of Hausdorff dimensions across dynamical spectra associated to conservative horseshoes, Moreira's theorem on the minima and Moreira-Romaña theorem on the interior of typical Markov and Lagrange spectra associated to horseshoes, and we discuss results of several authors as Ferenczi, Boshernitzan, Delecroix, Artigiani, Hubert, Marchese,

Ulcigrai, Parkkonen and Paulin for dynamical spectra associated to interval exchange transformations and to geodesic flows on negatively curved manifolds.

This book includes some new results, as:

• The precise description of portions of $M \setminus L$ in sections 2.3, 2.4 and 2.5.

• An improved lower estimate of $HD(M \setminus L)$, in subsection 2.5.5.

• An application of Cerqueira-Matheus-Moreira theorem to the Dirichlet spectrum, proving the continuity of the Hausdorff dimension of intersections of this spectrum with half-lines, in section 3.3.

We also include detailed expositions of the relation between the classical Markov spectrum and quadratic forms in Appendix C, of Freiman's rightmost gap of the classical spectra in Appendix D and of Vieira's proof of the closedness of dynamical spectra associated to Horseshoe's in Appendix G.

We would like to thank Jacob Palis for encouraging us to write this book, Yann Bugeaud for calling our attention to Dirichlet's spectrum and Sandoel Vieira for allowing us to use part of his Ph.D. thesis.

We also would like to thank Pierre Arnoux and Edmund Harriss for their collaboration with the cover design of this book.

Contents

Chapter 1

Classical Lagrange and Markov spectra

1.1 Diophantine approximations of real numbers

The study of rational approximations to the solutions of certain algebraic equations is a very old topic in Mathematics going back to the seminal works of Diophantus of Alexandria. For this reason, the subarea of Number Theory dedicated to the approximation of real numbers by rational numbers is a portion of the theory of *Diophantine approximations*.

Generally speaking, the quality $|\alpha - p/q|$ of a rational approximation $p/q \in \mathbb{Q}$ of a given a real number $\alpha \in \mathbb{R}$ is measured in comparison with the size $|q|$ of the denominator of p/q.

The most basic type of rational approximation comes from the fact that any real number lies between two consecutive integers: indeed, this says that for all $\alpha \in \mathbb{R}$ and $q \in \mathbb{N}$, there exists $p \in \mathbb{Z}$ such that $|q\alpha - p| \leq 1/2$, i.e.

$$\left| \alpha - \frac{p}{q} \right| \leq \frac{1}{2q}. \tag{1.1}$$

In 1842, Dirichlet [Di (1842)] invented his famous *pigeonhole principle* to largely improve over (1.1).

Theorem 1.1 (Dirichlet). *For any $\alpha \in \mathbb{R} \setminus \mathbb{Q}$ and any $Q \in \mathbb{N}$, there exists a rational number $p/q \in \mathbb{Q}$ with $0 < q \leq Q$ such that*

$$\left| \alpha - \frac{p}{q} \right| < \frac{1}{qQ}.$$

In particular, the inequality

$$\left| \alpha - \frac{p}{q} \right| < \frac{1}{q^2}$$

has infinitely many rational solutions $p/q \in \mathbb{Q}$.

Proof. Given $Q \in \mathbb{N}$, we consider the decomposition

$$[0,1) = \bigcup_{j=0}^{Q-1} \left[\frac{j}{Q}, \frac{j+1}{Q} \right)$$

of the interval $[0,1)$ into Q disjoint subintervals.

We have $Q+1$ distinct[1] numbers $\{j\alpha\} \in [0,1)$, $j = 0, \ldots, Q$, where $\{x\} := x - \lfloor x \rfloor \in [0,1)$ denotes the *fractional part* of x and $\lfloor x \rfloor$ is the *integer part* of x given by $\lfloor x \rfloor := \max\{n \in \mathbb{Z} : n \leq x\} \in \mathbb{Z}$. By the *pigeonhole principle*, some interval $\left[\frac{j}{Q}, \frac{j+1}{Q} \right)$ must contain two such numbers, say $\{n\alpha\}$ and $\{m\alpha\}$, $0 \leq n < m \leq Q$. It follows that

$$|\{m\alpha\} - \{n\alpha\}| < \frac{1}{Q},$$

i.e., $|q\alpha - p| < 1/Q$ where $0 < q := m - n \leq Q$ and $p := \lfloor m\alpha \rfloor - \lfloor n\alpha \rfloor \in \mathbb{Z}$. Therefore,

$$\left| \alpha - \frac{p}{q} \right| < \frac{1}{qQ} \leq \frac{1}{q^2}.$$

This proves the theorem. □

In 1891, Hurwitz [Hu (1891)] established[2] the optimality of Dirichlet's theorem:

Theorem 1.2 (Hurwitz). *For any $\alpha \in \mathbb{R} \setminus \mathbb{Q}$, the inequality*

$$\left| \alpha - \frac{p}{q} \right| \leq \frac{1}{\sqrt{5}q^2}$$

has infinitely many rational solutions $p/q \in \mathbb{Q}$.

Moreover, for all $\varepsilon > 0$, the inequality

$$\left| \frac{1+\sqrt{5}}{2} - \frac{p}{q} \right| \leq \frac{1}{(\sqrt{5} + \varepsilon)q^2}$$

has only finitely many rational solutions $p/q \in \mathbb{Q}$.

The first part of Hurwitz theorem is proved in Appendix A, while the second part of Hurwitz theorem is left as an exercise to the reader:

---

[1] $\alpha \notin \mathbb{Q}$ is used here.
[2] Actually, this statement was proved by Korkine and Zolotareff [KZ (1873)] in 1873, but it became widely known as *Hurwitz theorem* in the literature for some historical reason that we ignore.

Exercise 1.1

Show the second part of Hurwitz theorem. (Hint: use the identity $p^2 - pq - q^2 = \left(q\frac{1+\sqrt{5}}{2} - p\right)\left(q\frac{1-\sqrt{5}}{2} - p\right)$ relating $\frac{1+\sqrt{5}}{2}$ and its Galois conjugate $\frac{1-\sqrt{5}}{2}$).

Also, use your argument to give a bound on

$$\#\left\{\frac{p}{q} \in \mathbb{Q} : \left|\frac{1+\sqrt{5}}{2} - \frac{p}{q}\right| \leq \frac{1}{(\sqrt{5}+\varepsilon)q^2}\right\}$$

in terms of $\varepsilon > 0$.

1.2 Definition of the classical Lagrange spectrum

Hurwitz theorem leaves open the possibility that the statement "$\#\{p/q \in \mathbb{Q} : |\alpha - \frac{p}{q}| \leq \frac{1}{\sqrt{5}q^2}\} = \infty$" might admit an improvement for *certain* $\alpha \in \mathbb{R} \setminus \mathbb{Q}$. This fact suggests the following definition:

Definition 1.1. The *best constant of Diophantine approximation* of $\alpha \in \mathbb{R} \setminus \mathbb{Q}$ is

$$\ell(\alpha) := \limsup_{p,q \to \infty} \frac{1}{|q(q\alpha - p)|}.$$

Roughly speaking, $\ell(\alpha)$ represents the best constant ℓ such that $|\alpha - \frac{p}{q}| \leq \frac{1}{\ell q^2}$ has infinitely many rational solutions $p/q \in \mathbb{Q}$. By Hurwitz theorem, $\ell(\alpha) \geq \sqrt{5}$ for all $\alpha \in \mathbb{R} \setminus \mathbb{Q}$ and $\ell(\frac{1+\sqrt{5}}{2}) = \sqrt{5}$.

The Lagrange spectrum is the set of *finite* best constants of Diophantine approximations, i.e.:

Definition 1.2. The *Lagrange spectrum* is $L := \{\ell(\alpha) : \alpha \in \mathbb{R} \setminus \mathbb{Q}, \ell(\alpha) < \infty\} \subset \mathbb{R}$.

In 1926, Khinchin proved a famous theorem implying that $\ell(\alpha) = \infty$ for Lebesgue almost every $\alpha \in \mathbb{R} \setminus \mathbb{Q}$: see, e.g., his book [Kh (1964)] for more details. In particular, the Lagrange spectrum captures fine properties of certain real numbers beyond the (Lebesgue) typical behavior.

1.3 Definition of the Markov spectrum

Among the algebraic equations studied by Diophantus of Alexandria, one finds the problem of representing integers by sums of two squares. These studies were made systematic in Lagrange's works on quadratic forms. In this context, the analog of the Diophantine problem of finding rational approximations of real numbers for certain types of quadratic forms can be formulated in the following way.

Let $q(x, y) = ax^2 + bxy + cy^2$ be a *binary quadratic form* with real coefficients $a, b, c \in \mathbb{R}$. Suppose that q is *indefinite*[3] with positive *discriminant* $\Delta(q) := b^2 - 4ac$. What is the smallest value of $q(x, y)$ at non-trivial integral vectors $(x, y) \in \mathbb{Z}^2 \setminus \{(0, 0)\}$?

In a certain sense, the "smallest values" mentioned in the previous paragraph play the analogous role for real indefinite binary quadratic forms of the best constants of Diophantine approximation $\ell(\alpha)$ of irrational numbers $\alpha \in \mathbb{R} \setminus \mathbb{Q}$. We consider $\mathcal{Q} = \{q(x, y); q \text{ is an indefinite binary quadratic form with } \Delta(q) > 0\}$. In this setting, the Markov spectrum is defined by analogy with Lagrange spectrum in the following manner:

Definition 1.3. The *Markov spectrum* is

$$
M := \left\{ \frac{\sqrt{\Delta(q)}}{\inf\limits_{(x,y) \in \mathbb{Z}^2 \setminus \{(0,0)\}} |q(x, y)|} \in \mathbb{R} : q \in \mathcal{Q} \right\}.
$$

Remark 1.1. A similar Diophantine problem for *ternary* (and *n*-ary, $n \geq 3$) quadratic forms was proposed by Oppenheim in 1929. Oppenheim's conjecture was famously solved in 1987 by Margulis using *dynamics on homogeneous spaces*: the reader is invited to consult Witte Morris book [WM (2005)] for more details about this beautiful portion of Mathematics.

1.4   Beginning of the classical Lagrange and Markov spectra

In 1880, Markov [Ma (1880)] noticed a relationship between certain binary quadratic forms and rational approximations of certain irrational numbers. This allowed him to obtain the following statement about the initial portions of the classical spectra:

Theorem 1.3 (Markov). $L \cap (-\infty, 3) = M \cap (-\infty, 3) = \{k_1 < k_2 < k_3 < k_4 < \ldots\}$ *where* $k_1 = \sqrt{5}$, $k_2 = \sqrt{8}$, $k_3 = \frac{\sqrt{221}}{5}$, $k_4 = \frac{\sqrt{1517}}{13}$, $\ldots$ *is an explicit increasing sequence of quadratic surds*[4] *accumulating at 3.*

---

[3] I.e., q takes both positive and negative values.
[4] I.e., $k_n^2 \in \mathbb{Q}$ for all $n \in \mathbb{N}$.

In fact, $k_n = \sqrt{9 - \frac{4}{m_n^2}}$ where $m_n \in \mathbb{N}$ is the n-th Markov number, and a Markov number is the largest coordinate of a Markov triple (x, y, z), i.e., an integral solution of $x^2 + y^2 + z^2 = 3xyz$.

We refer the reader to Chapters 1 and 2 of Cusick–Flahive book [CF (1989)] and Bombieri's survey [B (2007)] for detailed expositions of the proof of Markov's theorem.

The objects introduced by Markov in his proof of Theorem 1.3 have extremely interesting behaviours. For example:

- all Markov triples can be deduced from $(1, 1, 1)$ by applying the so-called *Vieta involutions* V_1, V_2, V_3 given by

$$V_1(x, y, z) = (x', y, z)$$

where $x' = 3yz - x$ is the other solution of the second degree equation $X^2 - 3yzX + (y^2 + z^2) = 0$, etc. In other terms, all Markov triples appear in *Markov tree*[5]:

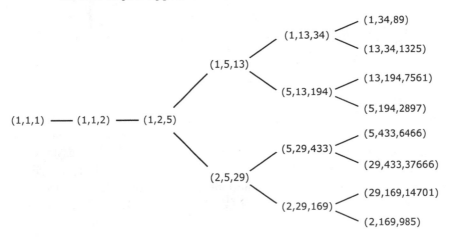

- in 1913, Frobenius formulated the so-called *Markov uniqueness conjecture* asserting that each Markov number z determines an *unique* Markov triple (x, y, z) with $x \leq y \leq z$. This fascinating conjecture remains open today and several of its aspects are described in Aigner's book [A (2013)].

---

[5]Namely, the tree where Markov triples (x, y, z) are displayed after applying permutations to put them in normalized form $x \leq y \leq z$, and two normalized Markov triples are connected if we can obtain one from the other by applying Vieta involutions.

- The set of Markov numbers is very sparse, e.g., Zagier [Za (1982)] showed that if $M(x) := \#\{m \text{ Markov number} : m \leq x\}$, then $M(x) = c(\log x)^2 + O(\log x(\log \log x)^2)$ for an *explicit* constant $c \simeq 0.18071704711507...$[6]

Remark 1.2. The reductions modulo prime numbers $p \in \mathbb{N}$ of Markov equation $x^2 + y^2 + z^2 = 3xyz$ produce an interesting family $\mathcal{G}_p$ of graphs whose vertices are solutions in $(\mathbb{Z}/p\mathbb{Z})^3 \setminus \{(0,0,0)\}$ of Markov equation and edges connect two solutions related by Vieta's involutions.

In fact, it was conjectured by Bourgain–Gamburd–Sarnak [BGSa (2016)] that these graphs are all connected and they remarked that some numerical experiments[7] suggest that $(\mathcal{G}_p)_{p \text{ prime}}$ is an *expander*[8] family. Moreover, Bourgain–Gamburd–Sarnak proved in [BGSb (2016)] that their conjecture is true for all primes outside a "small" exceptional set, the graphs $\mathcal{G}_p$ contain "giant components", and none of the connected components of $\mathcal{G}_p$ are "too small". Also, they announced that almost all Markov numbers are composite.

Remark 1.3. The Markov equation $x^2 + y^2 + z^2 = 3xyz$ is part of the *Markov–Hurwitz equations*

$$x_1^2 + \cdots + x_n^2 = ax_1 \ldots x_n + k$$

where $n \geq 3$, $a \geq 1$ and $k \in \mathbb{Z}$ are integer parameters. We mentioned above that the number of solutions of Markov equation in a ball of radius R grows like $(\log R)^2$, but Baragar [Ba (1998)] discovered that if the Markov–Hurwitz equation with parameters $n \geq 4$, $a \geq 1$ and $k = 0$ has non-trivial solutions, then the number of solutions in a ball of radius R grows as $(\log R)^{\beta(n)+o(1)}$ when $R \to \infty$, where $2.43 < \beta(4) < 2.477$ and, in general,

$$\frac{\log(n-1)}{\log 2} < \beta(n) < \frac{\log(n-1)}{\log 2} + o(n^{-0.58}).$$

Furthermore, Baragar's results were recently extended to the case $k > 0$ (in non-exceptional situations) by Gamburd–Magee–Ronan [GMR (2019)].

---

[6]Conjecturally, $M(x) = c(\log(3x))^2 + o(\log x)$, i.e., if m_n is the n-th Markov number (counted with multiplicity), then $m_n \sim \frac{1}{3}A^{\sqrt{n}}$ with $A = e^{1/\sqrt{c}} \simeq 10.5101504...$

[7]See also the recent article [CIL (2020)] for more informations.

[8]I.e., the two largest eigenvalues of the adjacency matrices of $\mathcal{G}_p$ stay *uniformly* away from each other.

1.5 Best rational approximations and continued fractions

The constant $\ell(\alpha)$ was defined in terms of rational approximations of $\alpha \in \mathbb{R} \setminus \mathbb{Q}$. In particular,

$$\ell(\alpha) = \limsup_{n \to \infty} \frac{1}{|s_n(s_n \alpha - r_n)|}$$

where $(r_n/s_n)_{n \in \mathbb{N}}$ is the sequence of best rational approximations of α. Here, p/q is called a *best rational approximation*[9] whenever

$$\left| \alpha - \frac{p}{q} \right| < \frac{1}{2q^2}.$$

The sequence $(r_n/s_n)_{n \in \mathbb{N}}$ of best rational approximations of α is produced by the so-called *continued fraction algorithm*.

Given $\alpha = \alpha_0 \notin \mathbb{Q}$, we define recursively $a_n = \lfloor \alpha_n \rfloor$ and $\alpha_{n+1} = \frac{1}{\alpha_n - a_n}$ for all $n \in \mathbb{N}$. We can write α as a *continued fraction*

$$\alpha = a_0 + \cfrac{1}{a_1 + \cfrac{1}{a_2 + \cfrac{1}{\ddots}}} =: [a_0; a_1, a_2, \dots]$$

and we denote

$$\mathbb{Q} \ni \frac{p_n}{q_n} := a_0 + \cfrac{1}{a_1 + \cfrac{1}{\ddots + \frac{1}{a_n}}} := [a_0; a_1, \dots, a_n].$$

Remark 1.4. Lévy's theorem [Le (1936)] (from 1936) says that $\sqrt[n]{q_n} \to e^{\pi^2/12 \log 2} \simeq 3.27582291872\dots$ for Lebesgue almost every $\alpha \in \mathbb{R}$. By elementary properties of continued fractions (recalled below), it follows from Lévy's theorem that $\sqrt[n]{|\alpha - \frac{p_n}{q_n}|} \to e^{-\pi^2/6 \log 2} \simeq 0.093187822954\dots$ for Lebesgue almost every $\alpha \in \mathbb{R}$.

Proposition 1.1. *p_n and q_n are recursively given by*

$$\begin{cases} p_{n+2} = a_{n+2} p_{n+1} + p_n, \, p_{-1} = 1, p_{-2} = 0 \\ q_{n+2} = a_{n+2} q_{n+1} + q_n, \, q_{-1} = 0, q_{-2} = 1 \end{cases}.$$

In other words, we have

$$\begin{pmatrix} p_{n+1} & p_n \\ q_{n+1} & q_n \end{pmatrix} \cdot \begin{pmatrix} a_{n+2} & 1 \\ 1 & 0 \end{pmatrix} = \begin{pmatrix} p_{n+2} & p_{n+1} \\ q_{n+2} & q_{n+1} \end{pmatrix} \quad (1.2)$$

or, more generally,

$$[a_0; a_1, \dots, a_{n-1}, z] = \frac{z p_{n-1} + p_{n-2}}{z q_{n-1} + q_{n-2}} \quad (1.3)$$

where z is a real (or complex) variable.

---

[9]This nomenclature will be justified later by Propositions 1.3 and 1.4 below.

Proof. Exercise.[10] $\square$

Corollary 1.1. $p_{n+1}q_n - p_n q_{n+1} = (-1)^n$ *for all* $n \geq 0$.

Proof. This follows from (1.2) because the matrix $\begin{pmatrix} * & 1 \\ 1 & 0 \end{pmatrix}$ has determinant -1. $\square$

Corollary 1.2. $\alpha = \frac{\alpha_n p_{n-1} + p_{n-2}}{\alpha_n q_{n-1} + q_{n-2}}$ *and* $\alpha_n = \frac{p_{n-2} - q_{n-2}\alpha}{q_{n-1}\alpha - p_{n-1}}$.

Proof. This is a consequence of (1.3) and the fact that $\alpha =: [a_0; a_1, \ldots, a_{n-1}, \alpha_n]$. $\square$

The relationship between $\frac{p_n}{q_n}$ and the sequence of best rational approximations is explained by the following two propositions:

Proposition 1.2. $\left| \alpha - \frac{p_n}{q_n} \right| \leq \frac{1}{q_n q_{n+1}} < \frac{1}{a_{n+1}q_n^2} \leq \frac{1}{q_n^2}$ *and, moreover, for all* $n \in \mathbb{N}$,

$$either \left| \alpha - \frac{p_n}{q_n} \right| < \frac{1}{2q_n^2} \quad or \quad \left| \alpha - \frac{p_{n+1}}{q_{n+1}} \right| < \frac{1}{2q_{n+1}^2}.$$

Proof. Note that α belongs to the interval with extremities p_n/q_n and p_{n+1}/q_{n+1} (by Corollary 1.2). Since this interval has size

$$\left| \frac{p_{n+1}}{q_{n+1}} - \frac{p_n}{q_n} \right| = \left| \frac{p_{n+1}q_n - p_n q_{n+1}}{q_n q_{n+1}} \right| = \left| \frac{(-1)^n}{q_n q_{n+1}} \right| = \frac{1}{q_n q_{n+1}}$$

(by Corollary 1.1), we conclude that $\left| \alpha - \frac{p_n}{q_n} \right| \leq \frac{1}{q_n q_{n+1}}$.

Furthermore, $\frac{1}{q_n q_{n+1}} = \left| \frac{p_{n+1}}{q_{n+1}} - \alpha \right| + \left| \alpha - \frac{p_n}{q_n} \right|$. Thus, if

$$\left| \alpha - \frac{p_n}{q_n} \right| \geq \frac{1}{2q_n^2} \quad and \quad \left| \alpha - \frac{p_{n+1}}{q_{n+1}} \right| \geq \frac{1}{2q_{n+1}^2},$$

then

$$\frac{1}{q_n q_{n+1}} \geq \frac{1}{2q_n^2} + \frac{1}{2q_{n+1}^2},$$

i.e., $2q_n q_{n+1} \geq q_n^2 + q_{n+1}^2$, i.e., $q_n = q_{n+1}$, a contradiction. $\square$

---

[10]Hint: Use induction and the fact that $[a_0; a_1, \ldots, a_{n-1}, a_n, z] = [a_0; a_1, \ldots, a_{n-1}, a_n + \frac{1}{z}]$.

In other terms, the sequence $(p_n/q_n)_{n \in \mathbb{N}}$ produced by the continued fraction algorithm contains best rational approximations with frequency at least $1/2$.

Conversely, the continued fraction algorithm detects *all* best rational approximations:

Proposition 1.3. *If* $|\alpha - \frac{p}{q}| < \frac{1}{2q^2}$, *then* $p/q = p_n/q_n$ *for some* $n \in \mathbb{N}$.

Proof. Exercise.[11] $\qquad\qquad\qquad\qquad\qquad\qquad\qquad\qquad\qquad\qquad\qquad\square$

The terminology "best rational approximation" is motivated by the previous proposition and the following result:

Proposition 1.4. *For all* $q < q_n$, *we have* $|\alpha - \frac{p_n}{q_n}| < |\alpha - \frac{p}{q}|$.

Proof. If $q < q_{n+1}$ and $p/q \neq p_n/q_n$, then

$$\left| \frac{p}{q} - \frac{p_n}{q_n} \right| \geq \frac{1}{qq_n} > \frac{1}{q_n q_{n+1}} = \left| \frac{p_{n+1}}{q_{n+1}} - \frac{p_n}{q_n} \right|.$$

Hence, p/q does not belong to the interval with extremities p_n/q_n and p_{n+1}/q_{n+1}, and so

$$\left| \alpha - \frac{p_n}{q_n} \right| < \left| \alpha - \frac{p}{q} \right|$$

because α lies between p_n/q_n and p_{n+1}/q_{n+1}. $\qquad\qquad\qquad\qquad\qquad\square$

In fact, the approximations (p_n/q_n) of α are usually quite impressive:

Example 1.1. $\pi = [3; 7, 15, 1, 292, 1, 1, 1, 2, 1, 3, 1, 14, 2, 1, \dots]$ so that

$$\frac{p_0}{q_0} = 3, \quad \frac{p_1}{q_1} = \frac{22}{7}, \quad \frac{p_2}{q_2} = \frac{333}{106}, \quad \frac{p_3}{q_3} = \frac{355}{113}, \quad \dots$$

The approximations p_1/q_1 and p_3/q_3 (called Yuelü and Milü in Wikipedia) are spectacular:

$$\left| \pi - \frac{22}{7} \right| < \frac{1}{700} < \left| \pi - \frac{314}{100} \right| \quad \text{and} \quad \left| \pi - \frac{355}{113} \right| < \frac{1}{3 \cdot 10^6} < \left| \pi - \frac{3141592}{10^6} \right|.$$

---

[11] Hint: Take $q_n \leq q < q_{n+1}$, suppose that $p/q \neq p_n/q_n$ and derive a contradiction in each case $q_{n+1}/2 \leq q < q_{n+1}$ and $q < q_{n+1}/2$ by analysing $|\alpha - \frac{p}{q}|$ and $|\frac{p}{q} - \frac{p_n}{q_n}|$ like in the proof of Proposition 1.4.

1.6   Perron characterization of Lagrange and Markov spectra

In 1921, Perron interpreted $\ell(\alpha)$ in terms of Dynamical Systems as follows.

Proposition 1.5. $\alpha - \frac{p_n}{q_n} = \frac{(-1)^n}{(\alpha_{n+1}+\beta_{n+1})q_n^2}$ *where* $\beta_{n+1} := \frac{q_{n-1}}{q_n} = [0; a_n, a_{n-1}, \ldots, a_1].$

Proof. Recall that $\alpha_{n+1} = \frac{p_{n-1}-q_{n-1}\alpha}{q_n\alpha-p_n}$ (cf. Corollary 1.2). Hence, $\alpha_{n+1} + \beta_{n+1} = \frac{p_{n-1}q_n-p_nq_{n-1}}{q_n(q_n\alpha-p_n)} = \frac{(-1)^n}{q_n(q_n\alpha-p_n)}$ (by Corollary 1.1). This proves the proposition. □

Therefore, the proposition says that $\ell(\alpha) = \limsup\limits_{n\to\infty}(\alpha_n + \beta_n)$. From the dynamical point of view, we consider the *symbolic space* $\Sigma = (\mathbb{N}^*)^{\mathbb{Z}} =: \Sigma^- \times \Sigma^+ = (\mathbb{N}^*)^{\mathbb{Z}^-} \times (\mathbb{N}^*)^{\mathbb{N}}$ equipped with the left *shift dynamics* $\sigma : \Sigma \to \Sigma$, $\sigma((a_n)_{n\in\mathbb{Z}}) := (a_{n+1})_{n\in\mathbb{Z}}$ and the *height function* $f : \Sigma \to \mathbb{R}$, $f((a_n)_{n\in\mathbb{Z}}) = [a_0; a_1, a_2, \ldots] + [0; a_{-1}, a_{-2}, \ldots]$. Then, the proposition above implies that
$$\ell(\alpha) = \limsup\limits_{n\to+\infty} f(\sigma^n(\underline{\theta}))$$
where $\alpha = [a_0; a_1, a_2, \ldots]$ and $\underline{\theta} = (\ldots, a_{-1}, a_0, a_1, \ldots)$. In particular,
$$L = \{\ell(\underline{\theta}) : \underline{\theta} \in \Sigma, \ell(\underline{\theta}) < \infty\} \tag{1.4}$$
where $\ell(\underline{\theta}) := \limsup\limits_{n\to+\infty} f(\sigma^n(\underline{\theta}))$.

As it turns out, the Markov spectrum has a *similar description*:

Proposition 1.6. *The Markov spectrum is*
$$M = \{m(\underline{\theta}) : \underline{\theta} \in \Sigma, m(\underline{\theta}) < \infty\} \tag{1.5}$$
where $m(\underline{\theta}) := \sup\limits_{n\in\mathbb{Z}} f(\sigma^n(\underline{\theta}))$.

The proof of this proposition relies on the classical reduction theory of binary quadratic forms and it is detailed in Appendix C below.

Remark 1.5. A *geometrical interpretation* of $\sigma : \Sigma \to \Sigma$ is provided by the so-called *Gauss map*[12]:
$$G(x) = \left\{\frac{1}{x}\right\} \tag{1.6}$$
for $0 < x \le 1$. Indeed, $G([0; a_1, a_2, \ldots]) = [0; a_2, \ldots]$, so that $\sigma : \Sigma \to \Sigma$ is a symbolic version of the *natural extension* of G. Furthermore, the identification $(\ldots, a_{-1}, a_0, a_1, \ldots) \simeq ([0; a_{-1}, a_{-2}, \ldots], [a_0; a_1, a_2, \ldots]) = (y, x)$ allows us to write the height function as $f((a_n)_{n\in\mathbb{Z}}) = x + y$.

---

[12]From Number Theory rather than Differential Geometry.

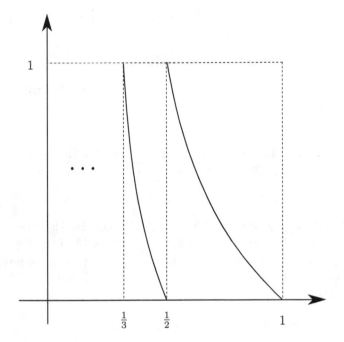

Fig. 1.1   The Gauss map.

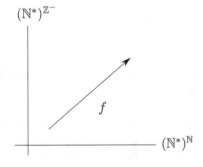

Fig. 1.2 $f(x, y) = x + y$.

Perron's dynamical interpretation of the Lagrange and Markov spectra is the starting point of many results about L and M which are not so easy to guess from their definitions:

Exercise 1.2

Show that $L \subset M$ are closed subsets of $\mathbb{R}$.

Remark 1.6. It was discovered by Freiman [Fr68 (1968)] that $M \setminus L \neq \emptyset$: indeed, he proved in 1968 that[13]

$$s = \overline{2212211221122112222} \in (\mathbb{N}^*)^{\mathbb{Z}}$$

has the property that $3.118120178 \simeq m(s) \in M \setminus L$. We will discuss in details the structure of the complement $M \setminus L$ of the Lagrange spectrum in the Markov spectrum in Chapter 2.

Proposition 1.7 (Perron). *The interval* $(\sqrt{12}, \sqrt{13})$ *is a gap of the Markov spectrum* M. *Moreover,* $m(\underline{\theta}) \leq \sqrt{12}$ *if and only if* $\underline{\theta} \in \{1, 2\}^{\mathbb{Z}}$.

Proof. Since $\sqrt{13} < 4$ and $\sup_{n \in \mathbb{Z}} \theta_n < m(\underline{\theta})$, we see that $m(\underline{\theta}) \leq \sqrt{13}$ implies that $\underline{\theta} \in \{1, 2, 3\}^{\mathbb{Z}}$. Moreover, if $\underline{\theta} \in \{1, 2, 3\}^{\mathbb{Z}}$ contains the strings 31, 13, 32 or 23, then $m(\underline{\theta}) \geq [3; 2, \overline{1, 3}] + [0; \overline{3, 1}] > 3.62 > \sqrt{13}$. Therefore, we have that $m(\underline{\theta}) \leq \sqrt{13}$ implies $\underline{\theta} \in \{1, 2\}^{\mathbb{Z}}$ or $\underline{\theta} = \overline{3}$. This proves the proposition because $m(\overline{3}) = 3 + 2[0; \overline{3}] = \sqrt{13}$ and $m(\underline{\theta}) \leq [2; \overline{1, 2}] + 2[0; \overline{1, 2}] = \sqrt{12}$ whenever $\underline{\theta} \in \{1, 2\}^{\mathbb{Z}}$. $\qquad\square$

Remark 1.7. In the previous argument, we used the following elementary result (whose proof is left as an exercise to the reader): if $\alpha = [a_0; a_1, \ldots, a_n, z, \ldots]$, $\beta = [a_0; a_1, \ldots, a_n, w, \ldots]$ and $z \neq w$, then $\alpha > \beta$ if and only if $(-1)^{n+1}(z - w) > 0$. This fact will be systematically (and often implicitly) invoked in our subsequent discussions of the classical Lagrange and Markov spectra.

1.7 Digression: Lagrange spectrum and cusp excursions on the modular surface

The Lagrange spectrum is related to the values of a certain height function H along the orbits of the geodesic flow g_t on the modular surface. Indeed, if we write cot vector by mean a cotangent vector, we will prove in the sequel that

$$L = \left\{ \limsup_{t \to +\infty} H(g_t(x)) < \infty : x \text{ is a unit cot vector to the modular surface} \right\}.$$

Remark 1.8. This fact is not surprising to experts: the Gauss map appears naturally by quotienting out the weak-stable manifolds of g_t as observed by Artin, Series, Arnoux, ... (see, e.g., [Ar (1994)]).

---

[13]Here $\overline{\theta_1 \ldots \theta_n}$ means infinite repetition of the block $\theta_1 \ldots \theta_n$.

An *unimodular lattice* in $\mathbb{R}^2$ has the form $g(\mathbb{Z}^2)$, $g \in SL(2, \mathbb{R})$, and the stabilizer in $SL(2, \mathbb{R})$ of the standard lattice $\mathbb{Z}^2$ is $SL(2, \mathbb{Z})$. In particular, the space of unimodular lattices in $\mathbb{R}^2$ is $SL(2, \mathbb{R})/SL(2, \mathbb{Z})$. As it turns out, this space is the unit cotangent bundle to the *modular surface* $\mathbb{H}/SL(2, \mathbb{Z})$, where $\mathbb{H} = \{z \in \mathbb{C} : \mathrm{Im}(z) > 0\}$ is the usual hyperbolic (Poincaré) upper-half plane and $\begin{pmatrix} a & b \\ c & d \end{pmatrix} \in SL(2, \mathbb{R})$ acts on $z \in \mathbb{H}$ via $\begin{pmatrix} a & b \\ c & d \end{pmatrix} \cdot z = \frac{az+b}{cz+d}$.

The *geodesic flow* of the modular surface is the action of $g_t = \begin{pmatrix} e^t & 0 \\ 0 & e^{-t} \end{pmatrix}$ on $SL(2, \mathbb{R})/SL(2, \mathbb{Z})$. The *stable* and *unstable* *manifolds* of g_t are the orbits of the *stable* and *unstable horocycle flows* $h_s = \begin{pmatrix} 1 & 0 \\ s & 1 \end{pmatrix}$ and $u_s = \begin{pmatrix} 1 & s \\ 0 & 1 \end{pmatrix}$: indeed, this follows from the definition of stable and unstable manifolds[14] and the facts that $g_t h_s = h_{se^{-2t}} g_t$ and $g_t u_s = u_{se^t} g_t$.

The set of *holonomy* (or *primitive*) *vectors* of $\mathbb{Z}^2$ is

$$\mathrm{Hol}(\mathbb{Z}^2) := \{(p, q) \in \mathbb{Z}^2 : \gcd(p, q) = 1\}.$$

In general, the set $\mathrm{Hol}(X)$ of holonomy vectors of $X = g(\mathbb{Z}^2)$, $g \in SL(2, \mathbb{Z})$, is

$$\mathrm{Hol}(X) := g(\mathrm{Hol}(\mathbb{Z}^2)) \subset \mathbb{R}^2.$$

The *systole* $\mathrm{sys}(X)$ of $X = g(\mathbb{Z}^2)$ is

$$\mathrm{sys}(X) := \min\{\|v\|_{\mathbb{R}^2} : v \in \mathrm{Hol}(X)\}.$$

Remark 1.9. By Mahler's compactness criterion [Mahler (1946)], $X \mapsto \frac{1}{\mathrm{sys}(X)}$ is a proper function on the non-compact space $SL(2, \mathbb{R})/SL(2, \mathbb{Z})$.

Remark 1.10. For later reference, we write $\mathrm{Area}(v) := |\mathrm{Re}(v)| \cdot |\mathrm{Im}(v)|$ for the area of the rectangle in $\mathbb{R}^2$ with diagonal $v = (\mathrm{Re}(v), \mathrm{Im}(v)) \in \mathbb{R}^2$.

Proposition 1.8. *The geodesic flow orbit of $X \in SL(2, \mathbb{R})/SL(2, \mathbb{Z})$ does not go straight to infinity (i.e., $\mathrm{sys}(g_t(X)) \to 0$ as $t \to +\infty$) if and only if there is no vertical vector in $\mathrm{Hol}(X)$. In this case, there are (unique) parameters $s, t, \alpha \in \mathbb{R}$ such that*

$$X = h_s g_t u_{-\alpha}(\mathbb{Z}^2).$$

---

[14]The reader is invited to consult the book of Hasselblatt–Katok [HK (1995)] for more explanations about the dynamics of geodesic and horocycle flows.

Proof. By unimodularity, any $X = g(\mathbb{Z}^2)$ has a single *short* holonomy vector. Since g_t contracts vertical vectors and expands horizontal vectors for $t > 0$, we have that $\text{sys}(g_t(X)) \to 0$ as $t \to +\infty$ if and only if $\text{Hol}(X)$ contains a vertical vector.

By Iwasawa decomposition, there are (unique) parameters $s, t, \theta \in \mathbb{R}$ such that $X = h_s g_t r_\theta$, where $r_\theta = \begin{pmatrix} \cos\theta & -\sin\theta \\ \sin\theta & \cos\theta \end{pmatrix}$. Since $\cos\theta \neq 0$ when $\text{Hol}(X)$ contains no vertical vector and, in this situation,

$$r_\theta = h_{\tan\theta} g_{\log\cos\theta} u_{-\tan\theta},$$

we see that $X = h_{s + e^{-2t}\tan\theta} \cdot g_{t + \log\cos\theta} \cdot u_{-\tan\theta}(\mathbb{Z}^2)$ (because $h_s g_t r_\theta = h_s g_t h_{\tan\theta} g_{\log\cos\theta} u_{-\tan\theta} = h_{s + e^{-2t}\tan\theta} \cdot g_{t + \log\cos\theta} \cdot u_{-\tan\theta}$). This ends the proof of the proposition. $\qquad\square$

Proposition 1.9. *Let $X = h_s g_t u_{-\alpha}(\mathbb{Z}^2)$ be an unimodular lattice without vertical holonomy vectors. Then,*

$$\ell(\alpha) = \limsup_{\substack{|Im(v)| \to \infty \\ v \in Hol(X)}} \frac{1}{Area(v)} = \limsup_{T \to +\infty} \frac{2}{sys(g_T(X))^2}.$$

Remark 1.11. This proposition says that the dynamical quantity $\limsup\limits_{T \to +\infty} \frac{2}{\text{sys}(g_T(X))^2}$ does *not* depend on the "weak-stable part" $h_s g_t$ (but only on α) and it can be computed *without* dynamics by simply studying almost vertical holonomy vectors in X.

Proof. Note that $\text{Area}(g_t(v)) = \text{Area}(v)$ for all $t \in \mathbb{R}$ and $v \in \mathbb{R}^2$. Since $\text{Area}(v) = \frac{\|g_{t(v)}(v)\|^2}{2}$ for $t(v) := \frac{1}{2}\log\frac{|Im(v)|}{|Re(v)|}$, the equality $\limsup\limits_{\substack{|Im(v)| \to \infty \\ v \in Hol(X)}} \frac{1}{\text{Area}(v)} = \limsup\limits_{T \to +\infty} \frac{2}{\text{sys}(g_T(X))^2}$ follows.

The relation $g_T h_s = h_{se^{-2T}} g_T$ and the continuity of the systole function imply that $\limsup\limits_{T \to +\infty} \frac{2}{\text{sys}(g_T(X))^2}$ depends only on α. Because any $v \in \text{Hol}(u_{-\alpha}(\mathbb{Z}^2))$ has the form $v = (p - q\alpha, q) = u_{-\alpha}(p, q)$ with $(p, q) \in \text{Hol}(\mathbb{Z}^2)$, the equality $\limsup\limits_{\substack{|Im(v)| \to \infty \\ v \in Hol(X)}} \frac{1}{\text{Area}(v)} = \ell(\alpha)$. $\qquad\square$

In summary, the previous propositions imply the following theorem:

Theorem 1.4. *The Lagrange spectrum L coincides with*

$$\left\{ \limsup_{T \to +\infty} H(g_T(x)) < \infty : x \in SL(2, \mathbb{R})/SL(2, \mathbb{Z}) \right\}$$

where $H(y) = \frac{2}{sys(y)^2}$ is a (proper) height function and g_t is the geodesic flow on $SL(2, \mathbb{R})/SL(2, \mathbb{Z})$.

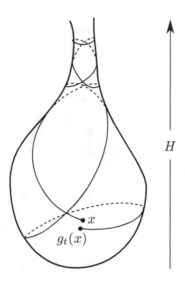

Fig. 1.3   Cusp excursion on the modular surface.

Remark 1.12. Several number-theoretical problems translate into dynamical questions on the modular surface: for example, Zagier [Za (1981)] showed that the Riemann hypothesis is equivalent to a certain speed of equidistribution of u_s-orbits on $SL(2, \mathbb{R})/SL(2, \mathbb{Z})$.

1.8   Final portion of the classical spectra: Hall's ray and Freiman's constant

In 1947, Hall [Hall (1947)] proved that:

Theorem 1.5 (Hall). *The half-line* $[6, +\infty)$ *is contained in* L.

This result motivates the following nomenclature: the biggest half-line $[c_F, +\infty) \subset L(\subset M)$ is called *Hall's ray*.

In 1975, Freiman [Fr75 (1975)] determined Hall's ray:

Theorem 1.6 (Freiman). *The Lagrange spectrum contains the half-line* $[4 + \frac{253589820 + 283798\sqrt{462}}{491993569}, +\infty)$ *and the Markov spectrum doesn't intersect* *the interval*

$$\left(\frac{19033619 - 2\sqrt{462}}{72101381} + \frac{594524 - \sqrt{243542}}{139318}, 4 + \frac{253589820 + 283798\sqrt{462}}{491993569} \right).$$

In particular, $c_F = 4 + \frac{253589820 + 283798\sqrt{462}}{491993569} \simeq 4.527829566...$

The constant c_F is called *Freiman's constant*. Unfortunately, the proof of the first part of Freiman's theorem is beyond the scope of this book: it takes almost 100 pages of technical calculations in the original article and, thus, it is not reasonable to discuss it here. On the other hand, the second part of Freiman's theorem is relatively short: see Appendix D below.

Let us now sketch the proof of Hall's theorem based on the following lemma:

Lemma 1.1 (Hall). *Denote by* $C(4) := \{[0; a_1, a_2, \ldots] \in \mathbb{R} : a_i \in \{1, 2, 3, 4\}\ \forall\, i \in \mathbb{N}\}$. *Then,*

$$C(4) + C(4) := \{x + y \in \mathbb{R} : x, y \in C(4)\} = [\sqrt{2} - 1, 4(\sqrt{2} - 1)],$$

that is, $C(4) + C(4) = [0.414\ldots, 1.656\ldots]$ *is an interval of length greater than 1.*

Remark 1.13. The reader can find a proof of this lemma in Cusick–Flahive's book [CF (1989)]. Interestingly enough, some of the techniques in the proof of Hall's lemma were rediscovered much later (in 1979) in the context of Dynamical Systems by Newhouse [Nh (1979)] (in the proof of his *gap lemma*).

Remark 1.14. $C(4)$ is a *dynamical Cantor set*[15] whose Hausdorff dimension is $> 1/2$ (see Remark 2.9 below). In particular, $C(4) \times C(4)$ is a planar Cantor set of Hausdorff dimension > 1 and Hall's lemma says that its image $f(C(4) \times C(4)) = C(4) + C(4)$ under the the projection $f(x, y) = x + y$ contains an interval. Hence, Hall's lemma can be thought as a sort of "particular case" of *Marstrand's theorem* [Mars (1954)] (ensuring that typical projections of planar sets with Hausdorff dimension > 1 has positive Lebesgue measure).

Proof of Theorem 1.5. For our current purposes, the specific form $C(4) + C(4)$ is *not* important: the *key point* is that $C(4) + C(4)$ is an interval of length > 1.

---

[15]See §2.3.1 and §2.3.1.1 in Chapter 2 below.

Indeed, given $6 \le \ell < \infty$, Hall's lemma guarantees the existence of $c_0 \in \mathbb{N}$, $5 \le c_0 \le \ell$ such that $\ell - c_0 \in C(4) + C(4)$. Thus,

$$\ell = c_0 + [0; a_1, a_2, \ldots] + [0; b_1, b_2, \ldots]$$

with $a_i, b_i \in \{1, 2, 3, 4\}$ for all $i \in \mathbb{N}$.

Define

$$\alpha := \left[0; \underbrace{b_1, c_0, a_1, \ldots}_{1^{st} \text{ block}}, \underbrace{b_n, \ldots, b_1, c_0, a_1, \ldots, a_n}_{n^{th} \text{ block}}, \ldots \right].$$

Since $c_0 \ge 5 > 4 \ge a_i, b_i$ for all $i \in \mathbb{N}$, Perron's characterization of $\ell(\alpha)$ implies that

$$L \ni \ell(\alpha) = \lim_{n \to \infty} (c_0 + [0; a_1, a_2, \ldots, a_n] + [0; b_1, b_2, \ldots, b_n]) = \ell.$$

This proves Theorem 1.5. □

1.9 Intermediate portions of the Lagrange and Markov spectra

Our discussion so far about the extremal portions of the classical spectra can be summarized as follows:

- $L \cap (-\infty, 3) = M \cap (-\infty, 3) = \{k_1 < k_2 < \cdots < k_n < \ldots\}$ is an *explicit* discrete set;
- $L \cap [c_F, \infty) = M \cap [c_F, \infty)$ is an *explicit* ray.

In particular, the classical Lagrange and Markov spectra *coincide* outside $(3, c_F)$.

On the other hand, we already saw some hints to the fact that the intermediate portions $L \cap [3, c_F]$ and $M \cap [3, c_F]$ are harder to describe: indeed, we mentioned that Perron found that $(\sqrt{12}, \sqrt{13})$ is a gap of the Markov spectrum (cf. Proposition 1.7) and Freimain found an element of $M \setminus L$ near 3.118120178 (cf. Remark 1.6).

The reader is invited to consult Chapter 5 of Cusick–Flahive book [CF (1989)] for a nice account on the list of maximal gaps in the classical spectra found before 1987. In this context, it is important to know the structure of the sequences $\underline{\theta} \in \Sigma$ such that $\ell(\underline{\theta})$ or $m(\underline{\theta})$ produces an endpoint of a gap in L or M. In Chapter 2, we discuss recent results of Gayfulin [Gay (2017)], [Gay2 (2019)] (correcting previous statements and/or arguments by Malyshev [Maly (1977)] and Dietz [Dietz (1985)]).

Besides the presence of many gaps, the intermediate portions of the classical spectra are difficult to describe because they represent the transition from the initial portion $\{k_1 < k_2 < \ldots\}$ of *zero* Lebesgue measure to the final portion $[c_F, \infty)$ of *positive* Lebesgue measure. In Chapter 6 of Cusick–Flahive book [CF (1989)], the reader finds a compilation of several results aiming to pinpoint the exact place

$$t_+ := \inf\{3 \leq t \leq c_F : M \cap [\sqrt{5}, t] \text{ has positive Lebesgue measure}\}$$

where the Markov spectrum goes from zero to positive Lebesgue measure. In particular, it is mentioned that Bumby [Bumby (1982)] made some computer-assisted calculations showing that $t_+ > 3.3343$. Also, they observe that it is not known whether the Markov spectrum contains a nontrivial open interval below $\sqrt{13}$.

Remark 1.15. To the best of our knowledge, despite certain heuristic arguments with arithmetic sums of Cantor sets (in the spirit of Remark 1.14 above), it is actually not known whether the interior of $M \cap [3, c_F]$ is not empty. In particular, a conjecture of Berstein [Be2 (1973), p. 77] from 1973 asserting that $[4.1, 4.52] \subset M$ is still open.

The belief that the intermediate portions of the classical spectra consist of intricate fractal sets was definitely confirmed by the following recent theorem of Moreira [Mor1 (2018)]:

Theorem 1.7 (Moreira). *For each $t \in \mathbb{R}$, the sets $L \cap (-\infty, t)$ and $M \cap (-\infty, t)$ have the same Hausdorff dimension, say $d(t) \in [0, 1]$.*

Moreover, the function $t \mapsto d(t)$ is continuous, $d(3 + \varepsilon) > 0$ for all $\varepsilon > 0$ and $d(\sqrt{12}) = 1$ (even though $\sqrt{12} = 3.4641\ldots < 4.5278\ldots = c_F$).

The proof of this theorem is presented in Chapter 3, where we also recall the notion of Hausdorff dimension. In a nutshell, the Hausdorff dimension is a measurement of "size" of fractal sets, so that Moreira's theorem says that the classical spectra become larger immediately after 3 and they reach the maximal size much before the beginning of Hall's ray.

Interestingly enough, we will also see in Chapter 3 that the techniques used in the proof of Theorem 1.7 have the following topological consequence for the Lagrange spectrum:

Theorem 1.8. *The set L' of accumulation points of L is perfect, i.e., $L' = L''$.*

Remark 1.16. It is not known whether $M' = M''$.

We conclude this introductory chapter with the following important remark about the nature of several theorems about the Lagrange and Markov spectra.

Remark 1.17. Many results about L and M are *dynamical*.[16] In particular, it is not surprising that many facts about L and M have counterparts for *dynamical Lagrange and Markov spectra*[17]: for example, Hall ray or intervals in dynamical Lagrange spectra were found by Parkkonen–Paulin [PP (2010)], Hubert–Marchese–Ulcigrai [HMU (2015)] and Moreira–Romaña [MR (2016)], and the continuity result in Moreira's theorem 1.7 was extended by Cerqueira–Matheus–Moreira in [CMM (2018)]. We will discuss some aspects of dynamical Lagrange and Markov spectra in Chapters 3, 4 and 5.

---

[16]I.e., they involve Perron's characterization of L and M, the study of Gauss map and/or the geodesic flow on the modular surface, etc.

[17]I.e., the collections of "records" of height functions along orbits of certain types of Dynamical Systems.

Chapter 2

Some results about the intermediate portions of the classical spectra

In this chapter, we explain a few recent results about the intermediate portions of the classical Lagrange and Markov spectra. More concretely, we start by explaining the proofs of Gayfulin's results in [Gay (2017)], [Gay2 (2019)] on attainable and admissible elements of the Lagrange spectrum and, afterwards, we focus on the study of the complement $M \setminus L$ of the Lagrange spectrum in the Markov spectrum. Then, we conclude this chapter with the discussion of a computer program allowing to rigorously draw pictures of small neighborhoods (in Hausdorff topology) of L and M.

In the sequel, besides the notations in Chapter 1 (e.g., $\ell(\alpha)$ is the best constant of Diophantine approximation for α, etc.), given $\underline{a} := (a_n)_{n \in \mathbb{Z}} \in (\mathbb{N}^*)^{\mathbb{Z}}$, we set

$$\lambda_k(\underline{a}) := [a_k; a_{k+1}, a_{k+2}, \dots] + [0; a_{k-1}, a_{k-2}, \dots],$$

so that its Markov value is $m(\underline{a}) = \sup_{k \in \mathbb{Z}} \lambda_k(\underline{a})$ and its Lagrange value is $\ell(\underline{a}) = \limsup_{k \to \infty} \lambda_k(\underline{a})$. Also, $\underline{a}_n = (\underline{a}, \underline{a}, ..., \underline{a})$ will stand for the n-fold concatenation of the finite string $\underline{a} = (a_1, a_2, ..., a_j)$. Moreover, the infinite concatenation of a string $\underline{a}$ to the right, resp. left of a string $\underline{b}$ is denoted by $(\underline{b}, \overline{\underline{a}}) = (\underline{b}, \underline{a}, \underline{a}, ...)$, resp. $(\overline{\underline{a}}, \underline{b}) = (..., \underline{a}, \underline{a}, \underline{b})$. Furthermore, we often point out the zeroth position of a string $\underline{a}$ with an asterisk, i.e., $\underline{a} = (a_{-m}, \dots, a_0^*, \dots, a_n)$. Finally, the transpose of a string $\underline{c} = (c_1, \dots, c_n)$ is denoted by $\underline{c}^T := (c_n, \dots, c_1)$.

2.1 Non-attainable numbers and non-admissible elements of the Lagrange spectrum

Definition 2.1. An irrational number $\alpha \in \mathbb{R} \setminus \mathbb{Q}$ which satisfies

$$\left| \alpha - \frac{p}{q} \right| \leq \frac{1}{\ell(\alpha)q^2} \tag{2.1}$$

for infinitely many rational numbers $\dfrac{p}{q}$ is called *attainable*.

By Propositions 1.4 and 1.5 in Chapter 1, if α is attainable if and only if

$$\lambda_n(\alpha) := \alpha_n + \beta_n \geq \ell(\alpha) \tag{2.2}$$

for infinitely many values of n.

It was claimed by Malyshev [Maly (1977)] in 1977 that for any $\ell \in L$ there exists an irrational α such that $\ell(\alpha) = \ell$ and α is attainable. As it turns out, Gayfulin [Gay (2017)] discovered a counterexample to Malyshev's claim: in what follows, we provide a detailed discussion of Gayfulin's theorem.

Consider $\theta = \overline{21}123^*3_221\overline{12}$ and let $\ell_G := m(\theta)$ be its Markov value. Since $[3; 3, 3, ...] > [3; 3, 2, ...]$ and $[0; 2, 1, ...] > [0; 3, 2, ...]$ we have that

$$\lambda_0(\theta) = [3; 3, 3, 2, 1, \overline{1, 2}] + [0; 2, 1, \overline{1, 2}] = \lambda_2(\theta) > [3; 3, 2, 1, \overline{1, 2}] + [0; 3, 2, 1, \overline{1, 2}]$$

that is, $\lambda_0(\theta) = \lambda_2(\theta) > \lambda_1(\theta)$. On the other hand, if $i \notin \{0, 1, 2\}$ then $\lambda_i(\theta) \leq 2 + 2[0; \overline{1, 3}] < 3.6 < \lambda_0(\theta)$. In particular, we have that $\ell_G = m(\theta) = \lambda_0(\theta) = \lambda_2(\theta) = 3.69147080089\ldots$

Lemma 2.1. *We have that $\ell_G = m(\theta) \in L$.*

Proof. Consider $\underline{b}_n = (2, 1)_n 1233321(1, 2)_n$ and $\theta_n = \overline{b}_n(2, 1)_n$ $123^*3321(1, 2)_n \overline{\underline{b}_n} = \overline{\underline{b}_n} \underline{b}_n^* \overline{\underline{b}_n}$. Then we have that $\ell_n = m(\theta_n) \in L$ and $\ell_n \to \ell$. Since L is closed, we have that $\ell \in L$. $\square$

Next, we will show that if $\ell(\alpha) = \ell_G$, then α is not attainable. First, we obtain obstructions in the continued fraction expansion of α with $\ell(\alpha) \leq \ell_G$.

Lemma 2.2. *Consider an irrational number $\alpha = [a_0; a_1, a_2, ...]$ with $\ell(\alpha) \leq \ell_G$ and let*

(1) $\mathcal{S}_0 = \{n \in \mathbb{N}; a_n = 4\} \cup \{n \in \mathbb{N}; (a_n, a_{n+1}) \in \{(1, 3), (3, 1)\}\}$,
(2) $\mathcal{S}_1 = \{n \in \mathbb{N}; (a_n, a_{n+1}, a_{n+2}) \in \{(3, 2, 2), (2, 2, 3)\}\}$,
(3) $\mathcal{S}_2 = \{n \in \mathbb{N}; (a_n, a_{n+1}, a_{n+2}) = (3, 2, 3)\}$
(4) $\mathcal{S}_3 = \{n \in \mathbb{N}; (a_{n-2}, a_{n-1}, a_n, a_{n+1}, a_{n+2}) = (1, 2, 3, 2, 1)\}$.

Then, $\mathcal{S} = \mathcal{S}_0 \cup \mathcal{S}_1 \cup \mathcal{S}_2 \cup \mathcal{S}_3$ is finite.

Proof. Since $\ell(\alpha) < 4$, it is not difficult to see that if $\alpha = [a_0; a_1, a_2, ...]$ then there exists $n_0 \in \mathbb{N}$ such that $a_n \leq 3, n > n_0$. Hence, without loss of generality, we suppose $a_n \leq 3$ for all n. This reduces the proof of the lemma to establish that the string $(a_0, a_1, \ldots)$ has the following properties:

(i) Neither $(1, 3)$ nor $(3, 1)$ occur infinitely many times.

(ii) Neither $(2, 2, 3)$ nor $(3, 2, 2)$ occur infinitely many times.

(iii) $(3, 2, 3)$ does not occur.

(iv) $(1, 2, 3, 2, 1)$ does not occur.

To prove (i), it is enough to show that $(a_n, a_{n+1}) = (3, 1)$ does not occur for infinitely many values of n (because the case of $(1, 3)$ is entirely analogous). If $(a_n, a_{n+1}) = (3, 1)$, then

$$\lambda_n(\alpha) = [3; 1, a_{n+2}, ...] + [0; a_{n-1}, ..., a_1] > [3; 1, \overline{1, 3}] + [0; 3, 1] > 3.8 > \ell_G + 10^{-1}.$$

Therefore, if $(3, 1)$ occurred infinitely often, one would conclude $\ell_G \geq \ell(\alpha) \geq \ell_G + 10^{-1}$, a contradiction. In particular, there is no loss of generality in assuming in the sequel that the patterns $(3, 1)$ and $(1, 3)$ do not occur in $(a_0, a_1, \dots)$.

Next we prove (ii). Assume that the set $\mathcal{S}_1 = \{n \in \mathbb{N}; (a_n, a_{n+1}, a_{n+2}) = (3, 2, 2)\}$ is infinite. For each $n \in \mathcal{S}_1$, we have

$$\lambda_n(\alpha) = [3; 2_2, a_{n+3}, ...] + [0; a_{n-1}, ..., a_1] \geq [3; 2_2, \overline{3, 2}] + [0; 3, 2] > 3.696$$
$$> \ell_G + 10^{-3},$$

because $(1, 3)$ and $(3, 1)$ not occur in α. In particular, $\ell_G \geq \ell(\alpha) > \ell_G + 10^{-3}$, a contradiction. By symmetry, $(2, 2, 3)$ also does not occur infinitely often. Hence, there is no loss of generality in assuming also that $(3, 2, 2)$ and $(2, 2, 3)$ do not occur in the continued fraction expansion of α.

Analogously, to prove (iii) we suppose that $\mathcal{S}_2 = \{n \in \mathbb{N}; (a_n, a_{n+1}, a_{n+2}) = (3, 2, 3)\}$ is infinite. For each $n \in \mathcal{S}_2$, we have that

$$\lambda_n(\alpha) = [3; 2, 3, a_{n+3}, ...] + [0; a_{n-1}, a_{n-2}, ..., a_1] > [3; 2, 3, \overline{3, 1}] + [0; 3, 2] > 3.71.$$

In particular, if $\mathcal{S}_2$ is infinite, then $\ell_G \geq \ell(\alpha) > 3.71$, a contradiction.

Finally, to prove (iv) suppose that $(a_{n-2}, a_{n-1}, a_n, a_{n+1}, a_{n+2}) = (1, 2, 3, 2, 1)$ infinitely many times. For each such n, we have

$$\lambda_n(\alpha) = [3; 2, 1, a_{n+3}, ...] + [0; 2, 1, a_{n-3}, ...].$$

Since we are supposing that $(1, 3)$ and $(3, 1)$ do not occur in the continued fraction expansion of α, we must have

$$\lambda_n(\alpha) > [3; 2, 1, \overline{2, 1}] + [0; 2, 1] > 3.699.$$

Thus, if $\mathcal{S}_3$ is infinite, then $\ell_G \geq \ell(\alpha) > 3.699$, a contradiction. This completes the proof. $\qquad\square$

Next, we give a necessary condition on the continued fraction expansion of an irrational number α with $\ell(\alpha) = \ell_G$.

Lemma 2.3. *Consider* $\alpha = [0; a_1, a_2, a_3, ...]$ *with* $\ell(\alpha) = \ell_G$. *Then there exist* N *such that for any integer* $n > N$ *the inequality*

$$\lambda_n(\alpha) > 3.691 \qquad (2.3)$$

holds only if

$$(a_{n-2}, a_{n-1}, a_n, a_{n+1}, a_{n+2}, a_{n+3}, a_{n+4}) = (1, 2, 3, 3, 3, 2, 1)$$

or

$$(a_{n-4}, a_{n-3}, a_{n-2}, a_{n-1}, a_n, a_{n+1}, a_{n+2}) = (1, 2, 3, 3, 3, 2, 1).$$

Proof. By Lemma 2.2 there exists an integer N such that in the infinite sequence $a_{N-10}, a_{N-9}, ...$ no element exceeds 3 and no pattern from the list in the set S appears. Consider $n > N$ such that (2.3) holds. By Lemma 2.2 we have three possibilities

(1) $(a_{n-1}, a_n, a_{n+1}) = (3, 3, 3)$
(2) $(a_{n-1}, a_n, a_{n+1}) = (2, 3, 3)$
(3) $(a_{n-1}, a_n, a_{n+1}) = (3, 3, 2)$.

In fact, if $(2, 3, 2)$ appears the only possible continuation is $(1, 2, 3, 2, 1)$, but the hypothesis on n does not allow this.

Moreover, in the case $a_n = 3 = a_{n-1} = a_{n+1}$, we see that

$$\lambda_n(\alpha) = [3; 3, a_{n+2}, ...] + [0; 3, a_{n-2}, ...] \le [3; 3, \overline{3, 2}] + [0; 3_2, 1] < 3.62 < 3.691,$$

a contradiction with (2.3).

Thus, we have $(a_{n-1}, a_{n+1}) \in \{(3, 2), (2, 3)\}$. By symmetry, it is enough to treat $(a_{n-1}, a_{n+1}) = (3, 2)$, that is, $(a_{n-1}, a_n, a_{n+1}) = (3, 3, 2)$. Since $(3, 2, 3)$ and $(3, 2, 2)$ do not occur, we must have $a_{n+2} = 1$. On the other hand, $a_{n-2} = 2$ or $a_{n-2} = 3$ because $(1, 3)$ does not occur.

If $a_{n-2} = 2$, then we would have $a_{n-3} = 1$ because $(2, 2, 3)$ does not occur. Thus,

$$\lambda_n(\alpha) = [3; 2, 1, a_{n+3}, ...] + [0; 3, 2, 1, a_{n-4}, ..., a_1] \le [3; 2, 1, \overline{1, 2}] + [0; 3, 2, 1, 3]$$

that is, $\lambda_n(\alpha) < 3.686$, because $[0; 3, 2, 1, a_{n-4}, ...] < [0; 3, 2, 1, 3]$ (as $a_{n-4} < 3$). This contradiction with (2.3) means that $a_{n-2} = 3$, that is, $(a_{n-2}, a_{n-1}, a_n, a_{n+1}, a_{n+2}) = (3, 3, 3, 2, 1)$.

Suppose that $a_{n-3} = 3$. Since $(1, 3)$ and $(3, 1)$ do not occur, one has

$$\lambda_n(\alpha) = [3; 2, 1, a_{n+3}, ...] + [0; 3_3, a_{n-4}, ...] \le [3; 2, 1, \overline{1, 2}] + [0; 3_4, 2, 2] < 3.69078.$$

Here, the inequality $[0; 3_3, a_{n-4}, ...] \leq [0; 3_4, 2, 2]$ holds because $a_{n-4} \leq 3$ and, if $a_{n-4} = 3$, then $a_{n-5} = 2$ implies $a_{n-6} = 1$. This is again a contradiction with (2.3), so that we must have $a_{n-3} = 2$. Since $(3, 2, 3)$ and $(2, 2, 3)$ do not occur, we conclude that $a_{n-4} = 1$, that is, $(a_{n-4}, a_{n-3}, a_{n-2}, a_{n-1}, a_n, a_{n+1}, a_{n+2}) = (1, 2, 3, 3, 3, 2, 1)$. This completes the argument. $\qquad\square$

The next lemma is the crucial step towards Gayfulin's counterexample to Malyshev's claim.

Lemma 2.4. *If $\alpha \in \mathbb{R} \setminus \mathbb{Q}$ and $\ell(\alpha) = \ell_G$, then there exists N such that*

$$\lambda_n(\alpha) < \ell_G \qquad (2.4)$$

for all $n > N$.

Proof. Suppose that there exists a sequence $\{k_j\}_{j \in \mathbb{N}}$ such that $\lambda_{k_j}(\alpha) \geq \ell_G$. By Lemma 2.3, there is J such that if $j > J + 2$, then a_{k_j} is equals to the leftmost or rightmost 3 in the string $(1, 2, 3, 3, 3, 2, 1)$. Moreover, we know that if $(1, 3)$ appears infinitely many times, then $\ell(\alpha) > \ell_G + 10^{-1}$ (see the proof of Lemma 2.2). In particular we can suppose that $(1, 3)$ does not occur. We can (and do) assume that a_{k_j} is the last 3 in this string (as the other case is analogous). In this context,

$$\lambda_{k_j}(\alpha) = [3; 3, 3, 2, 1, a_{k_j-5}, ..., a_1] + [0, 2, 1, ...].$$

Claim: $[3; 3, 3, 2, 1, a_{k_j-5}, ..., a_1] < [3; 3, 3, 2, 1, \overline{1, 2}]$. In fact, we write $[3; 3, 3, 2, 1, \overline{1, 2}] = [b_0; b_1, b_2, ...]$ and

$$4 \leq r = \max\{t \in \mathbb{N}; a_{k_j - t} = b_t\} \leq k_j - 1.$$

Case 1: $r < k_j - 1$ and r even. We have $b_{r+1} = 1$ and by definition of r we must have $a_{k_j-(r+1)} > 1$. But, r is even and then

$$[3; 3, 3, 2, 1, ..., a_{k_j-r}, a_{k_j-(r+1)}, ...] < [b_0; b_1, ..., b_r, 1, ...].$$

Case 2: $r < k_j - 1$ and r odd. In this case, $b_{r+1} = 2$ and then $a_{k_j-(r+1)} = 1$ or $a_{k_j-(r+1)} = 3$. This implies $(b_r, b_{r+1}) = (1, 2)$ and then $(a_{k_j-r}, a_{k_j-(r+1)}) = (1, 1)$ because of $(1, 3)$ does not occur. Using the fact that r is odd we have that

$$[3; 3_2, 2, 1, ..., a_{k_j-r}, a_{k_j-(r+1)}, ...] = [3; 3_2, 2, 1, ..., a_{k_j-r+1}, 1, 1, ...],$$

that is,

$$[3; 3_2, 2, 1, ..., a_{k_j-r}, a_{k_j-(r+1)}, ...] < [b_0; b_1, ..., b_r, 2, ...].$$

Case 3: $r = k_j - 1$. In this case, $[a_{k_j}; a_{k_j-1}, ..., a_1]$ would be a convergent of $[3; 3, 3, 2, 1, \overline{1, 2}]$. But, since $a_{k_j-(k_j-k_{j-1})} = a_{k_{j-1}} = 3$ and $b_{k_j-k_{j-1}} \in \{1, 2\}$ we have a contradiction.

These three cases prove the claim. Therefore

$$\lambda_{k_j}(\alpha) = [3; 3_2, 2, 1, a_{k_j-5}, ..., a_1] + [0, 2, 1, ...] < [3; 3_2, 2, 1, \overline{1, 2}] + [0; 2, 1, \overline{1, 2}]$$
$$= \ell_G.$$

This contradiction proves the Lemma.   □

A direct consequence of the previous lemma is the following result (originally proved in [Gay (2017)]):

Theorem 2.1 (Gayfulin). *Any irrational number α with $\ell(\alpha) = \ell_G$ is not attainable.*

Proof. By Lemma 2.4,

$$\left| \alpha - \frac{p_n}{q_n} \right| = \frac{1}{\lambda_{n+1}(\alpha)q_n^2} > \frac{1}{\ell_G q_n^2} = \frac{1}{\ell(\alpha)q_n^2} \qquad (2.5)$$

for every n sufficiently large.   □

Similarly to the definition of attainable irrational numbers, we have the following notion:

Definition 2.2. We say that $\ell \in L$ is *admissible* if $\ell = \ell(\alpha)$ for some attainable number α.

In this language, an immediate corollary of Theorem 2.1 is the fact that ℓ_G is a counterexample to Malyshev's claim:

Corollary 2.1 (Gayfulin). *The number $\ell_G = [3; 3, 3, 2, 1, \overline{1, 2}] + [0; 2, 1, \overline{1, 2}] \in L$ is not admissible.*

2.2 Left endpoints of maximal gaps of the Lagrange spectrum

The Lagrange spectrum is a closed subset of the real line $\mathbb{R}$ (see Exercise 1.2 in Chapter 1). Hence, its complement $\mathbb{R} \backslash L$ is a countable union of intervals. A *maximal gap* of L is a connected component of $\mathbb{R} \setminus L$.

It is known that ℓ_G is the left endpoint of a gap in the Markov spectrum M (see [CF (1989)], Table 2, p. 62). As $L \subset M$, Lemma 2.1 implies that $\ell_G \in L$ is the left endpoint of a maximal gap in the Lagrange spectrum.

The next theorem shows that each counterexample to Malyshev's claim is also the left endpoint of some gap in the Lagrange spectrum.

Remark 2.1. In what follows, we will use the fact (shown in Theorem 2 of Chapter 3 of Cusick–Flahive book [CF (1989)] for instance) that the Lagrange spectrum is related to the periodic points of the shift map $\sigma :$ $\Sigma \to \Sigma$, $\sigma((a_n)_n) = (a_{n+1})_n$, on the shift space $\Sigma = \{1, 2, 3, ...\}^{\mathbb{Z}}$ via

$$L = \overline{\{\ell(\theta); \theta \in \mathrm{Per}(\sigma)\}}$$

where $\mathrm{Per}(\sigma) = \{\theta \in \Sigma; \exists\, n > 0 \text{ with } \sigma^n(\theta) = \theta\}$ and $\ell(\theta) = \limsup_{n\to\infty} f(\sigma^n(\theta))$ with $f : \Sigma \to \mathbb{R}$ given by $f((a_n)_n) = [a_0; a_1, a_{-2}, ...] + [0; a_{-1}, a_{-2}, ...]$.

Remark 2.2. Recall from Proposition 1.5 that if $\alpha = [a_0; \underline{a}, b_1, b_2, ...]$ and $\beta = [a_0; \underline{a}, c_1, c_2, ...]$ with $\underline{a} = (a_1, ..., a_n)$ and $b_1 \neq c_1$, then

$$\alpha - \beta = \frac{(-1)^{n+1}([b_1; b_2, ...] - [c_1; c_2, ...])}{q(\underline{a})^2([b_1; b_2, ...] + [0; \underline{a}^T])([c_1; c_2, ...] + [0; \underline{a}^T])}.$$

In particular, $|\alpha - \beta| < \epsilon_n := 2^{-(n-1)}$ and, moreover, if $(a_1, ..., a_n, b_1, b_2, ...)$ and $(a_1, ..., a_n, c_1, c_2, ...)$ are sequences in $\{1, 2, 3, 4\}^{\mathbb{N}}$, then $|\alpha - \beta| > \delta_n := 5^{-(2n+4)}$.

Theorem 2.2. *If $\lambda \in L$ is not the left endpoint of a maximal gap of the Lagrange spectrum, then there exists an attainable α such that $\ell(\alpha) = \lambda$, i.e., λ is admissible.*

Proof. Firstly, we suppose that $\lambda = \ell(\alpha)$, where $\alpha = [0; \overline{\underline{c}}]$ and $\underline{c} = (c_1, c_2, ..., c_n)$. By definition of $\ell(\alpha)$ and Dirichlet's pigeonhole principle, there is $1 \leq j \leq n$ such that

$$\lambda = \lim_{m\to\infty} \lambda_{j+mn}(\alpha) = [c_j; c_{j+1}, c_{j+2}, ..., c_n, \overline{\underline{c}}] + [0; c_{j-1}, c_{j-2}, ..., c_1, \overline{\underline{c}^t}].$$

Next, we fix a sequence $\underline{d} = (d_1, d_2, ..., d_k)$ such that $[0; c_{j-1}, c_{j-2}, ..., c_1, \underline{d}^t] > [0; c_{j-1}, c_{j-2}, ..., c_1, \overline{\underline{c}^t}]$. Observe that, by definition, $\ell(\alpha) = \ell([0; \underline{d}, \overline{\underline{c}}])$. On the other hand, for each $m \in \mathbb{N}$, one has

$$\lambda_{j+mn+k}([0; \underline{d}, \overline{\underline{c}}]) = [c_j; c_{j+1}, c_{j+2}, ..., c_n, \overline{\underline{c}}] + [0; c_{j-1}, c_{j-2}, ..., c_1, \underline{c}_m^t, \underline{d}^t].$$

This implies that

$$[0; c_{j-1}, c_{j-2}, ..., c_1, \underline{c}_{2m}^t, \underline{d}^t] > [0; c_{j-1}, c_{j-2}, ..., c_1, \overline{\underline{c}^t}]$$

for every $m \in \mathbb{N}$. In particular,

$$\lambda_{j+2mn+k}([0; \underline{d}, \overline{\underline{c}}]) > \lambda = \ell(\alpha) = \ell([0; \underline{d}, \overline{\underline{c}}])$$

for every $m \in \mathbb{N}$. Thus, $[0; \underline{d}, \overline{c}]$ is attainable (as it satisfies (2.2)) and, *a fortiori*, λ is admissible.

Secondly, we assume that $\lambda \neq \ell(\alpha)$ for all $\alpha = [0; \overline{a}]$ and $\underline{a} = (a_1, a_2, ..., a_k)$. Since λ is not the left endpoint of a maximal gap of L, we can apply Remark 2.1 in order to select $\alpha_n = [0; \overline{c^n}]$, $\underline{c}^n = (c_1^n, ..., c_{t_n}^n)$ such that $\ell(\alpha_n) \searrow \lambda$ as $n \to \infty$. Of course, we can (and do) assume that all entries of all strings $\underline{c}^n$ do not surpass $\lambda + 1$. By Remark 2.2, given $\epsilon > 0$, there exists $N = N(\epsilon)$ such that, for any finite or infinite sequences $\underline{d}, \underline{e}$, we have

$$[c_j^n; c_{j+1}^n, ..., c_{t_n}^n, (\underline{c}^n)_N, \underline{d}] + [0; c_{j-1}^n, ..., c_1^n, (\underline{c}^n)_N, \underline{e}] > \ell(\alpha_n) - \epsilon \quad (2.6)$$

for all $n \in \mathbb{N}$.

We affirm that the lengths t_n of the strings $\underline{c}^n$ go to infinity. Otherwise, there would exist $T' \in \mathbb{N}$ such that $t_n < T'$ for all $n \in \mathbb{N}$. In particular, by Dirichlet's pigeonhole principle, one would be able to select $1 \leq T < T'$ such that $t_n = T$ for infinitely many $n \in \mathbb{N}$. Since the entries of the strings $\underline{c}^n$ are natural numbers below $\lambda + 1$, we can choose $\underline{c}^{n_0} = \underline{a}$ such that $\underline{c}^n = \underline{a}$ for infinitely many $n \in \mathbb{N}$. In particular, one would have $\lambda = \ell([0; \overline{a}])$, a contradiction.

By combining the facts that $t_n \to \infty$ as $n \to \infty$, the entries of all strings $\underline{c}^n$ are bounded by $\lambda + 1$, and the compactness of the space $\{1, ..., \lceil \lambda + 1 \rceil\}^{\mathbb{N}}$, we deduce that there exists an infinite subsequence $\tilde{\underline{c}}^n$ of $\underline{c}^n$ such that the first n elements of the periods $\tilde{\underline{c}}^n, \tilde{\underline{c}}^{n+1}, ...$ coincide.

Let us now show that λ is also admissible in this case. For this sake, let $\alpha_n := [0; \overline{\tilde{\underline{c}}^n}]$ and define $\rho_n = \dfrac{\ell(\alpha_n) - \lambda}{3} > 0$. Next, we select $N(n) := N(\rho_n)$ so that (2.6) holds. It follows that the irrational number

$$\gamma' := [0; \tilde{\underline{c}}_{2N(1)+1}^1, \tilde{\underline{c}}_{2N(2)+1}^2, ..., \tilde{\underline{c}}_{2N(n)+1}^n, ...]$$

satisfies $\lambda_j(\gamma') > \lambda$ for infinitely many values of j and $\ell(\gamma') = \lambda$, that is, λ is admissible.

Since the two cases above cover all possibilities for $\lambda \in L$ which is not a left endpoint of a maximal gap of L, the proof of the theorem is complete. $\qquad\qquad\qquad\qquad\qquad\qquad\qquad\qquad\qquad\qquad\qquad\qquad\qquad\quad$ $\square$

The compactness argument used in the proof of the previous argument can be easily adapted to derive the following lemma:

Lemma 2.5. *Suppose that $\alpha = [0; a_1, a_2, ...]$ with $\sup\limits_{n \in \mathbb{N}} a_i < \infty$. Then, there exists $\theta \in (\mathbb{N}^*)^{\mathbb{Z}}$ strongly associated to α in the sense that $\ell(\alpha) = \lambda_0(\theta) =$*

$m(\theta)$ *and, for each* $i \in \mathbb{N}$*, the string* $(\theta_{-i}, \ldots, \theta_0, \ldots, \theta_i)$ *appears infinitely often in the sequence* $(a_1, a_2, \ldots)$.

Proof. Exercise. $\qquad\qquad\qquad\qquad\qquad\qquad\qquad\qquad\qquad\qquad\qquad\square$

Our next goal is to study admissible left endpoints of maximal gaps of the Lagrange spectrum. In this direction, we will need the next four lemmas.

Lemma 2.6. *Let* $\alpha = [0; c_1, c_2, \ldots, c_N, \ldots] = [0; \underline{c}]$ *and* $\alpha' = [0; c'_1, c'_2, \ldots, c'_N, \ldots] = [0; \underline{c}']$, *with* $\underline{c}, \underline{c}' \in \{1, 2, 3, 4\}^{\mathbb{N}}$. *If any string of length* $2n+1$ *which appears in* $\underline{c}'$ *appears infinitely many times in* $\underline{c}$, *then*

$$\ell(\alpha') \le \ell(\alpha) + \frac{1}{2^{n-2}} := \ell(\alpha) + 2\epsilon_n.$$

Proof. Consider an increasing sequence t_j such that $\ell(\alpha') = \lim_{j \to \infty} \lambda_{t_j}(\alpha')$. Since $\underline{c}' \in \{1, 2, 3, 4\}^{\mathbb{N}}$, by compactness there exists a subsequence $\{\tilde{t}_j\}_j$ of $\{t_j\}_j$ such that $c'_{\tilde{t}_{j_1}+i} = c'_{\tilde{t}_{j_2}+i}$ for every $j_1, j_2 \in \mathbb{N}$ and $-n \le i \le n$. We still have

$$\ell(\alpha') = \lim_{j \to \infty} \lambda_{\tilde{t}_j}(\alpha')$$
$$= \lim_{j \to \infty} [c'_{\tilde{t}_j}; c'_{\tilde{t}_j+1}, c'_{\tilde{t}_j+2}, \ldots, c'_{\tilde{t}_j+n}, \ldots] + [0; c'_{\tilde{t}_j-1}, \ldots, c'_{\tilde{t}_j-n}, \ldots, c'_1].$$

Note that, by definition, the sequence of strings

$$(c'_{\tilde{t}_j-n}, \ldots, c'_{\tilde{t}_j-1}, c'_{\tilde{t}_j}, c'_{\tilde{t}_j+1}, c'_{\tilde{t}_j+2}, \ldots, c'_{\tilde{t}_j+n})$$

is constant as j varies, say $\underline{d} = (c'_{\tilde{t}_j-n}, \ldots, c'_{\tilde{t}_j-1}, c'_{\tilde{t}_j}, c'_{\tilde{t}_j+1}, c'_{\tilde{t}_j+2}, \ldots, c'_{\tilde{t}_j+n})$ for every j. Observe that $|\underline{d}| = 2n+1$, so that, by hypothesis, $\underline{d}$ occurs infinitely many times in $\underline{c}$. In other terms, there is an increasing sequence r_j such that

$$c'_{\tilde{t}_j+i} = c_{r_j+i}, \ \forall j \in \mathbb{N} \text{ and } -n \le i \le n.$$

In particular, by Remark 2.2, we have

$$\limsup_{j \to \infty} |\lambda_{\tilde{t}_j}(\alpha') - \lambda_{r_j}(\alpha)| < 2\epsilon_n.$$

By noting that $\limsup_j \lambda_{r_j}(\alpha) \le \ell(\alpha)$, we conclude $\ell(\alpha') \le \ell(\alpha) + 2\epsilon_n$. $\quad\square$

Lemma 2.7. *Given* $n \in \mathbb{N}$, *let* $N = N(n) = (2n+2)(4^{2n+2}+1)$. *If* $b_1, b_2, \ldots, b_N$ *is an arbitrary integer sequence of length* N *such that* $1 \le b_i \le 4$, $i = 1, 2, \ldots, N$, *then there exist* $i \equiv j \pmod{2}$ *such that* $b_{i+k} = b_{j+i}$ *for all* $k = 0, 1, 2, \ldots, 2n+1$.

The importance of the fact $i \equiv j \pmod{2}$ is that the sequence $(b_i, b_{i+1}, ..., b_{j-1})$ has even length.

Proof. For $0 \leq s \leq 4^{2n+2}$ we define

$$\underline{b}^{s+1} = (b_{s \cdot (2n+2)+1}, b_{s \cdot (2n+2)+2}, ..., b_{(s+1) \cdot (2n+2)}).$$

We note that $\underline{b}^{s+1} \in \{1, 2, 3, 4\}^{2n+2}$. Since $\#\{1, 2, 3, 4\}^{2n+2} = 4^{2n+2}$ and we have $4^{2n+2} + 1$ sequences $\underline{b}^{s+1}$, Dirichlet's pigeonhole principle implies that there are s, t such that $\underline{b}^{s+1} = \underline{b}^{t+1}$. In particular, if we write $i = s(2n+2) + 1$ and $j = t(2n+2) + 1$, we have that $b_{i+k} = b_{j+k}$ for every $0 \leq k \leq 2n+1$ and $i \equiv j \pmod{2}$. $\qquad \square$

Lemma 2.8. *Consider $\underline{b}$ a non-empty sequence of even length. If $\underline{a}$ is any finite sequence and $\underline{c}$ is any infinite non-periodic sequence, then*

$$\min([0; \underline{a}, \underline{b}, \underline{b}, \underline{c}], [0; \underline{a}, \underline{c}]) < [0; \underline{a}, \underline{b}, \underline{c}] < \max([0; \underline{a}, \underline{b}, \underline{b}, \underline{c}], [0; \underline{a}, \underline{c}]). \quad (2.7)$$

Proof. Since $\underline{c}$ is a non-periodic infinite sequence, the continued fractions in (2.7) are not equal. There is no loss of generality in assuming that the sequence $\underline{a}$ is empty. Suppose that $[0; \underline{b}, \underline{c}] > [0; \underline{b}, \underline{b}, \underline{c}]$. In this case, since $\underline{b}$ is non-empty and has even length, we have $[0; \underline{b}, \underline{c}] < [0; \underline{c}]$ (cf. Remark 1.7). The case when $[0; \underline{b}, \underline{c}] < [0; \underline{b}, \underline{b}, \underline{c}]$ is treated in exactly the same way. $\qquad \square$

Lemma 2.9. *Let $\alpha = [0; b_1, b_2, ..., b_n, ...]$ be an irrational number associated to an infinite sequence $(b_1, b_2, ...)$ which is not periodic. Consider the strings $\underline{b}^N = (b_1, b_2, ..., b_N)$ and the integers $i < j$ from Lemma 2.7. Define two new sequences*

$$\underline{a}^N = (b_1, b_2, ..., b_{i-1}, b_j, b_{j+1}, ..., b_N)$$

and

$$\underline{c}^N = (b_1, b_2, ..., b_{i-1}, b_i, ..., b_{j-1}, b_i, b_{i+1}, ..., b_N).$$

If $\tilde{a} = [0; \underline{a}^N, b_{N+1}, ...]$ and $\alpha' = [0; \underline{c}^N, b_{N+1}, ...]$, then

$$\max\{\tilde{a}, \alpha'\} > \alpha.$$

Proof. This is an easy application of Lemma 2.8: the details are left as an exercise for the reader. $\qquad \square$

At this stage, we are ready to prove the following theorem originally established in [Gay2 (2019)]:

Theorem 2.3 (Gayfulin). *If (a, b) is a maximal gap in L and α such that $\ell(\alpha) = a$. If θ is strongly associated to α, then θ is an eventually periodic sequence in the sense that $\theta = \overline{\underline{a}} \, \underline{b} \, \overline{\underline{c}}$, where $\underline{a}, \underline{b}$ and $\underline{c}$ are finite strings.*

Proof. Suppose θ is not periodic to the right side, the other case is analogous. Since θ is strongly associated to α, we can take a sequence $k(t) \to \infty$ such that

$$(a_{k(t)-t}, ..., a_{k(t)}, ..., a_{k(t)+t}) = (b_{-t}, ..., b_0, ..., b_t).$$

Then, $\lim_{t\to\infty} \lambda_{k(t)}(\alpha) = \lambda_0(\theta) = \ell(\alpha) = a$. Without loss of generality, we can assume that $\lim_{t\to\infty}(k(t+1) - k(t)) = +\infty$. Take n even such that $2^{-(n-1)} < (b-a)/2$. We consider $N = N(n)$ and $i < j$ for the sequence $(b_1, b_2, ...)$ as in Lemma 2.7. Since θ is not periodic to the right side, $r = \min\{m \in \mathbb{N}^*; b_{i+m} \neq b_{j+m}\}$ is well defined. We define $\theta_N^1 = b^N$ and $\theta_N^2 = c^N$ as in Lemma 2.9 and $\alpha^1 = [0; \theta_N^1, b_{N+1}, ...]$ and $\alpha^2 = [0; \theta_N^2, b_{N+1}, ...]$. By Lemma 2.9 we have $\max\{\alpha^1, \alpha^2\} > \gamma = [0; b_1, b_2, ...]$. We may assume that $\alpha^1 > \gamma$ as the other case is analogous. In this situation, set $\tilde{\theta} = (..., b_{-n}, ..., b_{-1}, b_0, \theta_N^1, b_{N+1}, ...)$. Note that

$$a = \lambda_0(\theta) < \lambda_0(\tilde{\theta}) < a + 2^{-(n-1)} < b. \qquad (2.8)$$

Consider the corresponding continued fraction α' which is obtained from the continued fraction α by replacing every segment $(a_{k(t)}, ..., a_{k(t)+N})$ by the segment $(a_{k(t)}, \theta_N^1)$ for every $t \geq j + r$. One can easily see that α' and α satisfy the condition of Lemma 2.6 and hence $\ell(\alpha') \leq \ell(\alpha) + 2^{-(n-2)} < b$. Since (a, b) is a maximal gap in L, we have

$$\ell(\alpha') \leq \ell(\alpha) = a. \qquad (2.9)$$

On the other hand, it is not hard to verify that for all $i \in \mathbb{N}$, the string $(\tilde{\theta}_{-i}, ..., \tilde{\theta}_0, ..., \tilde{\theta}_i)$ appears infinitely often in α': we leave the proof of this fact as an exercise for the reader. In any event, this implies that $\ell(\alpha') \geq \lambda_0(\tilde{\theta})$, so that (2.8) means that

$$a = \lambda_0(\theta) < \lambda_0(\tilde{\theta}) \leq \ell(\alpha') < \ell(\alpha) = a,$$

a contradiction. This proves the desired result. $\qquad\square$

Remark 2.3. By putting together Lemma 2.5 and Theorem 2.3, we obtain that any left endpoint a of a maximal gap of the Lagrange spectrum has the form $a = \lambda_0(A) = m(A)$ where A is an eventually periodic sequence. This statement was first claimed by Dietz [Dietz (1985)] in 1985, but it was pointed out in Chapter 5 of Cusick–Flahive book [CF (1989)] that Dietz's arguments contained a mistake. Fortunately, Gayfulin was able to come out with the arguments presented above in order to give the first complete proof of Dietz's claim.

Let us now pursue the presentation of Gayfulin's works [Gay (2017)] and [Gay2 (2019)] in order to construct an infinite sequence of non-admissible left endpoints of maximal gaps of the Lagrange spectrum.

Theorem 2.4. *A left endpoint a of a maximal gap of the Lagrange spectrum is admissible if and only if there exists $\alpha = [0; \underline{a}]$, $\underline{a}$ finite string, such that $\ell(\alpha) = a$.*

Proof. The first part of the proof of Theorem 2.2 is the sufficiency part of this theorem. In order to establish the necessity part, let a be an admissible left endpoint of the Lagrange spectrum. Consider $\alpha = [a_0; a_1, ...]$ such that $\ell(\alpha) = a$. Suppose that α is attainable, but $(a_1, a_2, ...)$ is not of the form $\underline{c}\,\overline{\underline{d}}$ for some finite strings $\underline{c}$, $\underline{d}$. Let n_j be a growing sequence of indices such that

$$\lambda_{n_j}(\alpha) \geq \ell(\alpha) \tag{2.10}$$

and

$$\lim_{j \to \infty} \lambda_{n_j}(\alpha) = \ell(\alpha).$$

Consider a sequence $\theta = (..., b_{-n}, ..., b_{-1}, b_0, b_1, ..., b_n, ...)$ strongly associated with α having the following property

$$(b_{-i}, ..., b_{-1}, b_0, b_1, ..., b_i) = (a_{n_j - i}, ..., a_{n_j - 1}, a_{n_j}, a_{n_j + 1}, ..., a_{n_j + i})$$

for infinitely many j's. By Theorem 2.3, θ is eventually periodic, that is, there exist a positive integer m and two finite sequences $\underline{a}$ and $\underline{c}$ such that

$$\theta = (\overline{\underline{a}}, b_{-m}, ..., b_0, ..., b_m, \overline{\underline{c}}).$$

It follows from (2.10) either

$$[a_{n_j}; a_{n_j + 1}, ...] \geq [b_0; b_1, ..., b_m, \overline{\underline{c}}] \text{ or } [a_{n_j}; a_{n_j - 1}, ..., a_1] \geq [0; b_{-1}, ..., b_{-m}, \overline{\underline{a}^T}] \tag{2.11}$$

for infinitely many j's. Our assumption on α implies that $[a_{n_j}; a_{n_j + 1}, ...] \neq [b_0, b_1, ..., b_m, \overline{\underline{c}}]$. Moreover, since $[0; a_{n_j - 1}, ..., a_1] \in \mathbb{Q}$, we have $[0; a_{n_j - 1}, ..., a_1] \neq [0; b_{-1}, ..., b_{-m}, \overline{\underline{a}^T}]$. Suppose that

$$[a_{n_j}; a_{n_j + 1}, ...] > [b_0; b_1, ..., b_m, \overline{\underline{c}}]. \tag{2.12}$$

Denote by p the length of period $\underline{c}$. Denote by r_j the minimal positive number such that $a_{n_j + r_j} \neq b_{r_j}$. Without loss of generality, we can assume that:

1. $\lim_{j \to \infty} (n_{j+1} - (n_j + r_j)) = \infty$

2. $[a_{n_j}; a_{n_j+1}, ...] > [b_0; b_1, ..., b_m, \underline{\bar{c}}]$ for every $j \in \mathbb{N}$.

3. $[a_{n_j}; a_{n_j+1}, ..., a_{n_j+m}] = [b_0; b_1, ..., b_m]$ for every $j \in \mathbb{N}$.

4. The sequence $(a_{n_j-j}, ..., a_{n_j}, ..., a_{n_j+j})$ coincides with the sequence $(b_{-j}, ..., b_0, ..., b_j)$ for every $j \in \mathbb{N}$.

5. The period length p is even.

Denote by t_j the number of times that $\underline{c}$ appears in the sequence $(b_{m+1}, ..., b_{r_j-1})$, that is, $t_j = \left\lfloor \dfrac{r_j - m - 1}{p} \right\rfloor$ and $\lim_{j \to \infty} t_j = \infty$. Denote by $\alpha^{(n)}$ the continued fraction obtained from $\alpha = [a_0; a_1, ...]$ as follows: if $t_j > n$, then every pattern

$$a_{n_j}, a_{n_j+1}, ..., a_{n_j+m}, \underline{c}_{t_j}, ..., a_{n_j+r_j}$$

is replaced by the pattern

$$a_{n_j}, a_{n_j+1}, ..., a_{n_j+m}, \underline{c}_n, ..., a_{n_j+r_j}.$$

By (2.12) and Remark 2.2, one has

$$[a_{n_j}; a_{n_j+1}, ..., a_{n_j+m}, \underline{c}_n, ..., a_{n_j+r_j}] > [b_0; b_1, ..., b_m, \underline{c}] + \delta_{m+(n+1)p}. \quad (2.13)$$

Since $\lim_{j \to \infty}(n_{j+1} - (n_j + r_j)) = \infty$ and the sequence $(a_{n_j-j}, ..., a_{n_j}, ..., a_{n_j+j})$ coincides with the sequence $(b_{-j}, ..., b_0, ..., b_j)$ for every $j \in \mathbb{N}$, one can easily see that $\ell(\alpha^{(n)}) \geq \ell(\alpha) + \delta_{m+(n+1)p}$. Furthermore, $\lim_{n \to \infty} \ell(\alpha^{(n)}) = \ell(\alpha) = a$. Indeed, every pattern of length np which occurs in the sequence of partial quotients of α infinitely many times must also occur in the sequence of partial quotients of $\alpha^{(n)}$ infinitely many times. Similarly, every pattern of length np which occurs in the sequence of partial quotients of $\alpha^{(n)}$ infinitely many times must also occur in the sequence of partial quotients of α infinitely many times too. Then, by Lemma 2.6 we have

$$|\ell(\alpha) - \ell(\alpha^{(n)})| \leq 2\epsilon_{np} = 2^{-(np-2)} \to 0$$

as $n \to \infty$. We affirm that this is a contradiction. In fact, recall that $\ell(\alpha^{(n)}) > \ell(\alpha)$ for every $n \in \mathbb{N}$ thanks to (2.13), so that $\lim_{n \to \infty} \ell(\alpha^{(n)}) = \ell(\alpha) = a$ would imply that there exists a positive integer N with $a = \ell(\alpha) < \ell(\alpha^{(n)}) < b$ for any $n > N$, and, *a fortiori*, (a, b) would not be a gap of the Lagrange spectrum.

Finally, if the inequality (2.12) does not hold infinitely often, then the reverse inequality

$$[0; a_{n_j-1}, ..., a_1] > [0; b_{-1}, ..., b_{-m}, \overline{\underline{a}^T}]$$

holds infinitely often and, thus, we can proceed exactly in the same way as above. $\square$

The previous theorem is equivalent to the following statement:

Corollary 2.2. *A left endpoint a of a gap in the Lagrange spectrum is not admissible if and only if there is no periodic sequence θ such that $a = m(\theta)$.*

This corollary helps in checking the non-admissibility of certain numbers in the Lagrange spectrum. More concretely, let

$$\alpha_n^* = [2; 1_{2n-2}, \overline{2_2, 1, 2}] + [0; 1_{2n-1}, 2, 1_{2n-2}, \overline{2_2, 1, 2}] \text{ and } \beta_n = 2 + 2[0; \overline{1_{2n}, 2}].$$

Gbur in [G (1976)] showed that (α_n^*, β_n) is a maximal gap in the Markov spectrum and Gayfulin showed in [Gay2 (2019)] that (α_n^*, β_n) is in fact a maximal gap in the Lagrange spectrum such that α_n^* is not admissible whenever $n \geq 2$. Before proving this, we shall show that $\beta_n, \alpha_n^* \in L$. Since $\beta_n = \ell([0; \overline{1_{2n}, 2}])$, we have $\beta_n \in L$. Given n, define $\underline{a}_k = ((212_2)_k, 1_{2n-2}, 2, 1_{2n-1}, 2, 1_{2n-2}, (2_2, 1, 2)_k)$. We leave as an exercise for the reader to show that

$$\ell([0; \overline{\underline{a}_k}]) = 2 + [0; 1_{2n-2}, (2_2 12)_k, ...] + [0; 1_{2n-1}, 2, 1_{2n-2}, (2_2 12)_k, ...].$$

In particular, $\lim_{k \to \infty} \ell([0; \overline{\underline{a}_k}]) = \alpha_n^*$. Since L is closed, we get $\alpha_n^* \in L$.

Next, we affirm that $\alpha_n^* < \beta_n$ for every $n \geq 1$. Indeed, if $u, v, w \in C(2) = \{[0; a_1, a_2, a_3, ...], a_j \in \{1, 2\}, \forall j \geq 1\}$, then $[2; 1_{2n-2}, 2 + u] + [0; 1_{2n-1}, 2 + v] \leq [2; 1_{2n-2}, 2, \overline{2, 1}] + [0; 1_{2n-2}, \overline{1, 2}] \leq 2[0; 1_{2n-2}, 1, \overline{1, 2}] \leq 2[0; 1_{2n}, 2 + w]$ (we leave the verification of these inequalities for the reader).

Clearly we have $\beta_n < \alpha_{n+1}^*$ for all $n \geq 1$, so, for every $n \geq 1$ we have $\alpha_n^* < \beta_n < \alpha_{n+1}^* < \beta_{n+1} < 2 + 2[0; \overline{1}] = 1 + \sqrt{5}$.

Lemma 2.10. *Consider a doubly infinite sequence $\theta = (b_i)_{i \in \mathbb{Z}}$ such that $m(\theta) < 1 + \sqrt{5}$. Then, $\theta \in \{1, 2\}^{\mathbb{Z}}$ and $(b_{i-1}, b_i, b_{i+1}, b_{i+2}) \notin \{(2, 1, 2, 1), (1, 2, 1, 2)\}$ for all $i \in \mathbb{Z}$.*

Proof. Exercise. ☐

Lemma 2.11. *Let $w = [0; a_1, a_2, ...]$ be a continued fraction with elements equal to 1 or 2. Suppose that the sequence $(a_1, a_2, ...)$ does not contain the pattern $(2, 1, 2, 1)$. Then $w \geq w_0 := [0; \overline{2, 1, 2_2}]$.*

Proof. We write $w_0 = [0; a_1', a_2', ...]$ and we consider $r = \min\{j \in \mathbb{N}; a_j \neq a_j'\}$. Suppose that $w < w_0$. Then $a_j = a_j'$ for $1 \leq j \leq 3$, and so $r > 3$. Note that $a_i' = 1$ iff $i \equiv 2 \pmod 4$. In particular, r cannot be odd because in this case, since $w < w_0$, we should have $a_r' = 1$. Then, r must be even. Since $w < w_0$ we must have $a_r' = 2$ and $a_r = 1$, so $r \equiv 0 \pmod 4$. Since r is even we have $a_{r-1}' = 2 = a_{r-1}$. Since $r - 2 \equiv 2 \pmod 4$,

$a'_{r-2} = 1 = a_{r-2}$ and finally $a_{r-3} = 2 = a'_{r-3}$. But this implies that $(a_{r-3}, a_{r-2}, a_{r-1}, a_r) = (2, 1, 2, 1)$, a contradiction. $\qquad\square$

The following lemma corresponds to property $P2$ of [G (1976)]:

Lemma 2.12. *Suppose that $m(A) = \lambda_0(A)$, where $A = (a_n)_{n \in \mathbb{Z}}$ is a sequence such that, for a certain positive integer m, $a_k = a_{m-k}$ for $0 \le k \le m$. Then, $[0; a_{m+1}, a_{m+2}, \dots] \le [0; a_{-1}, a_{-2}, \dots]$.*

Proof. Let $t = [0; a_{m+1}, a_{m+2}, \dots]$ and $y = [0; a_{-1}, a_{-2}, \dots]$, and let $f : [0, 1] \to [0, 1]$ given by $f(w) = [0; a_1, a_2, \dots, a_m + w]$. Then f is a contraction and $a_0 + y + f(t) = \lambda_0(A) = m(A) \ge \lambda_m(A) = a_m + t + f(y) = a_0 + t + f(y)$, so $y - f(y) \ge t - f(t)$, and thus $y \ge t$, since $h(w) = w - f(w)$ is an increasing function. $\qquad\square$

The result below is a version of Theorem 4, (i) of [G (1976)].

Theorem 2.5. *Let $\theta = (b_i)_{i \in \mathbb{Z}}$ be a sequence such that $b_0 = 2$ and $m(\theta) = \lambda_0(\theta) < 1 + \sqrt{5}$. Let x, y be such that $\{x, y\} = \{[0; b_1, b_2, b_3, \dots], [0; b_{-1}, b_{-2}, b_{-3}, \dots]\}$ and $x \le y$. Then, for some $n \ge 0$, we have $x = [0; 1_{2n}, 2, \dots]$, $y = [0; 1_{2n}, \dots]$ and $\beta_n \le m(\theta) = 2 + x + y \le \alpha^*_{n+1}$. Moreover, $m(\theta) = \alpha^*_{n+1}$ if, and only if, $x = [0; 1_{2n}, \overline{2_2 12}]$ and $y = [0; 1_{2n+1}, 2, 1_{2n}, \overline{2_2 12}]$.*

Proof. Since $x \le y$, we have $x \le \frac{x+y}{2} < \frac{\sqrt{5}-1}{2} = [0; \overline{1}]$, and thus the continued fraction of x starts with $[0; 1_{2n}, 2]$ for some $n \ge 0$. If $y \ge \frac{\sqrt{5}-1}{2}$, then the continued fraction of y should begin with $[0; 1_{2n+1}]$. Indeed, otherwise, $n \ge 1$ and $y \ge [0; 1_{2n-1}, 2, \overline{2, 1}]$ and $x + y \ge [0; 1_{2n}, \overline{2, 1}] + [0; 1_{2n-1}, 2, \overline{2, 1}] > 2 \cdot [0; \overline{1}] = \sqrt{5} - 1$ (the verification of these inequalities is left to the reader), a contradiction. On the other hand, if $y < \frac{\sqrt{5}-1}{2}$, then, since $y \ge x$, the continued fraction of y should start with $[0; 1_{2k}, 2]$ for some $k \ge n$.

We have $x = [0; 1_{2n}, 2 + t]$, for some $t \in [0, 1]$ which satisfies $t \le y$, by Lemma 2.12. So, if $f(w) = [0; 1_{2n}, 2 + w]$ then f is a decreasing contraction, and $y \ge x = f(t) \ge f(y)$, thus $y \ge x_0$, where $x_0 = [0; \overline{1_{2n}, 2}]$ is the fixed point of f, and we have $m(\theta) = 2 + y + x = 2 + y + f(t) \ge 2 + y + f(y) \ge 2 + x_0 + f(x_0) = 2 + 2x_0 = \beta_n$. On the other hand, since x starts with $[0; 1_{2n}, 2]$ and the strings 1212 and 2121 are forbidden, $x \le [0; 1_{2n}, \overline{2_2 12}]$, and the maximum value of y is of the type $y = [0; 1_{2n+1}, 2 + s]$, where, by Lemma 2.12, $s \le x \le [0; 1_{2n}, \overline{2_2 12}]$, so $y \le [0; 1_{2n+1}, 2, 1_{2n}, \overline{2_2 12}]$, and thus $m(\theta) = 2 + y + x \le 2 + [0; 1_{2n}, \overline{2_2 12}] + [0; 1_{2n+1}, 2, 1_{2n}, \overline{2_2 12}] = \alpha^*_{n+1}$. Notice that the equality holds only if $x = [0; 1_{2n}, \overline{2_2 12}]$ and $y =$

$[0; 1_{2n+1}, 2, 1_{2n}, \overline{2_2 12}]$. Since the intervals $[\beta_n, \alpha_{n+1}^*]$ are disjoint, we are done. $\qquad\square$

Corollary 2.3. *For each $n \geq 1$, (α_n^*, β_n) is a maximal gap of the Lagrange Spectrum.*

Theorem 2.6. *For all $n > 1$, α_n^* is not admissible.*

Proof. Suppose that α_{n+1}^* is admissible for some $n \geq 1$. Take α attainable such that $k(\alpha) = \alpha_{n+1}^*$. By Corollary 2.2 there exists $\theta = (b_i)_{i \in \mathbb{Z}}$ purely periodic strongly associated to α, and so $b_0 + [0; b_1.b_2, \dots] + [0; b_{-1}, b_{-2}, \dots] = \alpha_{n+1}^*$. We should then have $b_0 = 2$ and it follows from Theorem 2.5 that θ is equal to $(\overline{2, 1, 2_2}, 1_{2n}, 2, 1_{2n+1}, 2, 1_{2n}, \overline{2_2, 1, 2})$ or its transpose, which are not purely periodic, a contradiction. This concludes the proof of the Theorem. $\qquad\square$

2.3   The complement of the Lagrange spectrum in the Markov spectrum I

In 1968, Freiman found that the inclusion $L \subset M$ between the classical spectra is strict: more precisely, he proved in [Fr68 (1968)] that
$$\sigma := [\overline{2_4, 1_2, 2_2, 1}] + [0; 1, 2_2, 1_2, 2_4, 1, 2_2, 1_2, 2_2, 1_2, \overline{2_2, 1_2, 2_2, 1, 2_2}]$$
which is equals to 3.1181201781599... $\in M \setminus L$. and, in the same article, he showed that σ is actually an element of a *countable* infinite collection $\mathcal{F}$ of isolated points of M which are not in L.

Subsequently, Berstein showed in 1973 (cf. Theorem 1 at page 1 of [Bel (1973)]) that the collection $\mathcal{F}$ is contained in a maximal gap (c_∞, C_∞) of the Lagrange spectrum,[1] but the structure of this portion of $M \setminus L$ remained elusive until Matheus–Moreira [MM1 (2019)] proved in 2019 that this portion of $(M \setminus L) \cap (c_\infty, C_\infty)$ has a rich *fractal* structure:

Theorem 2.7. *The Hausdorff dimension of $(M \setminus L) \cap (c_\infty, C_\infty)$ satisfies:*
$$0.2628 < HD((M \setminus L) \cap (c_\infty, C_\infty)).$$

Remark 2.4. Actually, our proof of Theorem 2.7 allows us to give *complete* description of all elements of $(M \setminus L) \cap (c_\infty, C_\infty)$: see Remark 2.13 below. In particular, this provides some simplifications and complements to the original article [MM1 (2019)].

---

[1]Unfortunately, Berstein's original formulas for c_∞ and C_∞ contained mistakes. The correct expressions for c_∞ and C_∞ were derived in [MM1 (2019)] and they are presented in the sequel.

In a nutshell, the proof of Theorem 2.7 is based on a comparison between $(M \setminus L) \cap (c_\infty, C_\infty)$ and a Cantor set of real numbers whose continued fraction expansions satisfy some constraints.

More precisely, let Y be the Cantor set

$$Y := \{[0; \gamma] : \gamma \in \{1, 2\}^{\mathbb{N}} \text{ not containing the subwords in } P\} \qquad (2.14)$$

where $P = \{121, 212_31, 12_312, 2_312_3, 2_512_21_2, 1_22_212_5, 21_22_412_21_22_3, 2_31_22_212_41_22\}$.

Also, let

$$c_\infty := \overline{[2_4, 1_2, 2_2, 1]} + [0; \overline{1, 2_2, 1_2, 2_4}] = 3.11812017814369\ldots$$

and

$$\begin{aligned} C_\infty := &[2; 1, 2_2, 1_2, 2_4, 1, 2_2, 1_2, 2_4, 1, 2_2, 1_2, 2_2, 1, 2_4, 1_2, \overline{2_3, 1_3}] \\ &+ [0; 2_3, 1_2, 2_2, 1, 2_4, 1, \overline{1_3, 2_3}] \\ = &\, 3.118120178328746016\ldots \end{aligned}$$

Our general plan is to reduce Theorem 2.7 to the following result:

Theorem 2.8. $HD((M \setminus L) \cap (c_\infty, C_\infty)) = HD(Y)$ *(where Y is the Cantor set in (2.14)).*

Remark 2.5. Since $C_\infty < \sqrt{12}$, it follows from Proposition 1.7 that we can restrict our attention to the Markov values of sequences in $\{1, 2\}^{\mathbb{Z}}$ during our discussion of the proof of Theorem 2.7.

Before entering into the detailed proofs of these results, let us first review the notions of Hausdorff dimension and (dynamical) Cantor sets.

2.3.0.1 *Hausdorff dimension*

The *s-Hausdorff measure* $m_s(X)$ of a subset $X \subset \mathbb{R}^n$ is

$$m_s(X) := \lim_{\delta \to 0} \inf_{\substack{\bigcup_{i \in \mathbb{N}} U_i \supset X, \\ \text{diam}(U_i) \leq \delta \ \forall i \in \mathbb{N}}} \sum_{i \in \mathbb{N}} \text{diam}(U_i)^s.$$

The *Hausdorff dimension* of X is

$$HD(X) := \sup\{s \in \mathbb{R} : m_s(X) = \infty\} = \inf\{s \in \mathbb{R} : m_s(X) = 0\}.$$

Remark 2.6. There are many notions of dimension in the literature: for example, the *box-counting dimension* $d(X)$ of X is

$$d(X) = \lim_{\delta \to 0} \frac{\log N_X(\delta)}{\log(1/\delta)}$$

where $N_X(\delta)$ is the smallest number of boxes of side lengths $\leq \delta$ needed to cover X. As an exercise, the reader is invited to show that the Hausdorff dimension is always smaller than or equal to the box-counting dimension.

The following exercise (whose solution can be found in Falconer's book [F (1986)]) describes several elementary properties of the Hausdorff dimension:

Exercise 2.1

Show that:

(a) if $X \subset Y$, then $HD(X) \leq HD(Y)$;
(b) $HD(\bigcup_{i \in \mathbb{N}} X_i) = \sup_{i \in \mathbb{N}} HD(X_i)$; in particular, $HD(X) = 0$ whenever X is a countable set (such as $X = \{p\}$ or $X = \mathbb{Q}^n$);
(c) if $f : X \to Y$ is α-Hölder continuous,[2] then $\alpha \cdot HD(f(X)) \leq HD(X)$;
(d) $HD(\mathbb{R}^n) = n$ and, more generally, $HD(X) = m$ when $X \subset \mathbb{R}^n$ is a smooth m-dimensional submanifold.

Example 2.1. Cantor's middle-third set $C = \left\{ \sum_{i=1}^{\infty} \frac{a_i}{3^i} : a_i \in \{0, 2\} \forall i \in \mathbb{N} \right\}$ has Hausdorff dimension $\frac{\log 2}{\log 3} \in (0, 1)$: see Falconer's book [F (1986)] for more details.

2.3.1 *Dynamical Cantor sets*

A *dynamically defined Cantor set* $K \subset \mathbb{R}$ is

$$K = \bigcap_{n \in \mathbb{N}} \psi^{-n}(I_1 \cup \cdots \cup I_k)$$

where $I_1, \ldots, I_k$ are pairwise disjoint compact intervals, and $\psi : I_1 \cup \cdots \cup I_k \to I$ is a C^r-map from $I_1 \cup \cdots \cup I_k$ to its convex hull I such that:

(1) ψ is *uniformly expanding*: $|\psi'(x)| > 1$ for all $x \in I_1 \cup \cdots \cup I_k$;
(2) ψ is a (full) *Markov map*: $\psi(I_j) = I$ for all $1 \leq j \leq k$.

Remark 2.7. Dynamical Cantor sets are usually defined with a weaker Markov condition called *topological mixing*,[3] but we stick for now to this definition. (See [PT (1993)] for more details.)

---

[2]I.e., for some constant $C > 0$, one has $|f(x) - f(x')| \leq C|x - x'|^\alpha$ for all $x, x' \in X$.
[3]This means that $\psi^n(K \cap I_j) = K$ for all n sufficiently large and for all $1 \leq j \leq k$.

Example 2.2. Cantor's middle-third set $C = \left\{ \sum\limits_{i=1}^{\infty} \frac{a_i}{3^i} : a_i \in \{0,2\} \forall i \in \mathbb{N} \right\}$ is

$$C = \bigcap_{n \in \mathbb{N}} \psi^{-n}([0,1/3] \cup [2/3,1])$$

where $\psi : [0,1/3] \cup [2/3,1] \to [0,1]$ is given by

$$\psi(x) = \begin{cases} 3x, & \text{if } 0 \le x \le 1/3 \\ 3x - 2, & \text{if } 2/3 \le x \le 1 \end{cases}.$$

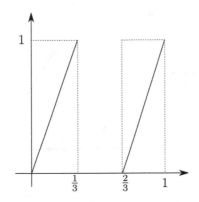

Dynamical construction of the standard middle-third Cantor set

Remark 2.8. A dynamical Cantor set is called *affine* when $\psi|_{I_j}$ is affine for all j. In this language, Cantor's middle-third set is an *affine dynamical Cantor set*.

Example 2.3. Given $A \ge 2$, let $C(A) := \{[0; a_1, a_2, \dots] : 1 \le a_i \le A \; \forall i \in \mathbb{N}\}$. This is a dynamical Cantor set associated to Gauss map: for example,

$$C(2) = \bigcap_{n \in \mathbb{N}} G^{-n}(I_1 \cup I_2)$$

where $I_1 = [\frac{\sqrt{3}-1}{2}, \frac{3-\sqrt{3}}{3}]$ and $I_2 = [\frac{\sqrt{3}}{3}, \sqrt{3} - 1]$ are the intervals depicted below.

Remark 2.9. Hensley [He (1992)] showed that

$$HD(C(A)) = 1 - \frac{6}{\pi^2 A} - \frac{72 \log A}{\pi^4 A^2} + O\left(\frac{1}{A^2}\right) = 1 - \frac{1 + o(1)}{\zeta(2)A}$$

and Jenkinson-Pollicott [JePo1 (2001)], [JePo2 (2018)] used thermodynamical formalism methods to obtain that

$$HD(C(2)) = 0.531280506277205141624468647368471785493059109018 39 \dots,$$
$$HD(C(3)) \simeq 0.705 \dots, \quad HD(C(4)) \simeq 0.788 \dots$$

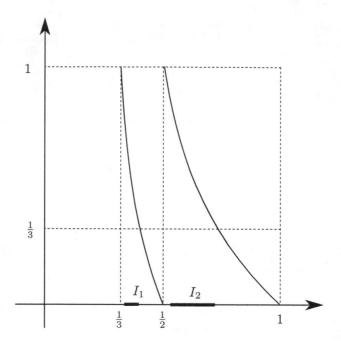

$$C(2) = \cap_{n \in \mathbb{N}} G^{-n}(I_1 \cup I_2)$$

2.3.1.1   *Gauss-Cantor sets*

The set $C(A)$ above is a particular case of *Gauss-Cantor set*:

Definition 2.3. Given $B = \{\beta_1, \ldots, \beta_l\}$, $l \geq 2$, a finite, primitive[4] alphabet of finite words $\beta_j \in (\mathbb{N}^*)^{r_j}$, the Gauss-Cantor set $K(B) \subset [0,1]$ associated to B is

$$K(B) := \{[0; \gamma_1, \gamma_2, \ldots] : \gamma_i \in B \ \forall i\}.$$

Example 2.4. $C(A) = K(\{1, \ldots, A\})$.

Exercise 2.2

Show that any Gauss-Cantor set $K(B)$ is dynamically defined.[5]

From the *symbolic* point of view, $B = \{\beta_1, \ldots, \beta_l\}$ as above induces a

---

[4]I.e., β_i doesn't begin by β_j for all $i \neq j$.
[5]Hint: For each word $\beta_j \in (\mathbb{N}^*)^{r_j}$, let $I(\beta_j) = I_j =$ the convex hull of $\{[0; \beta_j, a_1, \ldots] : a_i \in B \ \forall i\}$ and $\psi|_{I_j} := G^{r_j}$ where $G(x) = \{1/x\}$ is the Gauss map.

subshift

$$\Sigma(B) = \{(\gamma_i)_{i\in\mathbb{Z}} : \gamma_i \in B \; \forall i\} \subset \Sigma = (\mathbb{N}^*)^{\mathbb{Z}} = \Sigma^- \times \Sigma^+ := (\mathbb{N}^*)^{\mathbb{Z}^-} \times (\mathbb{N}^*)^{\mathbb{N}}$$

and the corresponding G-invariant version of $K(B)$ is $K^+(B) = \{[0,\gamma] : \gamma \in \Sigma^+(B)\}$, where $\Sigma^+(B) = \pi^+(\Sigma(B))$ and $\pi^+ : \Sigma \to \Sigma^+$ is the natural projection.[6]

2.3.2   Soft bounds on the Hausdorff dimension of dynamical Cantor sets

By definition, a Gauss-Cantor set K can be inductively constructed as follows. We start with the Markov partition $\mathcal{R}_1 = \{I_1, \ldots, I_r\}$ given by the connected components of the domain of the expanding map Ψ defining K. For each $n \geq 2$, we define $\mathcal{R}_n$ as the collection of connected components of $\Psi^{-1}(J)$, $J \in \mathcal{R}_{n-1}$. In this way, $K = \bigcap_{n\in\mathbb{N}} \bigcup_{R\in\mathcal{R}_n} R$.

As it was remarked in pages 68 to 70 of Palis–Takens book [PT (1993)], this description of K allows to derive some bounds on its Hausdorff dimension:

Proposition 2.1. *For all $n \in \mathbb{N}$, one has $\alpha_n \leq HD(K) \leq \beta_n$ where*

$$\sum_{R\in\mathcal{R}_n} \left(\frac{1}{\sup|(\Psi^n)'|_R|}\right)^{\alpha_n} = 1$$

and

$$\sum_{R\in\mathcal{R}_n} \left(\frac{1}{\inf|(\Psi^n)'|_R|}\right)^{\beta_n} = 1.$$

The proof of this proposition is presented in Appendix E below.

2.3.3   Forbidden strings for Markov values in (c_∞, C_∞)

After these preliminaries on Hausdorff dimension and Gauss-Cantor sets, we are ready to start the proof of Theorem 2.7. For this sake, our first step is to identify a list of finite strings such that the Markov value of any sequence in $\{1,2\}^{\mathbb{Z}}$ containing those strings must surpass C_∞.

Lemma 2.13. *If $B \in \{1,2\}^{\mathbb{Z}}$ contains any of the strings*

*(1) 12*1*

---

[6]This projection is related to local unstable manifolds of the left shift map on Σ.

(2) $2_2 12^* 2_2 1$

(3) $2_3 12^* 2_2$

(4) $2_4 2^* 12_2 1_2$

(5) $2_3 1_2 2_3 2^* 12_2 1_2 2_3$

(6) $2_2 1_2 2_3 2^* 12_2 1_2 2_3 1$

(7) $1_2 2_2 1_2 2_3 2^* 12_2 1_2 2_3$

(8) $2_2 1_2 2_1 2_2 2_3 2^* 12_2 1_2 2_5$

(9) $12_2 1_2 2_1 2_2 1_2 2_3 2^* 12_2 1_2 2_4 1$

(10) $2_4 12_2 1_2 2_3 2^* 12_2 1_2 2_4 1_2$

(11) $2_4 12_2 1_2 2_3 2^* 12_2 1_2 2_4 12_2 12$

(12) $12_4 12_2 1_2 2_3 2^* 12_2 1_2 2_4 12_2 1_3$

(13) $212_4 12_2 1_2 2_3 2^* 12_2 1_2 2_4 12_2 1_2 2_2$

then $\lambda_j(B) > 3.1181201786$ or $\lambda_{j+11}(B) > 3.15$ where j is the position in asterisk.

Proof. If B contains (1), then $\lambda_j(B) \geq [2; 1, \overline{1,2}] + [0; 1, \overline{1,2}] > 3.15$. If B contains (2), then $\lambda_j(B) \geq [2; 2_2, 1, \overline{1,2}] + [0; 1, 2_2, \overline{2,1}] > 3.12$.

If B contains (3), then $\lambda_j(B) \geq [2; 2_2, \overline{2,1}] + [0; 1, 2_3, \overline{2,1}] > 3.119$. If B contains (4), then $\lambda_j(B) \geq [2; 1, 2_2, 1_2, \overline{1,2}] + [0; 2_4, \overline{2,1}] > 3.1182$.

If B contains (5), then $\lambda_j(B) \geq [2; 1, 2_2, 1_2, 2_3, \overline{2,1}] + [0; 2_3, 1_2, 2_3, \overline{2,1}] > 3.118125$. If B contains (6), then $\lambda_j(B) \geq [2; 1, 2_2, 1_2, 2_3, 1, \overline{1,2}] + [0; 2_3, 1_2, 2_2, \overline{1,2}] > 3.118121$.

If B contains (7), then $\lambda_j(B) \geq [2; 1, 2_2, 1_2, 2_3, \overline{2,1}] + [0; 2_3, 1_2, 2_2, 1_2, \overline{1,2}] > 3.118121$. If B contains (8), then $\lambda_j(B) \geq [2; 1, 2_2, 1_2, 2_5, \overline{2,1}] + [0; 2_3, 1_2, 2_2, 1, 2_2, \overline{2,1}] > 3.1181206$.

If B contains (9) and $\lambda_{j+11}(B) \leq 3.15$, then the discussion of (1) above implies that $\lambda_j(B) \geq [2; 1, 2_2, 1_2, 2_4, 1, 2, 2, \overline{2,1}] + [0; 2_3, 1_2, 2_2, 1, 2_2, 1, \overline{1,2}] > 3.1181202$.

If B contains (10), then $\lambda_j(B) \geq [2; 1, 2_2, 1_2, 2_4, 1_2, \overline{1,2}] + [0; 2_3, 1_2, 2_2, 1, 2_4, \overline{2,1}] > 3.1181202$. If B contains (11), then $\lambda_j(B) \geq [2; 1, 2_2, 1_2, 2_4, 1, 2_2, 1, 2, \overline{2,1}] + [0; 2_3, 1_2, 2_2, 1, 2_4, \overline{2,1}] > 3.1181201787$.

If B contains (12), then $\lambda_j(B) \geq [2; 1, 2_2, 1_2, 2_4, 1, 2_2, 1_3, \overline{1,2}] + [0; 2_3, 1_2, 2_2, 1, 2_4, 1, \overline{1,2}] > 3.1181201786$. If B contains (13), then $\lambda_j(B) \geq [2; 1, 2_2, 1_2, 2_4, 1, 2_2, 1_2, 2_2, \overline{2,1}] + [0; 2_3, 1_2, 2_2, 1, 2_4, 1, 2, \overline{2,1}] > 3.1181201789$. $\square$

2.3.3.1 *Allowed strings for Markov values in* (c_∞, C_∞)

Our second step towards the proof of Theorem 2.7 is to identify a list of finite strings contributing a value below c_∞ to the calculation of Markov values of sequences in $\{1, 2\}^{\mathbb{Z}}$.

Lemma 2.14. *If* $B \in \{1, 2\}^{\mathbb{Z}}$ *contains any of the strings*

(15) $1_2 2^* 2$
(16) $2_2 1 2^* 2 1$
(17) $2_3 2^* 1 2_2 1 2$
(18) $1 2_3 2^* 1 2_2 1_3$
(19) $1 2_3 2^* 1 2_2 1_2 2_2 1$
(20) $2 1 2_3 2^* 1 2_2 1_2 2_3$
(21) $1_3 2_3 2^* 1 2_2 1_2 2_3$

then $\lambda_j(B) < 3.118117$ *or* $\lambda_{j-6}(B) > 3.15$ *where* j *is the position in asterisk.*

Proof. If B contains (15), $\lambda_j(B) \leq [2; 2, \overline{2, 1}] + [0; 1_2, \overline{1, 2}] < 3.05$. If B contains (16), $\lambda_j(B) \leq [2; 2, 1, \overline{1, 2}] + [0; 1, 2_2, \overline{2, 1}] < 3.09$. If B contains (17), $\lambda_j(B) \leq [2; 1, 2_2, 1, 2, \overline{2, 1}] + [0; 2_3, \overline{2, 1}] < 3.118$. If B contains (18), $\lambda_j(B) \leq [2; 1, 2_2, 1_3, \overline{1, 2}] + [0; 2_3, 1, \overline{1, 2}] < 3.118$.

If B contains (19) and $\lambda_{j-6}(B) \leq 3.15$, then Lemma 2.13 (1) implies that $\lambda_j(B) \leq [2; 1, 2_2, 1_2, 2_2, 1, \overline{1, 2}] + [0; 2_3, 1, 1, 2, 2, \overline{2, 1}] < 3.118117$.

If B contains (20), $\lambda_j(B) \leq [2; 1, 2_2, 1_2, 2_3, \overline{1, 2}] + [0; 2_3, 1, 2, \overline{2, 1}] < 3.118$. If B contains (21), $\lambda_j(B) \leq [2; 1, 2_2, 1_2, 2_3, \overline{1, 2}] + [0; 2_3, 1_3, \overline{1, 2}] < 3.11801$. $\square$

2.3.3.2 *Sequences in* $\{1, 2\}^{\mathbb{Z}}$ *with Markov values in* (c_∞, C_∞)

In what follows, we use Lemmas 2.13 and 2.14 to impose severe restrictions on sequences with Markov values in the interval (c_∞, C_∞).

Lemma 2.15. *Let* $B \in \{1, 2\}^{\mathbb{Z}}$ *such that* $3.118117 < \lambda_0(B)$ *and* $\lambda_n(B) < 3.1181201786$ *for* $n \in \{0, \pm 2, \pm 6, \pm 9, \pm 11, \pm 15\}$. *Then,* $B_{-14} \ldots B_{16}$ *or* $(B_{-16} \ldots B_{14})^T$ *equals to*

$$1_2 2_4 1 2_2 1 2_2 4 1 2_2 1 2_2 4 1 2_2 1 2_2 4 1 2_2 1 2_2.$$

Proof. After performing a transposition if necessary, we see that Lemma 2.13 (1), Lemma 2.14 (15) and our assumption on $\lambda_0(B)$ imply

$$B_{-1} B_0 B_1 B_2 = 2 2^* 1 2.$$

By Lemma 2.13 (1) and our assumption on $\lambda_{\pm 2}(B)$, we get

$$B_{-1}B_0B_1B_2B_3 = 22^*122.$$

Our hypothesis on $\lambda_0(B)$ allows to successively apply Lemma 2.14 (16), Lemma 2.13 (2), Lemma 2.13 (3), Lemma 2.14 (17), Lemma 2.13 (4), Lemma 2.14 (18) to deduce that

$$B_{-4}\ldots B_6 = 12_32^*12_21_22.$$

By Lemma 2.13 (1) and our assumption on $\lambda_{\pm 6}(B)$, we get

$$B_{-4}\ldots B_7 = 12_32^*12_21_22_2.$$

In view of our assumption on $\lambda_0(B)$, by successively applying Lemma 2.14 (19), (20), (21), we obtain that

$$B_{-6}\ldots B_8 = 21_22_32^*12_21_22_3.$$

By Lemma 2.13 (1) and our assumption on $\lambda_{\pm 6}(B)$, we get

$$B_{-7}\ldots B_8 = 2_21_22_32^*12_21_22_3.$$

In view of our assumption on $\lambda_0(B)$, by successively applying Lemma 2.13 (5), (6), (7), we derive that

$$B_{-9}\ldots B_9 = 212_21_22_32^*12_21_22_4.$$

By Lemma 2.13 (1) and our assumption on $\lambda_{\pm 9}(B)$, we obtain

$$B_{-10}\ldots B_9 = 2_212_21_22_32^*12_21_22_4.$$

In view of our assumption on $\lambda_0(B)$, by successively applying Lemma 2.13 (8), (9), we get

$$B_{-11}\ldots B_{10} = 2_312_21_22_32^*12_21_22_41.$$

In view of our assumption on $\lambda_{\pm 9}(B)$, by successively applying Lemma 2.13 (2), (4), we get

$$B_{-13}\ldots B_{10} = 12_412_21_22_32^*12_21_22_41.$$

By Lemma 2.13 (10) and our assumption on $\lambda_0(B)$, we obtain

$$B_{-13}\ldots B_{11} = 12_412_21_22_32^*12_21_22_412.$$

In view of our assumption on $\lambda_{\pm 11}(B)$, by successively applying Lemma 2.13 (1), (3), we get

$$B_{-13}\ldots B_{13} = 12_412_21_22_32^*12_21_22_412_21.$$

In view of our assumption on $\lambda_0(B)$, by successively applying Lemma 2.13 (11), (12), we deduce that

$$B_{-13}\ldots B_{15} = 1_2 4 12_2 1_2 2_3 2^* 12_2 1_2 2_4 12_2 1_2 2.$$

By Lemma 2.13 (1) and our assumption on $\lambda_{\pm 15}(B)$, we obtain

$$B_{-13}\ldots B_{16} = 1_2 4 12_2 1_2 2_3 2^* 12_2 1_2 2_4 12_2 1_2 2_2.$$

Finally, by Lemma 2.13 (13) and our assumption on $\lambda_0(B)$, we conclude that

$$B_{-14}\ldots B_{16} = 1_2 2_4 12_2 1_2 2_3 2^* 12_2 1_2 2_4 12_2 1_2 2_2.$$

$\square$

A careful inspection of the proof of the previous lemma reveals that we actually showed that:

Lemma 2.16. *Let $B \in \{1,2\}^{\mathbb{Z}}$ such that $B_{-6}\ldots B_8 = 21_2 2_4 12_2 1_2 2_3$ and $\lambda_n(B) < 3.1181201786$ for $n \in \{0, -6, -9, 11, 15\}$. Then, $B_{-14}\ldots B_{16}$ equals*

$$1_2 2_4 12_2 1_2 2_4 12_2 1_2 2_4 12_2 1_2 2_2.$$

In particular:

(1) either $B_{-15}\ldots B_{16} = 1_3 2_4 12_2 1_2 2_4 12_2 1_2 2_4 12_2 1_2 2_2$,

(2) or $B_{-15}\ldots B_{16} = 21_2 2_4 12_2 1_2 2_4 12_2 1_2 2_4 12_2 1_2 2_2$ and the vicinity of B_{-9} is $B_{-15}\ldots B_{-1} = 21_2 2_4 12_2 1_2 2_3$.

Using these facts, we can produce another forbidden string for sequences with Markov values in the interval (c_∞, C_∞):

Lemma 2.17. *Let $A \in \{1,2\}^{\mathbb{Z}}$ such that, for some $n \in \mathbb{Z}$ and $a \in \mathbb{N}$, one has $\lambda_k(A) < 3.1181201786$ for $k - n \in \{-21 - 6j, -19 - 6j : j = 0, \ldots, a\}$, $k = 9, 27, 33, 35, 37$, and $k - n \in \{39 + 6j, 41 + 6j, 43 + 6j : j = 0, \ldots, a\}$. If $A_{n-15}\ldots A_{n+16} = 1_3 2_4 12_2 1_2 2_4 12_2 1_2 2_4 12_2 1_2 2_2$, then*

$$\lambda_n(A) \geq \left[2; 1, 2_2, 1_2, 2_4, 1, 2_2, 1_2, 2_4, 1, 2_2, 1_2, 2_2, 1, 2_4, 1_2, \underbrace{2_3, 1_3, \ldots, 2_3, 1_3}_{a+2 \ times}, \ldots \right]$$

$$+ \left[0; 2_3, 1_2, 2_2, 1, 2_4, 1, \underbrace{1_3, 2_3, \ldots, 1_3, 2_3}_{a+1 \ times}, \ldots \right].$$

In particular, the subsequence $1_3 2_4 12_2 1_2 2_4 12_2 1_2 2_4 12_2 1_2 2_2$ or its transpose is not contained in a bi-infinite sequence $A \in \{1,2\}^{\mathbb{Z}}$ with $m(A) < C_\infty$.

Proof. If $A_{n-15}\ldots A_{n+16} = 1_3 2_4 12_2 1_2 2_4 12_2 1_2 2_4 12_2 1_2 2_2$, then

$$\lambda_n(A) = [0; 2_3, 1_2, 2_2, 1, 2_4, 1_3, \ldots] + [2; 1, 2_2, 1_2, 2_4, 1, 2_2, 1_2, 2_2, \ldots]$$
$$\geq [0; 2_3, 1_2, 2_2, 1, 2_4, 1_4, 2, \ldots] + [2; 1, 2_2, 1_2, 2_4, 1, 2_2, 1_2, 2_2, \ldots].$$

By Lemma 2.13 (1) and our assumption on $\lambda_{n-17}(A)$, we have

$$\lambda_n(A) \geq [0; 2_3, 1_2, 2_2, 1, 2_4, 1_4, 2, \ldots] + [2; 1, 2_2, 1_2, 2_4, 1, 2_2, 1_2, 2_3, \ldots]$$
$$\geq [0; 2_3, 1_2, 2_2, 1, 2_4, 1_4, 2_2, \ldots] + [2; 1, 2_2, 1_2, 2_4, 1, 2_2, 1_2, 2_3, \ldots]$$
$$\geq [0; 2_3, 1_2, 2_2, 1, 2_4, 1_4, 2_3, 1, \ldots] + [2; 1, 2_2, 1_2, 2_4, 1, 2_2, 1_2, 2_3, \ldots].$$

By Lemma 2.13 (1), (2) and our assumption on $\lambda_{n-21}(A)$, $\lambda_{n-19}(A)$, we get

$$\lambda_n(A) \geq [0; 2_3, 1_2, 2_2, 1, 2_4, 1_4, 2_3, 1, \ldots] + [2; 1, 2_2, 1_2, 2_4, 1, 2_2, 1_2, 2_3, \ldots]$$
$$\geq [0; 2_3, 1_2, 2_2, 1, 2_4, 1_4, 2_3, 1_2, \ldots] + [2; 1, 2_2, 1_2, 2_4, 1, 2_2, 1_2, 2_3, \ldots]$$
$$\geq [0; 2_3, 1_2, 2_2, 1, 2_4, 1_4, 2_3, 1_3, 2, \ldots] + [2; 1, 2_2, 1_2, 2_4, 1, 2_2, 1_2, 2_3, \ldots].$$

By Lemma 2.13 (1) and our assumption on $\lambda_{n-23}(A)$, we obtain

$$\lambda_n(A) \geq [0; 2_3, 1_2, 2_2, 1, 2_4, 1_4, 2_3, 1_3, 2, \ldots] + [2; 1, 2_2, 1_2, 2_4, 1, 2_2, 1_2, 2_3, \ldots]$$
$$\geq [0; 2_3, 1_2, 2_2, 1, 2_4, 1_4, 2_3, 1_3, 2_2, \ldots] + [2; 1, 2_2, 1_2, 2_4, 1, 2_2, 1_2, 2_3, \ldots]$$
$$\geq [0; 2_3, 1_2, 2_2, 1, 2_4, 1_4, 2_3, 1_3, 2_3, \ldots] + [2; 1, 2_2, 1_2, 2_4, 1, 2_2, 1_2, 2_3, \ldots].$$

By recursively applying the previous arguments at the positions $20+6j$, $1 \leq j \leq a$, we see that our assumptions on $\lambda_{n-19-6j}(A)$ and $\lambda_{n-21-6j}(A)$ for $1 \leq j \leq a$ together with Lemma 2.13 (1), (2) imply that

$$\lambda_n(A) \geq \left[0; 2_3, 1_2, 2_2, 1, 2_4, 1, \underbrace{1_3, 2_3, \ldots, 1_3, 2_3}_{a+1 \text{ times}}, \ldots\right]$$
$$+ [2; 1, 2_2, 1_2, 2_4, 1, 2_2, 1_2, 2_3, \ldots].$$

On the other hand, the fact that $A_n \ldots A_n = 21_2 2_4 12_2 1_2 2_3$ and our assumption on $\lambda_9(A)$ allow to apply Lemma 2.16 to get

$$\lambda_n(A) \geq \left[0; 2_3, 1_2, 2_2, 1, 2_4, 1, \underbrace{1_3, 2_3, \ldots, 1_3, 2_3}_{a+1 \text{ times}}, \ldots\right]$$
$$+ [2; 1, 2_2, 1_2, 2_4, 1, 2_2, 1_2, 2_4, 1, 2_2, 1_2, 2_2, \ldots].$$

By Lemma 2.13 (1), (2), (4) and our assumption on $\lambda_{27}(A)$, we have

$$\lambda_n(A) \geq \left[0; 2_3, 1_2, 2_2, 1, 2_4, 1, \underbrace{1_3, 2_3, \ldots, 1_3, 2_3}_{a+1 \text{ times}}, \ldots\right]$$
$$+ [2; 1, 2_2, 1_2, 2_4, 1, 2_2, 1_2, 2_4, 1, 2_2, 1_2, 2_2, 1, 2_4, 1_2, 2, \ldots].$$

By Lemma 2.13 (1) and our assumption on $\lambda_{33}(A)$, we have

$$\lambda_n(A) \geq \left[0; 2_3, 1_2, 2_2, 1, 2_4, 1, \underbrace{1_3, 2_3, \ldots, 1_3, 2_3}_{a+1 \text{ times}}, \ldots\right]$$
$$+[2; 1, 2_2, 1_2, 2_4, 1, 2_2, 1_2, 2_4, 1, 2_2, 1_2, 2_2, 1, 2_4, 1_2, 2_3, 1, \ldots].$$

By Lemma 2.13 (1) and our assumption on $\lambda_{37}(A)$, and by Lemma 2.13 (2) and our assumption on $\lambda_{35}(A)$, we have

$$\lambda_n(A) \geq \left[0; 2_3, 1_2, 2_2, 1, 2_4, 1, \underbrace{1_3, 2_3, \ldots, 1_3, 2_3}_{a+1 \text{ times}}, \ldots\right]$$
$$+[2; 1, 2_2, 1_2, 2_4, 1, 2_2, 1_2, 2_4, 1, 2_2, 1_2, 2_2, 1, 2_4, 1_2, 2_3, 1_3, \ldots].$$

By recursively applying Lemma 2.13 (1) at the positions $39 + 6j$, $39 + 4(j+1)$, $0 \leq j \leq a$, and Lemma 2.13 (2) at the positions $39 + 2(j+1)$, $0 \leq j \leq a$, we see that our assumptions on $\lambda_{n+39+6j}(A)$, $\lambda_{n+43+6j}(A)$ and $\lambda_{n+41+6j}(A)$ for $0 \leq j \leq a$ imply that

$$\lambda_n(A) \geq \left[0; 2_3, 1_2, 2_2, 1, 2_4, 1, \underbrace{1_3, 2_3, \ldots, 1_3, 2_3}_{a+1 \text{ times}}, \ldots\right]$$
$$+\left[2; 1, 2_2, 1_2, 2_4, 1, 2_2, 1_2, 2_4, 1, 2_2, 1_2, 2_2, 1, 2_4, 1_2, 2_3, 1_3, \underbrace{2_3, 1_3, \ldots, 2_3, 1_3}_{a+1 \text{ times}}, \ldots\right].$$

Finally, assume that $A \in \{1, 2\}^{\mathbb{Z}}$ is a bi-infinite sequence with $m(A) < C_{\infty}$ containing $1_3 2_4 1 2_2 1_2 2_4 1 2_2 1_2 2_4 1 2_2 1_2 2_2$ or its transpose, say $A_{l-15} \ldots A_{l+16}$ or $(A_{l-16} \ldots A_{l+15})^T$ equals $1_3 2_4 1 2_2 1_2 2_4 1 2_2 1_2 2_4 1 2_2 1_2 2_2$ for some $l \in \mathbb{Z}$. Our discussion above would then imply that

$$C_{\infty} > m(A) \geq \lambda_l(A)$$
$$\geq [2; 1, 2_2, 1_2, 2_4, 1, 2_2, 1_2, 2_4, 1, 2_2, 1_2, 2_2, 1, 2_4, 1_2, \overline{2_3, 1_3}]$$
$$+[0; 2_3, 1_2, 2_2, 1, 2_4, 1, \overline{1_3, 2_3}]$$
$$:= C_{\infty},$$

a contradiction. This proves the lemma. □

At this point, we are ready to give a preliminary characterization of the sequences in $\{1, 2\}^{\mathbb{Z}}$ giving rise to a Markov value in (c_{∞}, C_{∞}):

Proposition 2.2. *Let* $m \in M \cap (c_{\infty}, C_{\infty})$. *Then,* $m = m(B) = \lambda_0(B)$ *for a sequence* $B \in \{1, 2\}^{\mathbb{Z}}$ *with the following properties:*

(1) $\ldots B_{-14} \ldots B_0 B_1 \ldots B_{16} = \overline{1_2 2 \overline{1_2} 2_4} 1_2 2 1_2 2_4 1_2 2 1_2 2_2;$

(2) *there exists $N \geq 17$ such that $B_N B_{N+1} \ldots$ is a word on 1 and 2 satisfying:*

   (a) *it does not contain the subwords (1) to (13) and their transposes, and $2 1_2 2_4 1_2 2 1_2 2_3$,*

   (b) *if it contains the subword $2_3 1_2 2_2 1 2_4 1_2 2 = B_{n-8} \ldots B_{n+6}$, then*
$$B_n \ldots B_{n+9} \cdots = \overline{2_4 \overline{1_2} 2_2 1}.$$

Proof. Let $B \in \{1,2\}^{\mathbb{Z}}$ be a bi-infinite sequence such that $\lambda_0(B) = m(B) = m$. Since $3.118117 < c_\infty < m < C_\infty < 3.1181201786$, Lemma 2.15 says that $B_{-14} \ldots B_{16}$ or $(B_{-16} \ldots B_{14})^T$ equals to $1_2 2_4 1 2_2 1_2 2_4 1 2_2 1_2 2_4 1 2_2 1_2 2_2$.

Thus, by reversing B if necessary, we obtain a bi-infinite sequence $B \in \{1,2\}^{\mathbb{Z}}$ such that $m = m(B) = \lambda_0(B)$ and $B_{-14} \ldots B_{16} = 1_2 2_4 1 2_2 1_2 2_4 1 2_2 1_2 2_4 1 2_2 1_2 2_2$.

Because $m(B) = m < C_\infty$, we know from Lemma 2.17 that B does not contain the word $1_3 2_4 1 2_2 1_2 2_4 1 2_2 1_2 2_4 1 2_2 1_2 2_2$, and, thus, we can successively apply Lemma 2.16 at the positions $-9k$, $k \in \mathbb{N}$, to get that
$$\ldots B_{-14} \ldots B_0 B_1 \ldots B_{16} = \overline{1_2 \overline{1_2} 2_4} 1_2 2 1_2 2_4 1_2 2 1_2 2_2.$$

Moreover, Lemma 2.13 implies that the word $B_{17} \ldots$ does not contain the subwords (1) to (13) and their transposes.

Furthermore, the subword $2 1_2 2_4 1 2_2 1_2 2_3$ can not appear in $B_n \ldots$ for all $n \geq 17$. Indeed, if this happens, since $m(B) = m < C_\infty$, it would follow from Lemma 2.17 that B does not contain the subsequence $1_3 2_4 1 2_2 1_2 2_4 1 2_2 1_2 2_4 1 2_2 1_2 2_2$ and, hence, one could repeatedly apply Lemma 2.16 to deduce that $B = \overline{1_2 \overline{1_2} 2_4}$, a contradiction because this would mean that $c_\infty < m = m(B) = m(\overline{1_2 \overline{1_2} 2_4}) = c_\infty$.

In summary, we showed that there exists $N \geq 17$ such that the word $B_N \ldots$ does not contain the subwords (1) to (13) and their transposes, and $2 1_2 2_4 1 2_2 1_2 2_3$.

Finally, if the word $B_{17} \ldots$ contains the subword $2_3 1_2 2_2 1 2_4 1_2 2 = B_{n-8} \ldots B_{n+6}$, since B does not contain the transpose of $1_3 2_4 1 2_2 1_2 2_4 1 2_2 1_2 2_4 1 2_2 1_2 2_2$ (thanks to Lemma 2.17 and the fact that $m(B) = m < C_\infty$), then one can apply Lemma 2.16 at the positions $n + 9k$ for all $k \in \mathbb{N}$ to get that
$$B_n \ldots B_{n+9} \cdots = \overline{2_4 \overline{1_2} 2_2 1}.$$

This completes the argument. $\square$

Remark 2.10. The original version of the proposition above is Proposition 6.4 in [MM1 (2019)] and it contains a minor typographical error: in fact, the authors wrote "$B_n \ldots B_{n+9} \cdots = \overline{2_4 1 2_2 1 2}$" rather than "$B_n \ldots B_{n+9} \cdots = \overline{2_4 1 2_2 2_2 1}$". As it turns out, this little typographical mistake affects only Proposition 6.4 and, consequently, the description of the countable set $\mathcal{C}$ in Proposition 7.2 in [MM1 (2019)], but, of course, these typographical mistakes have no consequence to all other statements in [MM1 (2019)].

2.3.4 *The maximal gap of the Lagrange spectrum containing* σ

The ideas in the proof of the previous proposition can be recycled to establish the following result:

Proposition 2.3. *The interval* (c_∞, C_∞) *is a maximal gap of* L.

Proof. Recall from Remark 2.1 that the Lagrange spectrum L is the closure of the set of Lagrange values $\ell(\underline{\theta})$ of periodic sequences $\underline{\theta} = \overline{(\theta_1, \ldots, \theta_n)}$. In particular, $c_\infty = \ell(\overline{2_4, 1_2, 2_2, 1}) \in L$.

Using Lemma 2.14, we can also show that $C_\infty \in L$. In fact, our task is to exhibit a sequence $(P_a)_{a\in\mathbb{N}}$ of finite words in 1 and 2 such that $\lim\limits_{a\to\infty} \ell(\overline{P_a}) = C_\infty$. We claim that

$$P_a := \underbrace{2_3 1_3 \ldots 2_3 1_3}_{a \text{ times}} 1 2_4 1 2_2 1_2 2_3 2^* 1 2_2 1 2_4 1 2_2 1 2_4 1 2_2 1 2_2 1 2_4 1_2 \underbrace{2_3 1_3 \ldots 2_3 1_3}_{a \text{ times}}$$

satisfy $\lim\limits_{a\to\infty} \ell(\overline{P_a}) = C_\infty$. Indeed, we start by observing that $\lim\limits_{a\to\infty} \lambda_0(\overline{P_a}) = C_\infty$ (with the convention that the zeroth position corresponds to 2^*). Next, we notice that Lemma 2.14 (15), (16), (19) imply that $\lambda_j(\overline{P_a}) < 3.118117$ except possibly when the $j = \pm 9$ or 0 modulo the length P_a. Moreover,

$$\lim_{a\to\infty} \lambda_{-9}(\overline{P_a})$$
$$= [2; 1, 2_2, 1_2, 2_4, 1, 2_2, 1_2, 2_4, 1, 2_2, 1_2, 2_4, 1, 2_2, 1_2, 2_2, 1, 2_4, 1_2, \overline{2_3, 1_3}]$$
$$+ [0; 2_3, 1, \overline{1_3, 2_3}]$$
$$= 3.1180041084\ldots$$

and

$$\lim_{a\to\infty} \lambda_9(\overline{P_a}) = [2; 1, 2_2, 1_2, 2_4, 1, 2_2, 1_2, 2_2, 1, 2_4, 1_2, \overline{2_3, 1_3}]$$
$$+ [0; 2_3, 1_2, 2_2, 1, 2_4, 1_2, 2_2, 1, 2_4, 1, \overline{1_3, 2_3}]$$
$$= 3.11812017817071\ldots$$

In particular, $\ell(\overline{P_a}) = m(\overline{P_a}) = \lambda_0(\overline{P_a})$ for all $a \in \mathbb{N}$ sufficiently large. Thus, we have that $C_\infty = \lim\limits_{a \to \infty} \lambda_0(\overline{P_a}) = \lim\limits_{a \to \infty} \ell(\overline{P_a}) \in L$, as desired.

At this point, our task is reduced to prove that

$$L \cap (c_\infty, C_\infty) = \emptyset.$$

For this sake, we will adapt the idea of the proof of Proposition 2.2. By contradiction, suppose that $\alpha \in L \cap (c_\infty, C_\infty)$ and fix $B \in \{1,2\}^{\mathbb{Z}}$ with $\ell(B) = \alpha$. Since $3.118117 < c_\infty < \alpha < C_\infty < 3.1181201786$, we can choose $N \in \mathbb{N}$ such that

$$\lambda_n(B) < 3.1181201786$$

for all $|n| \geq N$, and we can fix a monotone sequence $\{n_k\}_{k \in \mathbb{N}}$ such that $|n_k| \geq N$ and $\lambda_{n_k}(B) > 3.118117$ for all $k \in \mathbb{Z}$. Moreover, by reversing B if necessary, we can assume that $n_k \to +\infty$ as $k \to \infty$ and $\limsup\limits_{n \to +\infty} \lambda_n(B) = \alpha$.

We affirm that $1_3 2_4 12_2 1_2 2_4 12_2 1_2 2_4 12_2 1_2 2_2$ or its transpose can not be contained in $B_n B_{n+1} \ldots$ for all $n \geq N$: otherwise, we would have a sequence $m_k \to +\infty$ as $k \to \infty$ such that $B_{m_k-15} \ldots B_{m_k+16}$ or $(B_{m_k-16} \ldots B_{m_k+15})^T$ equals $1_3 2_4 12_2 1_2 2_4 12_2 1_2 2_4 12_2 1_2 2_2$ for all $k \in \mathbb{N}$; hence, by Lemma 2.17, the fact that $\lambda_n(B) < 3.1181201786$ for all $n \geq N$ would imply that

$$\lambda_{m_k}(B)$$

$$\geq \left[2; 1, 2_2, 1_2, 2_4, 1, 2_2, 1_2, 2_4, 1, 2_2, 1_2, 2_2, 1, 2_4, 1_2, \underbrace{2_3, 1_3, \ldots, 2_3, 1_3}_{a_k+2 \text{ times}}, \ldots \right]$$

$$+ \left[0; 2_3, 1_2, 2_2, 1, 2_4, 1, \underbrace{1_3, 2_3, \ldots, 1_3, 2_3}_{a_k+1 \text{ times}}, \ldots \right]$$

where $a_k = \lfloor \frac{m_k - 43 - N}{6} \rfloor$; on the other hand, since $a_k \to \infty$ as $k \to \infty$, it would follow that $C_\infty > \alpha \geq \limsup\limits_{k \to \infty} \lambda_{m_k}(B) \geq C_\infty$, an absurd.

Thus, the discussion of the previous paragraph allows us to select $R \geq N$ such that $B_R B_{R+1} \ldots$ does not contain $1_3 2_4 12_2 1_2 2_4 12_2 1_2 2_4 12_2 1_2 2_2$ or its transpose.

Note that, by Lemma 2.15, our choices of $N \in \mathbb{N}$ and $\{n_k\}_{k \in \mathbb{N}}$ imply that

(1) either $B_{n_k-14} \ldots B_{n_k+16} = 1_2 2_4 12_2 1_2 2_4 12_2 1_2 2_4 12_2 1_2 2_2$

(2) or $(B_{n_k-16} \ldots B_{n_k+14})^T = 1_2 2_4 12_2 1_2 2_4 12_2 1_2 2_4 12_2 1_2 2_2$

for each $k \in \mathbb{N}$ with $n_k \geq N + 15$.

If the first possibility occurs for all $n_k > R + 15$, then the facts that $\lambda_n(B) < 3.1181201786$ for all $n \geq N$ and the sequence $B_R B_{R+1} \ldots$ does not contain $1_3 2_4 12_2 1_2 2_4 12_2 1_2 2_4 12_2 1_2 2_2$ allow to apply $d_k := \lfloor \frac{n_k - 6 - R}{9} \rfloor$ times Lemma 2.16 at the positions $n_k - 9(j-1)$, $j = 1, \ldots, d_k$ to deduce that the sequence B has the form

$$\ldots B_{n_k - 9d_k} \ldots B_{n_k} \ldots B_{n_k + 16} \ldots$$
$$= \ldots \underbrace{212_2 1_2 2_3, \ldots, 212_2 1_2 2_3}_{d_k \text{ times}} 212_2 1_2 2_4 12_2 1_2 2_2 \ldots$$

Because $R - 15 \leq n_k - 9d_k \leq R + 16$ and $n_k \to +\infty$, we get that B has the form $\ldots \overline{212_2 1_2 2_3}$.

If the second possibility occurs for some $k_0 \in \mathbb{N}$ with $n_{k_0} \geq R + 15$, then the facts that $\lambda_n(B) < B_\infty < \alpha_\infty + 10^{-6}$ for all $n \geq N$ and the sequence $B_R B_{R+1} \ldots$ does not contain the subsequence $(1_3 2_4 12_2 1_2 2_4 12_2 1_2 2_4 12_2 1_2 2_2)^T$ allow to apply Lemma 2.16 at the positions $n_k + 9a$, $a \in \mathbb{N}$, to deduce that the sequence B has the form

$$\ldots B_{n_{k_0}} B_{n_{k_0}+1} \ldots B_{n_k+10} \cdots = \ldots \overline{2_4 1_2 2_2 1}.$$

In any case, each possibility above would imply that

$$c_\infty < \alpha = \limsup_{n \to +\infty} \lambda_n(B) = \ell(\overline{2_4 1_2 2_2 1}) = c_\infty,$$

a contradiction.

In summary, the existence of $\alpha \in L \cap (c_\infty, C_\infty)$ leads to a contradiction in any scenario. $\qquad \square$

2.3.5 *Proof of Theorem 2.8*

The description of $(M \setminus L) \cap (c_\infty, C_\infty) = M \cap (c_\infty, C_\infty)$ provided by Propositions 2.2 and 2.3 allows us to compare this piece of $M \setminus L$ with the Cantor set

$$Y := \{[0; \gamma] : \gamma \in \{1, 2\}^{\mathbb{N}} \text{ not containing the subwords in } P\}$$

where $P = \{121, 212_3 1, 12_3 12, 2_3 12_3, 2_5 12_2 1_2, 1_2 2_2 12_5, 21_2 2_4 12_2 1_2 2_3, 2_3 1_2 2_2 12_4 1_2 2\}$. This is the content of the next two propositions.

Remark 2.11. In principle, the set of prohibited strings should be the finite set of words consisting of the words (1) to (13) in Lemma 2.13 above and their transposes, and the words $21_2 2_4 12_2 1_2 2_3$ and its transpose (as described in the original article [MM1 (2019)]), but the words from (5) to (13) are redundant at this stage because they already contain the word $21_2 2_4 12_2 1_2 2_3$.

Remark 2.12. The set Y is a topologically mixing dynamically defined Cantor set (cf. Remark 2.7). In fact, since the set of prohibited words P consist of words whose size is bounded by 15, we can define Y in a dynamical way via the Markov partition determined by the set W of all words of size 14 not containing any factor in P and, for $\alpha, \beta \in W$, allow the transition $\alpha\beta$ if, and only if, $\alpha\beta$ does not contain any factor in P. This Markov partition is topologically mixing because any word (or factor of a word) in W can be continued up to connect with the symmetric block $[11, 222]$. Indeed, it is enough to verify this for the words obtained from the elements of P by removing their last entry and this is done as follows:

(1)$12 \to$ (1)$12\bar{2}$, (22)$12 \to$ (22)$1221\bar{2}$, $12221 \to 12221\bar{1}$,

(2212)$21222 \to$ (2212)$21222\bar{2}$, (1112)$21222 \to$ (1112)$2122221\bar{1}$, (122112)$21222 \to$ (122112)$21222\bar{2}$, (222112)$21222 \to$ (222112)$2122221122\bar{1}$, (2)$222122 \to$ (2)$22212212212\bar{2}$,

$222221221 \to 2222212212\bar{2}$, (1)$112212222 \to$ (1)$1122122221\bar{1}$,

(122)$112212222 \to$ (122)$112212222211\bar{1}$, (222)$112212222 \to$ (222)$1122122221122\bar{1}$,

$21122221221122 \to 21122221221122\bar{1}$ and $22211221222211 \to 22211221222211\bar{1}$

(where $(1), (22), (2212)$, etc. indicate possible prefixes).

Proposition 2.4. $(M \setminus L) \cap (3.118120178159, 3.118120178173)$ *contains the set* $\{[\overline{2_4, 1_2, 2_2, 1}] + [0; 1, 2_2, 1_2, 2_4, 1, 2_2, 1_2, 2_2, 1, \gamma] \ : \ 1_2 2_2 1\gamma \in \{1, 2\}^{\mathbb{N}}$ *does not contain the subwords in* $P\}$.

Proof. Consider $B = \overline{12_2 1_2 2_4}; 12_2 1_2 2_4 12_2 1_2 2_2 1\gamma$ where $1_2 2_2 1\gamma \in \{1, 2\}^{\mathbb{N}}$ does not contain subwords in P and ; serves to indicate the zeroth position.

Note that
$$\lambda_0(B) \le [\overline{2_4, 1_2, 2_2, 1}] + [0; 1, 2_2, 1_2, 2_4, 1, 2_2, 1_2, 2_2, 1, \overline{2, 1}] < 3.118120178173$$
and
$$\lambda_0(B) \ge [\overline{2_4, 1_2, 2_2, 1}] + [0; 1, 2_2, 1_2, 2_4, 1, 2_2, 1_2, 2_2, 1, \overline{1, 2}] > 3.118120178159,$$
and Lemma 2.14 (15), (16), (19) imply that $\lambda_n(B) < 3.118117$ for all positions $n \le 18$ except possibly for $n = -9k$ with $k \ge 0$.

On the other hand,
$$\lambda_{-9k}(B) = [\overline{2_4, 1_2, 2_2, 1}]$$
$$+ \left[0; \underbrace{1, 2_2, 1_2, 2_4, \ldots, 1, 2_2, 1_2, 2_4}_{k \text{ times}}, 1, 2_2, 1_2, 2_4, 1, 2_2, 1_2, 2_2, 1_2, \gamma \right]$$
$$< [\overline{2_4, 1_2, 2_2, 1}] + [0; 1, 2_2, 1_2, 2_4, 1, 2_2, 1_2, 2_2, 1_2, \gamma] = \lambda_0(B),$$
for all $k \ge 1$.

Furthermore, since $1_2 2_2 1_2 \gamma$ does not contain subwords in P, it follows from (the proof of) Lemma 2.15 that $\lambda_n(B) < 3.118117$ for all $n \geq 19$.

This shows that $m(B) = \lambda_0(B) = [\overline{2_4, 1_2, 2_2, 1}] + [0; 1, 2_2, 1_2, 2_4, 1, 2_2, 1_2, 2_2, 1, \gamma]$ belongs to $(M \setminus L) \cap (3.118120178159, 3.118120178173)$. $\qquad\square$

Proposition 2.5. $(M \setminus L) \cap (c_\infty, C_\infty)$ *is contained in the union of*

$$\mathcal{C} = \{[0; \overline{2_3, 1_2, 2_2, 1, 2}] + [2; 1, 2_2, 1_2, 2_4, 1, 2_2, 1_2, 2_2, \theta, \overline{2_4, 1_2, 2_2, 1, 2}] :$$
$$\theta \text{ is a finite word in 1 and 2}\}$$

and the sets

$$\mathcal{D}(\delta) = \{0; \overline{2_3, 1_2, 2_2, 1, 2}] + [2; 1, 2_2, 1_2, 2_4, 1, 2_2, 1_2, 2_2, \delta, \gamma] :$$
$$\text{no subword of } \gamma \in \{1, 2\}^{\mathbb{N}} \text{ belongs to } P\},$$

where δ is a finite word in 1 and 2.

Proof. By Proposition 2.2, if $m \in (M \setminus L) \cap (c_\infty, C_\infty)$, then $m = m(B) = \lambda_0(B)$ with

$$B = \overline{212_2 1_2 2_3} 2^* 1 2_2 1_2 2_4 1 2_2 1_2 2_2 \delta \gamma$$

where the asterisk indicates the zeroth position, δ is a finite word in 1 and 2, and the infinite word γ satisfies:

(1) γ does not contain the subwords (1) to (13) and their transposes, and $21_2 2_4 1 2_2 1_2 2_3$,
(2) if γ contains the subword $2_3 1_2 2_2 1 2_4 1_2 2$, then $\gamma = \mu \overline{2_4, 1_2, 2_2, 1, 2}$ with μ a finite word in 1 and 2.

Hence:

(1) if γ contains $2_3 1_2 2_2 1 2_4 1_2 2$, then

$$m(B) = \lambda_0(B) = [0; \overline{2_3, 1_2, 2_2, 1, 2}]$$
$$+ [2; 1, 2_2, 1_2, 2_4, 1, 2_2, 1_2, 2_2, \delta, \mu, \overline{2_4 1 2_2 1_2}]$$

where $\theta = \delta\mu$ is a finite word in 1 and 2, i.e., $m(B) \in \mathcal{C}$;
(2) otherwise,

$$m(B) = \lambda_0(B) = [0; \overline{2_3, 1_2, 2_2, 1, 2}] + [2; 1, 2_2, 1_2, 2_4, 1, 2_2, 1_2, 2_2, \delta, \gamma]$$

where γ does not contain the subwords (1) to (13) and their transposes, and $21_2 2_4 1 2_2 1_2 2_3$ and its transpose $2_3 1_2 2_2 1 2_4 1_2 2$, i.e., $m(B) \in \mathcal{D}(\delta)$.

This completes the argument. $\qquad\square$

Remark 2.13. A careful analysis of the proof of Theorem 2.5 actually gives that $(M \setminus L) \cap (c_\infty, C_\infty)$ is *equal* to the union of

$$\mathcal{C}' = \{[0; \overline{2_3, 1_2, 2_2, 1, 2}] + [2; 1, 2_2, 1_2, 2_4, 1, 2_2, 1_2, 2_2, 1, \theta,$$
$$2_2 1_2 2_2 1 \overline{2_4, 1_2, 2_2, 1, 2}] : \theta \text{ is a (possibly empty)}$$

finite word in 1 and 2 such that $[0; 1, \theta] \geq [0; \theta^T, 1]$ and no subword of $1_2 2_2 1 \theta 2_2 1_2$ belongs to $P\}$ and

$$\mathcal{D}' = \{0; \overline{2_3, 1_2, 2_2, 1, 2}] + [2; 1, 2_2, 1_2, 2_4, 1, 2_2, 1_2, 2_2, 1, \gamma] :$$
$$\text{no subword of } 1_2 2_2 1 \gamma \in \{1,2\}^{\mathbb{N}} \text{ belongs to } P\}.$$

At this stage, Theorem 2.8 follows directly from Propositions 2.4 and 2.5: on one hand, by Proposition 2.4, $(M \setminus L) \cap (c_\infty, C_\infty)$ contains a set bi-Lipschitz equivalent to Y and, hence,

$$HD((M \setminus L) \cap (c_\infty, C_\infty)) \geq HD(Y).$$

On the other hand, by Proposition 2.5, $(M \setminus L) \cap (c_\infty, C_\infty)$ is contained in

$$\mathcal{C} \cup \bigcup_{n \in \mathbb{N}} \left(\bigcup_{\delta \in \{1,2\}^n} \mathcal{D}(\delta) \right).$$

Since $\mathcal{C}$ is a countable set and $\{\mathcal{D}(\delta) : \delta \in \{1,2\}^n, n \in \mathbb{N}\}$ is a countable family of subsets bi-Lipschitz equivalent to Y, it follows that

$$HD((M \setminus L) \cap (c_\infty, C_\infty)) \leq HD(Y).$$

This proves Theorem 2.8.

2.3.6   *Proof of Theorem 2.7*

Note that the definition of Y in (2.14) implies that Y contains the Gauss-Cantor set $K(\{1_2, 2_2\})$. Thus, Theorem 2.8 implies that

$$HD((M \setminus L) \cap (c_\infty, C_\infty)) = HD(Y) \geq HD(K(\{1_2, 2_2\})).$$

Therefore, the proof of Theorem 2.7 is reduced to show that:

Proposition 2.6. *One has* $0.2628 < HD(K(\{1_2, 2_2\})) < 0.2646$.

Proof. The convex hull of $K(\{1_2, 2_2\})$ is the interval I with extremities $[0; \overline{1}]$ and $[0; \overline{2}]$. The images $I_{11} := \phi_{11}(I)$ and $I_{22} := \phi_{22}(I)$ of I under the inverse branches

$$\phi_{11}(x) := \cfrac{1}{1 + \cfrac{1}{1 + \frac{1}{x}}} \quad \text{and} \quad \phi_{22}(x) := \cfrac{1}{2 + \cfrac{1}{2 + \frac{1}{x}}}$$

of the first two iterates of the Gauss map $G(x) := \{1/x\}$ provide the first step of the construction of the Cantor set $K(\{1_2, 2_2\})$. In general, the collection $\mathcal{R}^n$ of intervals of the nth step of the construction of $K(\{1_2, 2_2\})$ is

$$\mathcal{R}^n := \{\phi_{x_1} \circ \cdots \circ \phi_{x_n}(I) : (x_1, \ldots, x_n) \in \{11, 22\}^n\}.$$

Given $R = \phi_{y_1} \circ \ldots \phi_{y_n}(I) \in \mathcal{R}^n$ associated to a string $(y_1, \ldots, y_n) = (x_1, \ldots, x_{2n}) \in \{11, 22\}^n$, let

$$\lambda_{n,R} := \inf_{x \in R} |(\Psi^n)'(x)|$$

$$= \min\left\{ \prod_{i=1}^{2n}\left(\frac{1}{[0; x_i, \ldots, x_{2n}, \overline{1}]}\right)^2, \prod_{i=1}^{2n}\left(\frac{1}{[0; x_i, \ldots, x_{2n}, \overline{2}]}\right)^2 \right\}$$

and

$$\Lambda_{n,R} := \sup_{y \in R} |(\Psi^n)'(y)|$$

$$= \max\left\{ \prod_{i=1}^{2n}\left(\frac{1}{[0; x_i, \ldots, x_{2n}, \overline{1}]}\right)^2, \prod_{i=1}^{2n}\left(\frac{1}{[0; x_i, \ldots, x_{2n}, \overline{2}]}\right)^2 \right\}.$$

Here, our computation of $\lambda_{n,R}$ and $\Lambda_{n,R}$ uses the fact that $(\Psi^n)|_R$ comes from a Möbius transformation induced by an integral matrix with determinant ± 1 (cf. Chapter 1, §1.5), so that $(\Psi^n)'|_R$ is monotone on R.

By Proposition 2.1, if we define $\alpha_n \in [0, 1]$, $\beta_n \in [0, 1]$ by

$$\sum_{R \in \mathcal{R}^n}\left(\frac{1}{\Lambda_{n,R}}\right)^{\alpha_n} = 1 = \sum_{R \in \mathcal{R}^n}\left(\frac{1}{\lambda_{n,R}}\right)^{\beta_n},$$

then $\alpha_n \leq HD(K(\{1_2, 2_2\})) \leq \beta_n$ for all $n \in \mathbb{N}$. Therefore, we can estimate on $K(\{1_2, 2_2\})$ by computing α_n and β_n for some particular values of $n \in \mathbb{N}$.

A quick computer search for the values of α_{12} and β_{12} reveals that

$$\alpha_{12} = 0.2628\ldots \quad \text{and} \quad \beta_{12} = 0.2645\ldots$$

In particular, $0.2628 < \alpha_{12} \leq HD(K(\{1_2, 2_2\})) \leq \beta_{12} < 0.2646$, so that the proof of Proposition 2.6 and, *a fortiori*, Theorem 2.7 is now complete. $\square$

2.4 The complement of the Lagrange spectrum in the Markov spectrum II

In 1973, Freiman [Fr73 (1973)] found that $M \setminus L$ also contains the number

$$\alpha_\infty := [2; \overline{1_2, 2_3, 1, 2}] + [0; 1, 2_3, 1_2, 2, 1, \overline{2}] = 3.2930442654\ldots$$

and Berstein computed[7] in [Bel (1973)] the extremities

$$b_\infty := [2; \overline{1_2, 2_3, 1, 2}] + [0; \overline{1, 2_3, 1_2, 2}] = 3.2930442439\ldots$$

and

$$B_\infty := [2; 1, \overline{1, 2_3, 1, 2, 1_2, 2, 1_2, 2}]$$
$$+ [0; 1, 2_3, 1_2, 2, 1, 2_3, 1_2, 2, 1, 2_2, \overline{1, 2_3, 1, 2, 1_2, 2, 1_2, 2}]$$
$$= 3.2930444814\ldots$$

of the maximal gap (b_∞, B_∞) of the Lagrange spectrum containing σ.

Contrary to the elements of $M \setminus L$ in the collection $\mathcal{F}$ mentioned in §2.3 above, the element α_∞ is *not* isolated: indeed, Flahive showed in 1977 that $\alpha_\infty = \lim_{n \to \infty} \alpha_n$, where

$$\alpha_n := [2; \overline{1_2, 2_3, 1, 2}] + [0; 1, 2_3, 1_2, 2, 1, 2_n, \overline{1, 2, 1_2, 2_3}] \in M \setminus L$$

for all $n \geq 4$ (see, e.g., Chapter 3 of Cusick–Flahive book [CF (1989)]).

The structure of $(M \setminus L) \cap (b_\infty, B_\infty)$ was determined by Matheus–Moreira [MM2 (2019)] in 2019. As a by-product of their statements, they showed that

Theorem 2.9. *The Hausdorff dimension of* $(M \setminus L) \cap (b_\infty, B_\infty)$ *satisfies*

$$0.353 < HD((M \setminus L) \cap (b_\infty, B_\infty)).$$

The proof of this result follows the same strategy from §2.3, namely, we will compare $(M \setminus L) \cap (b_\infty, B_\infty)$ with the Cantor set X given by

$$X := \{[0; \gamma] : \gamma \in \{1, 2\}^{\mathbb{N}} \text{ not containing the subwords in } Q\}, \quad (2.15)$$

where

$$Q := \{21212, 12121, 2121_3, 1_3212, 2_31_21_22_21, 12_21_2212_3, 12_3121_22_2, 2_21_22121_231\}$$

in order to reduce Theorem 2.9 to the following result:

Theorem 2.10. $HD((M \setminus L) \cap (b_\infty, B_\infty)) = HD(X)$ *(where X is the Cantor set in (2.15)).*

Since the arguments used to show Theorems 2.9 and 2.10 are really parallel to those involved in the proofs of Theorem 2.7 and 2.8, we will just sketch the main steps of the proof of Theorem 2.9.

---

[7]The formulas of b_∞ and B_∞ in [Bel (1973)] contained mistakes which were later corrected by Matheus–Moreira in [MM2 (2019)].

2.4.1 Forbidden and allowed strings for Markov values in (b_∞, B_∞)

We begin with the following elementary lemma.

Lemma 2.18. *Let* $B \in \{1,2\}^{\mathbb{Z}}$.

(xii') *If* B *contains* $1212_2121_22_312^*1_22_3121_22_3121_22$, *then* $\lambda_j(B) > B_\infty + 6 \times 10^{-9}$ *where* j *indicates the position in asterisk.*

(xii") *If* B *contains* $2_212_2121_22_312^*1_22_3121_22_3121_22$, *then* $\lambda_j(B) < B_\infty - 10^{-9}$ *where* j *indicates the position in asterisk.*

Proof. If (xii') occurs, then

$$\lambda_j(B) = [2; 1_2, 2_3, 1, 2, 1_2, 2_3, 1, 2, 1_2, 2, \dots] + [0; 1, 2_3, 1_2, 2, 1, 2_2, 1, 2, 1, \dots]$$
$$\geq [2; 1_2, 2_3, 1, 2, 1_2, 2_3, 1, 2, 1_2, 2, \overline{1, 2}] + [0; 1, 2_3, 1_2, 2, 1, 2_2, 1, 2, 1, \overline{1, 2}]$$
$$> B_\infty + 6 \times 10^{-9}.$$

If (xii") occurs, then

$$\lambda_j(B) = [2; 1_2, 2_3, 1, 2, 1_2, 2_3, 1, 2, 1_2, 2, \dots] + [0; 1, 2_3, 1_2, 2, 1, 2_2, 1, 2_2 \dots]$$
$$\leq [2; 1_2, 2_3, 1, 2, 1_2, 2_3, 1, 2, 1_2, 2, \overline{2, 1}] + [0; 1, 2_3, 1_2, 2, 1, 2_2, 1, 2_2, \overline{2, 1}]$$
$$< B_\infty - 10^{-9}.$$

$\square$

Next, we recall following result extracted from Lemma 2 in Chapter 3 of [CF (1989)]:

Lemma 2.19. *If* $B \in \{1,2\}^{\mathbb{Z}}$ *contains any of the subsequences*

(a) 1^*
(b) 22^*
(c) $1_22^*1_2$
(d) $2_212^*1_221$
(e) $12_212^*1_22$
(f) $2_412^*1_22_3$

then $\lambda_j(B) < \alpha_\infty - 10^{-5}$ *where* j *indicates the position in asterisk.*

The next three lemmas are similar to Lemmas 2.15, 2.16 and 2.17 above, and their proofs are left as exercises to the reader (who might consult [MM2 (2019)] for solutions).

Lemma 2.20. *Let $B = (B_m)_{m \in \mathbb{Z}} \in \{1,2\}^{\mathbb{Z}}$ be a bi-infinite sequence. Suppose that $\lambda_n(B) \leq \alpha_\infty + 10^{-6}$ at a certain position $n \in \mathbb{Z}$. Then, the sole possible situations are:*

(1) $B_n = 1$ and $\lambda_n(B) < \alpha_\infty - 10^{-5}$;
(2) $B_{n-1}B_n = 22$ and $\lambda_n(B) < \alpha_\infty - 10^{-5}$;
(3) $B_nB_{n+1} = 22$ and $\lambda_n(B) < \alpha_\infty - 10^{-5}$;
(4) $B_{n-2}B_{n-1}B_nB_{n+1}B_{n+2} = 11211$ and $\lambda_n(B) < \alpha_\infty - 10^{-5}$;
(5) $B_{n-3}\ldots B_{n+4} \in \{21121221, 22121121\}$ and $\lambda_n(B) < \alpha_\infty - 10^{-5}$;
(6) $B_{n-4}\ldots B_{n+3} \in \{12112122, 12212112\}$ and $\lambda_n(B) < \alpha_\infty - 10^{-5}$;
(7) $B_{n-5}\ldots B_{n+5} \in \{22211212222, 22221211222\}$ and $\lambda_n(B) < \alpha_\infty - 10^{-5}$;
(8) $B_{n-5}\ldots B_{n+4} = 1222121122$;
(9) $B_{n-4}\ldots B_{n+5} = 2211212221$.

In particular, the subwords 212^*12, 212^*1_3, 1_32^*12, 1212^*1_2, 1_22^*121, $2_312^*1_22_21$ and $12_21_22^*12_3$ are forbidden (where the asterisk indicates the nth position).

Lemma 2.21. *Let $B = (B_m)_{m \in \mathbb{Z}} \in \{1,2\}^{\mathbb{Z}}$ be a bi-infinite sequence. Suppose that $\lambda_{n-7}(B), \lambda_n(B), \lambda_{n+7}(B) \leq \alpha_\infty + 10^{-6}$ for some $n \in \mathbb{Z}$.*

(1) If $B_{n-5}\ldots B_{n+4} = 12_3121_22_2$, then:

    (a) either $B_{n-10}\ldots B_{n+11} = 2_2121_22_3121_22_3121_221$,
    (b) or $B_{n-10}\ldots B_{n+11} = 2_2121_22_3121_22_3121_22_2$ and, in particular, the vicinity of $B_{n+7} = 2$ is $B_{n+2}\ldots B_{n+11} = 12_3121_22_2$.

(2) If $B_{n-4}\ldots B_{n+5} = 2_21_221_23_1$, then:

    (a) either $B_{n-11}\ldots B_{n+10} = 121_221_23_1_221_23_1_221_2_2$,
    (b) or $B_{n-11}\ldots B_{n+10} = 2_21_221_23_1_221_23_1_221_2_2$ and, in particular, the vicinity of $B_{n-7} = 2$ is $B_{n-11}\ldots B_{n-2} = 2_21_221_23_1$.

Lemma 2.22. *Let $A \in \{1,2\}^{\mathbb{Z}}$ be a bi-infinite sequence. Suppose that, for some $n \in \mathbb{Z}$ and $a \in \mathbb{N}$, one has $\lambda_{n\pm7}(A) \leq B_\infty + 6 \times 10^{-9}$, $\lambda_{n\pm(17+6k)}(A) \leq \alpha_\infty + 10^{-6}$ for each $k = 1, \ldots, 2a$, and $\lambda_{n\pm(7+6j)}(A) \leq \alpha_\infty + 10^{-6}$ for each $j = 1, \ldots, 2a$.*

If $A_{n-10} \dots A_{n+11}$ or $(A_{n-11} \dots A_{n+10})^T$ equals $2_2 121_2 23 121_2 23 121_2 21$, then

$$\lambda_n(A) \geq \left[2; 1, \underbrace{1, 2_3, 1, 2, 1_2, 2, 1_2, 2, \dots, 1, 2_3, 1, 2, 1_2, 2, 1_2, 2}_{a+1 \; times}, \dots \right]$$

$$+ \left[0; 1, 2_3, 1_2, 2, 1, 2_3, 1_2, 2, 1, 2_2, \right.$$

$$\left. \underbrace{1, 2_3, 1, 2, 1_2, 2, 1_2, 2, \dots, 1, 2_3, 1, 2, 1_2, 2, 1_2, 2}_{a \; times}, \dots \right].$$

In particular, the subsequence $2_2 121_2 23 121_2 23 121_2 21$ or its transpose $121_2 212_3 1_2 212_3 1_2 212_2$ is not contained in a bi-infinite sequence $A \in \{1,2\}^{\mathbb{Z}}$ with $m(A) < B_\infty$.

2.4.2 The Lagrange spectrum does not meet the interval (b_∞, B_∞)

Proposition 2.7. *If $\alpha \in M \cap (b_\infty, B_\infty)$, then $\alpha \notin L$.*

Proof. Suppose that $\alpha \in L \cap (b_\infty, B_\infty)$. Let $B \in \{1,2\}^{\mathbb{Z}}$ be a bi-infinite sequence such that $\ell(B) := \limsup_{i \to \infty} \lambda_i(B) = \alpha$.

Since $\alpha_\infty - 10^{-5} < b_\infty < \alpha < B_\infty$, we can fix $N \in \mathbb{N}$ large enough such that

$$\lambda_n(B) < B_\infty$$

for all $|n| \geq N$, and we can select a monotone sequence $\{n_k\}_{k \in \mathbb{N}}$ such that $|n_k| \geq N$ and $\lambda_{n_k}(B) \geq \alpha_\infty - 10^{-5}$ for all $k \in \mathbb{Z}$. Moreover, by reversing B if necessary, we can assume that $n_k \to +\infty$ as $k \to \infty$ and $\limsup_{n \to +\infty} \lambda_n(B) = \alpha$.

We have two possibilities:

(1) either the sequence $B_n B_{n+1} \dots$ contains the subsequence $2_2 121_2 23 121_2 23 121_2 21$ or its transpose $121_2 212_3 1_2 212_3 1_2 212_2$ for all $n \geq N$,

(2) or there exists $R \geq N$ such that $B_R B_{R+1} \dots$ does not contain the subsequence $2_2 121_2 23 121_2 23 121_2 21$ or its transpose $121_2 212_3 1_2 212_3 1_2 212_2$.

In the first scenario, let $\{m_k\}_{k \in \mathbb{N}}$ be a monotone sequence such that $B_{m_k-10} \dots B_{m_k+11}$ or $(B_{m_k-11} \dots B_{m_k+10})^T$ equals $2_2 121_2 23 121_2 23 121_2 21$

for all $k \in \mathbb{N}$ and $m_k \to +\infty$ as $k \to \infty$. By Lemma 2.22, the fact that $\lambda_n(B) < B_\infty$ for all $n \geq N$ would imply that

$$
\lambda_{m_k}(B) \geq \left[2; 1, \underbrace{1, 2_3, 1, 2, 1_2, 2, 1_2, 2, \dots, 1, 2_3, 1, 2, 1_2, 2, 1_2, 2}_{a_k+1 \text{ times}} \dots \right]
$$

$$
+ \left[0; 1, 2_3, 1_2, 2, 1, 2_3, 1_2, 2, 1, 2_2, \right.
$$

$$
\left. \underbrace{1, 2_3, 1, 2, 1_2, 2, 1_2, 2, \dots, 1, 2_3, 1, 2, 1_2, 2, 1_2, 2}_{a_k \text{ times}}, \dots \right]
$$

where $a_k = \lfloor \frac{m_k - 17 - N}{6} \rfloor$. Since $a_k \to \infty$ as $k \to \infty$, it would follow that

$$
B_\infty > \alpha \geq \limsup_{k \to \infty} \lambda_{m_k}(B) \geq B_\infty,
$$

a contradiction.

In the second scenario, we note that, by Lemma 2.20, for each $k \in \mathbb{N}$, we have

(1) either $B_{n_k-5} \dots B_{n_k+4} = 1 2_3 1 2 1 2_2 2_2$
(2) or $B_{n_k-4} \dots B_{n_k+5} = 2_2 1_2 2 1 2_3 1$.

If the first possibility occurs for some $k_0 \in \mathbb{N}$ with $n_{k_0} \geq R + 10$, then the facts that $\lambda_n(B) < B_\infty < \alpha_\infty + 10^{-6}$ for all $n \geq N$ and the sequence $B_R B_{R+1} \dots$ does not contain the subsequence $2_2 1 2 1 2_2 2_3 1 2 1 2_2 2_3 1 2 1 2_2 1$ allow to repeatedly apply Lemma 2.21 at the positions $n_{k_0} + 7a$, $a \in \mathbb{N}$, to deduce that the sequence B has the form

$$
\dots B_{n_{k_0}} B_{n_{k_0}+1} B_{n_{k_0}+2} \dots = \dots 2 \overline{1_2 2_3 1 2}.
$$

If the second possibility occurs for all $n_k > R + 10$, then the facts that $\lambda_n(B) < B_\infty < \alpha_\infty + 10^{-6}$ for all $n \geq N$ and the sequence $B_R B_{R+1} \dots$ does not contain the subsequence $1 2 1_2 2 1 2_3 1_2 2 1 2_3 1_2 2 1 2_2$ allow to apply $d_k := \lfloor \frac{n_k - 4 - R}{7} \rfloor$ times Lemma 2.21 at the positions $n_k - 7(j - 1)$, $j = 1, \dots, d_k$, to deduce that the sequence B has the form

$$
\dots B_{n_k-7d_k} \dots B_{n_k} \dots B_{n_k+10} \dots = \dots \underbrace{2 1 2_3 1 2, \dots, 2 1 2_3 1_2}_{d_k \text{ times}} 2 1 2_3 1_2 2 1 2_2 \dots
$$

Because $R - 4 \leq n_k - 7d_k \leq R + 11$ and $n_k \to +\infty$, we deduce that B has the form $\dots \overline{2 1 2_3 1_2}$.

In any case, the second scenario would imply that

$$b_\infty < \alpha = \limsup_{n \to +\infty} \lambda_n(B) = \ell(\overline{1_2 2_3 12}) = b_\infty,$$

a contradiction.

In summary, the existence of $\alpha \in L \cap (b_\infty, B_\infty)$ leads to a contradiction in any scenario. This proves the proposition. $\square$

2.4.3   *Preliminary description of sequences with Markov values in (b_∞, B_∞)*

Proposition 2.8. *Let $m \in M \cap (b_\infty, B_\infty)$. Then, $m = m(B) = \lambda_0(B)$ for a sequence $B \in \{1,2\}^{\mathbb{Z}}$ with the following properties:*

(1) $B_{-10} \ldots B_0 B_1 \ldots B_7 \cdots = 2_2 121_2 2_3 12 1\overline{1_2 2_3 12}$;

(2) there exists $N \geq 11$ such that $\ldots B_{-N-1} B_{-N}$ is a word on 1 and 2 satisfying:

    (a) it does not contain the subwords 21212, 12121, 2121_3, $1_3 212$, $2_3 121_2 2_2 1$, $12_2 1_2 212_3$ and $12_3 121_2 2_2$,

    (b) if it contains the subword $2_2 1_2 212_3 1 = B_{n-4} \ldots B_{n+5}$, then

$$\ldots B_{n-7} \ldots B_{n+10} = \overline{212_3 1_2} 212_3 1_2 212_2.$$

Proof. Let $B \in \{1,2\}^{\mathbb{Z}}$ be a bi-infinite sequence such that $m = m(B)$. Since $m < B_\infty$, Proposition 2.7 implies that $\limsup_{i \to \infty} \lambda_i(B) = \ell(B) \leq b_\infty < m$.

Therefore, we can select N_0 large enough such that $\lambda_n(B) < \frac{b_\infty + m}{2} < m$ for all $|n| \geq N_0$. In particular, $m = m(B) = \lambda_{n_0}(B)$ for some $|n_0| < N_0$.

It follows that we can shift B in order to obtain a sequence — still denoted by B — such that $\lambda_0(B) = m(B) = m$. Since $m > b_\infty > \alpha_\infty - 10^{-5}$, Lemma 2.20 says that

$$B_{-5} \ldots B_4 = 1222121122 \quad \text{or} \quad B_{-4} \ldots B_5 = 2211212221.$$

Thus, by reversing B if necessary, we obtain a bi-infinite sequence $B \in \{1,2\}^{\mathbb{Z}}$ such that $m = m(B) = \lambda_0(B)$ and $B_{-5} \ldots B_4 = 1222121122$.

Because $\lambda_n(B) \leq m < B_\infty$ for all $n \in \mathbb{Z}$, we know from Lemma 2.22 that B does not contain the subsequence $2_2 121_2 2_3 121_2 2_3 121_2 21$, and, thus, we can successively apply Lemma 2.21 at the positions $7k$, $k \in \mathbb{N}$, to get that

$$B_{-10} \ldots B_0 B_1 \ldots B_7 \cdots = 2_2 121_2 2_3 12 \overline{1_2 2_3 12}.$$

Moreover, Lemma 2.20 implies that the word $\ldots B_{-11}$ does not contain the subwords 21212, 12121, 2121_3, 1_3212, $2_3121_22_21$ and 12_21_22123.

Furthermore, the subword $12_3121_22_2$ can not appear in $\ldots B_n$ for all $n \leq -11$. Indeed, if this happens, since $m(B) = m < B_\infty$, it would follow from Lemma 2.22 that B does not contain the subsequence $2_2121_22_3121_22_3121_221$ and, hence, one could repeatedly apply Lemma 2.21 to deduce that $B = \overline{1_22_312}$, a contradiction because this would imply $b_\infty < m = m(B) = m(\overline{1_22_312}) = b_\infty$.

In summary, we showed that there exists $N \geq 11$ such that the word $\ldots B_{-N}$ does not contain the subwords 21212, 12121, 2121_3, 1_3212, $2_3121_22_21$, 12_21_22123 and $12_3121_22_2$.

Finally, if the word $\ldots B_{-11}$ contains the subword $2_21_221123_31 = B_{n-4} \ldots B_{n+5}$, since B does not contain the subsequence $121_2212_31_2212_31_2212_2$ (thanks to Lemma 2.22 and the fact that $\lambda_n(B) < B_\infty$ for all $n \in \mathbb{Z}$), then one can apply Lemma 2.21 at the positions $n - 7k$ for all $k \in \mathbb{N}$ to get that

$$\ldots B_{n-7} \ldots B_{n+10} = \overline{212_31_2}212_31_2212_2.$$

This completes the proof of the proposition.   □

2.4.4   Comparison between $M \setminus L$ near α_∞ and the Cantor set X

The description of $(M \setminus L) \cap (b_\infty, B_\infty) = M \cap (b_\infty, B_\infty)$ provided by Propositions 2.7 and 2.8 allows us to compare this piece of $M \setminus L$ with the Cantor set

$$X := \{[0; \gamma] : \gamma \in \{1, 2\}^{\mathbb{Z}} \text{ not containing the subwords in } Q\}$$

where

$$Q := \{21212, 12121, 2121_3, 1_3212, 2_3121_22_21, 12_21_22123, 12_3121_22_2, 2_21_221123_31\}$$

introduced in (2.15) above.

Remark 2.14. Similarly to Remark 2.12, one has that X is a topologically mixing dynamically defined Cantor set. Indeed, since the set of prohibited words Q consist of words whose size is bounded by 10, we can consider a Markov partition determined by the set V of all words of size 9 not containing any factor in Q and, for $\alpha, \beta \in V$, allow the transition $\alpha\beta$ if, and only if, $\alpha\beta$ does not contain any factor in Q.

This Markov partition is topologically mixing because any word (or factor of a word) in V can be continued up to connect with the symmetric block $[1, 22]$. Indeed, it is enough to verify this for the words obtained from the elements of Q by removing their last entry:

$(2)2121 \rightarrow (2)2121\overline{1}, (21)1212 \rightarrow (21)1212\overline{2}, (2)21211 \rightarrow (2)212112\overline{1}$,

$11121 \rightarrow 11121\overline{1}, (2)222121122 \rightarrow (2)222121122\overline{2}$,

$122112122 \rightarrow 122112122\overline{1}, 122212112 \rightarrow 122212112\overline{1}$ and

$(2)221121222 \rightarrow (2)221121222\overline{2}$ (where $(2), (21)$, etc. indicate forced prefixes).

Proposition 2.9. $(M \setminus L) \cap (\alpha_\infty - 10^{-8}, \alpha_\infty + 10^{-8})$ *contains the set*

$$\{[2; \overline{1_2, 2_3, 1, 2}] + [0; 1, 2_3, 1_2, 2, 1, 2_4, \gamma] : 2_3\gamma \in \{1, 2\}^{\mathbb{N}}$$

does not contain the subwords in Q}.

Proof. Consider the sequence

$$B = \gamma^T, 2_4, 1, 2, 1_2, 2_3, 1, 2; \overline{1_2, 2_3, 1, 2}$$

where $2_3\gamma \in \{1, 2\}^{\mathbb{N}}$ does not contain subwords in P and ; serves to indicate the zeroth position.

On one hand,

$$\lambda_0(B) \leq [2; \overline{1_2, 2_3, 1, 2}] + [0; 1, 2_3, 1_2, 2, 1, 2_4, 1, 2, 1] < \alpha_\infty + 10^{-8}$$

and

$$\lambda_0(B) \geq [2; \overline{1_2, 2_3, 1, 2}] + [0; 1, 2_3, 1_2, 2, 1, 2_4, 2, 1] > \alpha_\infty - 10^{-8},$$

and items (a), (b) and (f) of Lemma 2.19 imply that

$$\lambda_n(B) < \alpha_\infty - 10^{-5}$$

for all positions $n \geq -12$ except possibly for $n = 7k$ with $k \geq 1$.

On the other hand,

$$\lambda_{7k}(B) = [2; \overline{1_2, 2_3, 1, 2}] + \left[0; \underbrace{1, 2_3, 1_2, 2, \ldots, 1, 2_3, 1_2, 2}_{k \text{ times}}, 1, 2_3, 1_2, 2, 1, 2_4, \ldots\right]$$

$$< [2; \overline{1_2, 2_3, 1, 2}] + [0; 1, 2_3, 1_2, 2, 1, 2_3, 1_2, 2, 1]$$

$$< [2; \overline{1_2, 2_3, 1, 2}] + [0; 1, 2_3, 1_2, 2, 1, 2_4]$$

$$< [2; \overline{1_2, 2_3, 1, 2}] + [0; 1, 2_3, 1_2, 2, 1, 2_4, \ldots] = \lambda_0(B),$$

so that $\lambda_0(B) - \lambda_{7k}(B) > [0; 1, 2_3, 1_2, 2, 1, 2_4] - [0; 1, 2_3, 1_2, 2, 1, 2_3, 1_2, 2, 1] > 10^{-9}$ for all $k \geq 1$.

Moreover, since $2_3\gamma$ does not contain subwords in Q, it follows from (the proof of) Lemma 2.20 that $\lambda_n(B) < \alpha_\infty - 10^{-5}$ for all $n \leq -13$.

This shows that $m(B) = \lambda_0(B) = [2; \overline{1_2, 2_3, 1, 2}] + [0; 1, 2_3, 1_2, 2, 1, 2_4, \gamma]$ belongs to $(M \setminus L) \cap (\alpha_\infty - 10^{-8}, \alpha_\infty + 10^{-8})$. $\qquad \square$

Proposition 2.10. $(M \setminus L) \cap (b_\infty, B_\infty)$ *is contained in the union of*

$$\mathcal{C}'' = \{[2; \overline{1_2, 2_3, 1, 2}] + [0; 1, 2_3, 1_2, 2, 1, 2_2, \theta, \overline{1_2, 2_3, 1, 2}] :$$
$$\theta \text{ is a finite word in } 1 \text{ and } 2\}$$

and the sets

$$\mathcal{D}''(\delta) = \{[2; \overline{1_2, 2_3, 1, 2}] + [0; 1, 2_3, 1_2, 2, 1, 2_2, \delta, \gamma] :$$
$$\text{no subword of } \gamma \in \{1, 2\}^{\mathbb{N}} \text{ belongs to } Q\},$$

where δ is a finite word in 1 and 2.

Proof. By Proposition 2.8, if $m \in (M \setminus L) \cap (b_\infty, B_\infty)$, then $m = m(B) = \lambda_0(B)$ with

$$B = \gamma^T \delta^T 2_2 121_2 2_3 12^* \overline{1_2 2_3 12}$$

where the asterisk indicates the zeroth position, δ is a finite word in 1 and 2, and the infinite word γ satisfies:

(1) γ^T does not contain the subwords 21212, 12121, 2121$_3$, 1$_3$212, 2$_3$121$_2$2$_2$1, 12$_2$1$_2$212$_3$ and 12$_3$121$_2$2$_2$
(2) if γ^T contains 2$_2$1$_2$212$_3$1, then $\gamma^T = \overline{212_3 1_2} \mu^T$ with μ a finite word in 1 and 2.

It follows that:

(1) if γ^T contains 2$_2$1$_2$212$_3$1, then

$$m(B) = \lambda_0(B) = [2; \overline{1_2, 2_3, 1, 2}] + [0; 1, 2_3, 1_2, 2, 1, 2_2, \delta, \mu, \overline{1_2, 2_3, 1, 2}]$$

where $\theta = \delta\mu$ is a finite word in 1 and 2, i.e., $m(B) \in \mathcal{C}''$;
(2) otherwise,

$$m(B) = \lambda_0(B) = [2; \overline{1_2, 2_3, 1, 2}] + [0; 1, 2_3, 1_2, 2, 1, 2_2, \delta, \gamma]$$

where γ does not contain the subwords 21212, 12121, 2121$_3$, 1$_3$212, 2$_3$121$_2$2$_2$1, 12$_2$1$_2$212$_3$, 12$_3$121$_2$2$_2$ and 2$_2$1$_2$212$_3$1, i.e., $m(B) \in \mathcal{D}''(\delta)$.

This completes the proof of the proposition. $\square$

2.4.5 *Proof of Theorem 2.10*

By putting together Propositions 2.9 and 2.10, we can derive Theorem 2.10. Indeed, by Proposition 2.4, $(M \setminus L) \cap (b_\infty, B_\infty)$ contains a set bi-Lipschitz equivalent to X and, hence,

$$HD((M \setminus L) \cap (b_\infty, B_\infty)) \geq HD(X).$$

By Proposition 2.5, $(M \setminus L) \cap (b_\infty, B_\infty)$ is contained in

$$\mathcal{C}'' \cup \bigcup_{n \in \mathbb{N}} \left(\bigcup_{\delta \in \{1,2\}^n} \mathcal{D}''(\delta) \right).$$

Since $\mathcal{C}''$ is a countable set and $\{\mathcal{D}''(\delta) : \delta \in \{1,2\}^n, n \in \mathbb{N}\}$ is a countable family of subsets bi-Lipschitz equivalent to X, it follows that

$$HD((M \setminus L) \cap (b_\infty, B_\infty)) \leq HD(X).$$

This proves Theorem 2.10.

2.4.6 *Proof of Theorem 2.9*

Note that the definition of X in (2.15) implies that X contains[8] the Gauss-Cantor set $K(\{1, 2_2\})$. Thus, Theorem 2.10 implies that the proof of Theorem 2.9 is reduced to show that:

Proposition 2.11. *One has $HD(K(\{1, 2_2\})) > 0.353$.*

Proof. The convex hull of $K(\{1, 2_2\})$ is the interval J with extremities $[0; \overline{2}]$ and $[0; 1, \overline{2}]$. The images $J_1 := \phi_1(J)$ and $J_{22} := \phi_{22}(J)$ of J under the inverse branches

$$\phi_1(x) := \frac{1}{1 + \frac{1}{x}} \quad \text{and} \quad \phi_{22}(x) := \frac{1}{2 + \frac{1}{2 + \frac{1}{x}}}$$

of the first two iterates of the Gauss map $G(x) := \{1/x\}$ provide the first step of the construction of the Cantor set $K(\{1, 2_2\})$. In general, given $n \in \mathbb{N}$, the collection $\mathcal{R}^n$ of intervals of the nth step of the construction of $K(\{1, 2_2\})$ is given by

$$\mathcal{R}^n := \{\phi_{x_1} \circ \cdots \circ \phi_{x_n}(I) : (x_1, \ldots, x_n) \in \{1, 22\}^n\}.$$

---

[8]Interestingly enough, after the publication of [MM2 (2019)], the authors became aware that the fact that $(M \setminus L) \cap (b_\infty, B_\infty)$ contains a bi-Lipschitz copy of the Gauss-Cantor set $K(\{1_2, 2_2\})$ was first established in the PhD thesis of Yu [Y (1976)] from 1976 in order to prove that $M \setminus L$ is uncountable.

Similarly to the argument used in the proof of Proposition 2.6, we have that $\alpha_n \leq HD(K(\{1, 2_2\})) \leq \beta_n$ for all $n \in \mathbb{N}$, where

$$\sum_{R \in \mathcal{R}^n} \left(\frac{1}{\Lambda_{n,R}} \right)^{\alpha_n} = 1 = \sum_{R \in \mathcal{R}^n} \left(\frac{1}{\lambda_{n,R}} \right)^{\beta_n}$$

and

$$\lambda_{n,R} := \inf_{x \in R} |(\Psi^n)'(x)|$$

$$= \min \left\{ \prod_{i=1}^{k} \left(\frac{1}{[0; x_i, \ldots, x_k, \overline{2}]} \right)^2, \prod_{i=1}^{k} \left(\frac{1}{[0; x_i, \ldots, x_k, 1, \overline{2}]} \right)^2 \right\},$$

and

$$\Lambda_{n,R} := \sup_{y \in R} |(\Psi^n)'(y)|$$

$$= \max \left\{ \prod_{i=1}^{k} \left(\frac{1}{[0; x_i, \ldots, x_k, \overline{2}]} \right)^2, \prod_{i=1}^{k} \left(\frac{1}{[0; x_i, \ldots, x_k, 1, \overline{2}]} \right)^2 \right\}$$

for $R = \phi_{y_1} \circ \ldots \phi_{y_n}(I) \in \mathcal{R}^n$ associated to a string $\{1, 22\}^n = (y_1, \ldots, y_n) = (x_1, \ldots, x_k) \in \{1, 2\}^k$.

A computer search for the values of α_{12} and β_{12} reveals that

$$\alpha_{12} = 0.353465... \quad \text{and} \quad \beta_{12} = 0.357917...$$

In particular, $0.353 < \alpha_{12} \leq HD(K(\{1, 2_2\})) \leq \beta_{12} < 0.35792$, so that the proof of Proposition 2.11 and, *a fortiori*, Theorem 2.9 is now complete. $\qquad\qquad\square$

2.4.7   *Full description of* $(M \setminus L) \cap (b_\infty, B_\infty)$

Closing our discussions related to Theorem 2.9, let us take the opportunity to complement the original article [MM2 (2019)] by giving a complete description of $(M \setminus L) \cap (b_\infty, B_\infty)$.

Theorem 2.11. *$(M \setminus L) \cap (b_\infty, B_\infty)$ is the union of*

$$\mathcal{C}_1 = \{ [2; \overline{1_2, 2_3, 1, 2}] + [0; 1, 2_3, 1_2, 2, 1, 2_4, \theta, 2_2, 1, 2, \overline{1_2, 2_3, 1, 2}] :$$
$$\theta \text{ is a finite word in } 1 \text{ and } 2$$

such that $[0; 2_2, \theta] \geq [0; \theta^T, 2_2]$ *and no subword of* $2_3 \theta 2_2 1$ *belongs to* $Q\}$,

$$\mathcal{C}_2 = \{ [2; \overline{1_2, 2_3, 1, 2}] + [0; 1, 2_3, 1_2, 2, 1, 2_2, 1_2, \theta, 2_2, 1, 2, \overline{1_2, 2_3, 1, 2}] :$$
$$\theta \text{ is a finite word in } 1 \text{ and } 2$$

such that $[0; 1_2, \theta] \geq [0; \theta^T, 1_2]$ *and no subword of* $1_2 2 1_2 \theta 2_2 1$ *belongs to* $Q\}$,

$$\mathcal{C}_3 = \{[2; \overline{1_2, 2_3, 1, 2}] + [0; 1, 2_3, 1_2, 2, 1, 2_2, 1, 2_2, \theta, 2_2, 1, 2, \overline{1_2, 2_3, 1, 2}] :$$

θ *is a finite word in 1 and 2*

such that $[0; 1, 2_2, \theta] \geq [0; \theta^T, 2_2, 1]$ *and no subword of* $21 2 \theta 2_2 1$ *belongs to* $Q\}$,

$$X_1 = \{[2; \overline{1_2, 2_3, 1, 2}] + [0; 1, 2_3, 1_2, 2, 1, 2_4, 1, 2, \overline{1_2, 2_3, 1, 2}]\},$$

$$X_2 = \{[2; \overline{1_2, 2_3, 1, 2}] + [0; 1, 2_3, 1_2, 2, 1, 2_2, 1, 2_2, 1, 2, \overline{1_2, 2_3, 1, 2}]\},$$

and the sets

$$D_1 = \{[2; \overline{1_2, 2_3, 1, 2}] + [0; 1, 2_3, 1_2, 2, 1, 2_4, \gamma] : 2_3\gamma \in \{1, 2\}^{\mathbb{N}}$$

does not contain the subwords in $Q\}$,

$$D_2 = \{[2; \overline{1_2, 2_3, 1, 2}] + [0; 1, 2_3, 1_2, 2, 1, 2_2, 1_2, \gamma] : 1_2 2 1_2 \gamma \in \{1, 2\}^{\mathbb{N}}$$

does not contain the subwords in $Q\}$

and

$$D_3 = \{[2; \overline{1_2, 2_3, 1, 2}] + [0; 1, 2_3, 1_2, 2, 1, 2_2, 1, 2_2, \gamma] : 21 2\gamma \in \{1, 2\}^{\mathbb{N}}$$

does not contain the subwords in $Q\}$.

Proof. It follows from Proposition 2.8 and the following remarks: the fact that $D_1 \subset (M \setminus L) \cap (b_\infty, B_\infty)$ is the statement of Proposition 2.4, and the fact that $D_2 \cup D_3 \cup X_1 \cup X_2 \subset (M \setminus L) \cap (b_\infty, B_\infty)$ can be proven in the same way, using the estimates of Lemma 2.18 (notice that X_1 and X_2 correspond to sequences θ of "negative sizes" in the cases of $\mathcal{C}_1$ and $\mathcal{C}_3$). For the discrete part $\mathcal{C}_1 \cup \mathcal{C}_2 \cup \mathcal{C}_3$, the conditions $[0; 2_2, \theta] \leq [0; \theta^T, 2_2]$ (resp $[0; 1_2, \theta] \leq [0; \theta^T, 1_2]$ and $[0; 1, 2_2, \theta] \leq [0; \theta^T, 2_2, 1]$) stand to guarantee that the Markov value is indeed attained in position 0, in the notation of Proposition 2.8. □

2.5 The complement of the Lagrange spectrum in the Markov spectrum III

The constructions of elements of $M \setminus L$ by Freiman and Flahive near $3.11812\ldots$ and $3.29304\ldots$ were conjectured by Cusick to cease to work beyond $\sqrt{12}$: in fact, one reads at page 516 of his article [C (1975)] from 1975 the phrase "I think it is likely that L and M coincide above $\sqrt{12} = 3.46410$".

In a recent work [MM3 (2020)], Matheus–Moreira gave a negative answer to Cusick's conjecture by showing the following result:

Theorem 2.12. *The intersection of* $M \setminus L$ *with the interval* $(3.7, 3.71)$ *has Hausdorff dimension* > 0.53128 *(and, a fortiori,* $(M \setminus L) \cap (3.7, 3.71) \neq \emptyset$*).*

Remark 2.15. Actually, we will explain below how to slightly modify the original arguments in [MM3 (2020)] to derive that $HD((M \setminus L) \cap (3.7, 3.71)) > 0.53224$.

The proof of Theorem 2.12 is based on some qualitative *dynamical* insights leading to a series of quantitative estimates with continued fractions.

More precisely,[9] we know that the Lagrange and Markov spectra are obtained by studying the height function $f((b_n)_{n \in \mathbb{Z}}) := \lambda_0((b_n)_{n \in \mathbb{Z}}) = [b_0; b_1, \dots] + [0; b_{-1}, \dots]$ along the orbits of the shift dynamics $\sigma : \Sigma \to \Sigma$ on the shift space $\Sigma = (\mathbb{N}^*)^{\mathbb{Z}}$. The natural map $\Sigma \to \mathbb{R} \times \mathbb{R}$ sending $(b_i)_{i \in \mathbb{Z}}$ to $([b_0; b_1, \dots], [0; b_{-1}, \dots])$ allows to transfer the shift dynamics $\sigma : \Sigma \to \Sigma$ and the height function $f : \Sigma \to \mathbb{R}$ to the plane $\mathbb{R}^2$: in this way, we obtain geometric realizations of σ and f, namely, σ becomes the natural extension $\widetilde{G} : (\mathbb{R} \setminus \mathbb{Q}) \times (\mathbb{R} \setminus \mathbb{Q}) \cap (0, 1)) \to (\mathbb{R} \setminus \mathbb{Q}) \times (\mathbb{R} \setminus \mathbb{Q}) \cap (0, 1))$ of the Gauss map and f becomes $\widetilde{f} : \mathbb{R}^2 \to \mathbb{R}$, $\widetilde{f}(x, y) := x + y$.

From this geometric point of view, the constructions of Freiman and Flahive of some elements of $M \setminus L$ can be thought as follows. First, they choose a periodic word $\dots \alpha \alpha \dots$, i.e., a periodic point $p_\alpha \in \mathbb{R}^2$ of $\widetilde{G}$ with a certain Markov value $\ell = \widetilde{f}(p_\alpha) = \max_{n \in \mathbb{Z}} \widetilde{f}(\widetilde{G}^n(p_\alpha))$. Secondly, they pick a Cantor set $\Sigma_\alpha \subset \Sigma$ such that the maximum of f on Σ_α is attained at a periodic word generating a periodic point $p_{21} \in \mathbb{R}^2$ of $\widetilde{G}$ with $\ell > \rho = \widetilde{f}(p_{21}) = \max_{n \in \mathbb{Z}} \widetilde{f}(\widetilde{G}^n(p_{21}))$.

In a certain sense, Freiman and Flahive found their examples of elements of $M \setminus L$ by investigating the intersections $W^u_{loc}(\Sigma_\alpha) \cap W^s_{loc}(p_\alpha)$ and $W^s_{loc}(\Sigma_\alpha) \cap W^u_{loc}(p_\alpha)$ between the local stable and unstable sets of Σ_α and $\dots \alpha \alpha \dots$. Here, they picked p_α so that its local invariant manifolds intersect the invariant manifolds of the subset $\Lambda_\alpha \subset \mathbb{R}^2$ related to Σ_α at *distinct* heights with respect to $\widetilde{f}(x, y) = x + y$. In fact, one can show that the height μ of the point $\{q_\alpha\} := W^u_{loc}(p_\alpha) \cap W^s_{loc}(p_{21})$ is *strictly smaller* than the minimal height ν of *any* point $r \in W^s_{loc}(p_\alpha) \cap W^u_{loc}(\Lambda_\alpha)$: this situation is depicted in Figure 2.1 below and we will call it *replication mechanism* in the sequel.

Moreover, the $\widetilde{G}$-orbit of q_α is *locally unique* in the sense that some portion of the $\widetilde{G}$-orbit of any point $z \in \mathbb{R}^2$ with $\sup_{n \in \mathbb{Z}} \widetilde{f}(\widetilde{G}^n(z))$ close to μ must stay close to the first few $\widetilde{G}$-iterates of q_α: this feature is called *local uniqueness* in what follows.

From the point of view of continued fractions, the previous paragraphs

---

[9]In the sequel, we assume some familiarity with basic aspects of the standard theory of hyperbolic sets (and we recommend the book [HK (1995)] for all necessary details).

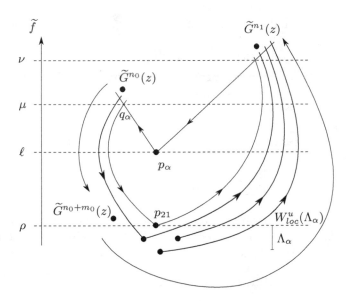

Fig. 2.1   Dynamics behind some elements in $M \setminus L$.

can be implemented as follows. Flahive introduced in her paper [Gb (1977)] the notion of *semi-symmetric* words[10] and, in our language, she proved that the conditions for the replication mechanism above are often met by the periodic points p_α associated to *non* semi-symmetric word α. In particular, it is not surprising *a posteriori* to realize that Freiman's elements σ and α_∞ in $M \setminus L$ described in §2.3 and §2.4 above are related to the non semi-symmetric words (of odd lengths[11]) 222211221 and 2112221.

For our current purpose of establishing Theorem 2.12, we have chosen the non semi-symmetric word (of odd length) $\alpha = 3322212$. Once this choice is made, we compute the Markov value ℓ of the periodic sequence $\ldots \alpha\alpha \ldots$, and we select a Cantor set $\Sigma_\alpha := \{1, 2\}^{\mathbb{Z}}$ because Proposition 1.7 ensures that Markov values of sequences in Σ_α are $\leq \sqrt{12} < \ell$. In this setting, the study of the heights of intersections of local invariant manifolds (related to the replication mechanism) translates into the question of getting *distinct* answers for the Markov values of sequences obtained by gluing

---

[10]A finite string is semi-symmetric if it is a palindrome or a concatenation of two palindromes.

[11]We insist on non semi-symmetric words of *odd* length because any modification of the associated infinite periodic sequence will force a definite increasing of the Markov value in one of two consecutive periods.

words in Σ_α to the left and to the right of $\ldots\alpha\alpha\ldots$. More concretely, since α is *not* semi-symmetric, if α decomposes as $\alpha = xy$, then the Markov values μ of $\ldots\alpha\alpha z = \ldots xyxyz$ with $z \in \Sigma_\alpha$ could be $\mu > \ell$ and *systematically* smaller than the Markov values ν of $w\alpha\alpha\cdots = wxyxy\ldots$ with $w \in \Sigma_\alpha$ (because the gluings of y and z is a different problem from the gluings of w and x). For example, if we try to glue the sequence $2121\cdots \in \Sigma_\alpha$ on the right of the periodic sequence $\ldots\alpha\alpha\cdots = \ldots 33222123322212\ldots$ without increasing too much the Markov value of the resulting sequence, we might go for

$$\ldots 33222123322212212121\ldots$$

whose Markov value μ is $3.70969985975\ldots$ On the other hand, if we try to glue $2121\cdots \in \Sigma_\alpha$ on the left of $\ldots\alpha\alpha\cdots = \ldots 33222123322212\ldots$ without increasing too much the Markov value, the best choice is

$$\ldots 21212122123322212332212\ldots$$

$$\ldots 212121221233222123322212\ldots$$

whose Markov value ν is $3.70969985982\ldots$: this replication mechanism is the content of Lemma 2.31 below.

In other words, the cost of gluing any $w \in \Sigma_\alpha$ and $\alpha\alpha\ldots$ is always higher than the cost of the sequence $\ldots\alpha\alpha z$. Hence, if we have *local uniqueness*, the Markov value μ of $\ldots\alpha\alpha z$ is likely to belong to the complement of L because any attempt to modify the left side of $\ldots\alpha\alpha z$ to reproduce big chunks of this sequence (in order to show that $\mu \in L$) would fail since it ends up producing a subword close to the sequence $z\alpha\alpha\ldots\alpha\alpha z$ whose Markov value would be $\nu > \mu$. The relevant local uniqueness property is established in Corollary 2.10 below.

After this discussion of the key ideas behind Theorem 2.12, let us now start the discussion of its proof. Since we are interested in the portion of $M \setminus L$ between 3.7 and 3.71, we deal exclusively with sequences $a = (a_n)_{n\in\mathbb{Z}} \in \{1, 2, 3\}^{\mathbb{Z}}$ in the sequel.

2.5.1 *Local uniqueness of candidate sequences*

We begin by proving the relevant local uniqueness property mentioned above. For this sake, we collect *without* giving proofs a series of elementary lemmas in the spirit of the "forbidden and allowed strings" discussions from §2.3 and §2.4 and we present only the proofs of a few corollaries: the reader is invited to consult the original article [MM3 (2020)] for all details skipped here.

Lemma 2.23.

  (i) $\lambda_0(\ldots 3^*1\ldots) > 3.822$
 (ii) $\lambda_0(\ldots 23^*2\ldots) > 3.7165$
(iii) $\lambda_0(\ldots 33^*3\ldots) < 3.61279.$

An immediate corollary of this lemma is:

Corollary 2.4. *If* $3.62 < \lambda_0(a) < 3.71$, *then* $a = \ldots 33^*2\ldots$ *up to transposition.*

Lemma 2.24.

 (iv) $\lambda_0(\ldots 33^*21\ldots) < 3.6973$
  (v) $\lambda_0(\ldots 33^*23\ldots) > 3.72$ *or* $\lambda_{-1}(\ldots 33^*23\ldots) > 3.822.$

Corollary 2.5. *If* $3.698 < \lambda_0(a) < 3.71$ *and* $\lambda_i(a) < 3.71$ *for* $|i| \leq 1$, *then* $a = \ldots 33^*22\ldots$ *up to transposition.*

Lemma 2.25.

  (vi) $\lambda_0(\ldots 333^*22\ldots) > 3.71$
 (vii) $\lambda_i(\ldots 233^*221\ldots) > 3.7099$ *for some* $i \in \{-3, 0, 5\}$
(viii) $\lambda_0(\ldots 233^*223\ldots) < 3.7087$
  (ix) $\lambda_0(\ldots 3233^*222\ldots) \leq \lambda_0(\ldots 2233^*222\ldots) < 3.7084.$

Corollary 2.6. *If* $3.7087 < \lambda_0(a) < 3.7099$ *and* $\lambda_i(a) < 3.7099$ *for* $|i| \leq 5$, *then* $a = \ldots 1233^*222\ldots$ *up to transposition.*

Lemma 2.26.

   (x) $\lambda_i(\ldots 1233^*2223\ldots) > 3.7099$ *for some* $i \in \{-5, 0\}$
  (xi) $\lambda_0(\ldots 11233^*2221\ldots) < 3.7096$
 (xii) $\lambda_0(\ldots 21233^*2222\ldots) > 3.71$
(xiii) $\lambda_0(\ldots 21233^*22211\ldots) > 3.7097$
(xiv) $\lambda_0(\ldots 111233^*22221\ldots) \geq \lambda_0(\ldots 111233^*22222\ldots) > 3.7097$
 (xv) $\lambda_i(\ldots 111233^*22223\ldots) > 3.7097$ *for some* $i \in \{-7, 0, 5\}$
(xvi) $\lambda_0(\ldots 211233^*22223\ldots) \leq \lambda_0(\ldots 211233^*22222\ldots) < 3.70957$
(xvii) $\lambda_i(\ldots 121233^*22212\ldots) > 3.7097$ *for some* $i \in \{-7, 0, 7\}$
(xviii) $\lambda_0(\ldots 321233^*22212\ldots) < 3.709604.$

Corollary 2.7. *If* $3.709604 < \lambda_0(a) < 3.7097$ *and* $\lambda_i(a) < 3.7097$ *for* $|i| \leq 7$, *then* $a = \ldots 221233^*22212\ldots$ *or* $\ldots 211233^*22221\ldots$ *up to transposition.*

Proof. By Corollary 2.6, $a = \ldots 1233^*222\ldots$. By Lemma 2.26 (x), $a = \ldots 1233^*2221\ldots$ or $\ldots 1233^*2222\ldots$. By Lemma 2.23 (i), Lemma 2.26 (xi), (xii), $a = \ldots 21233^*2221\ldots$ or $\ldots 11233^*2222\ldots$

By Lemma 2.23 (i), Lemma 2.26 (xiii), (xiv), (xv), (xvi), $a = \ldots 21233^*22212\ldots$ or $\ldots 11233^*22221\ldots$

By Lemma 2.23 (i), Lemma 2.26 (xiv), (xvii), (xviii), $a = \ldots 221233^*22212\ldots$ or $\ldots 211233^*22221\ldots$ $\square$

Lemma 2.27.

(xix) $\lambda_0(\ldots 221233^*222121\ldots) < 3.709642$

(xx) $\lambda_0(\ldots 1221233^*222122\ldots) \leq \lambda_0(\ldots 2221233^*222122\ldots) < 3.709693$

(xxi) $\lambda_0(\ldots 1221233^*2221233\ldots) < 3.70968$

(xxii) $\lambda_0(\ldots 3221233^*2221233\ldots) > 3.7097$

(xxiii) $\lambda_0(\ldots 1211233^*222211\ldots) \leq \lambda_0(\ldots 2211233^*222211\ldots) < 3.70969$

(xxiv) $\lambda_0(\ldots 1211233^*222212\ldots) < 3.70969$

(xxv) $\lambda_0(\ldots 3211233^*222212\ldots) > 3.7097$.

Corollary 2.8. *If* $3.709693 < \lambda_0(a) < 3.7097$ *and* $\lambda_i(a) < 3.7097$ *for* $|i| \leq 9$, *then* $a = \ldots 2221233^*2221233\ldots$ *or* $\ldots 33211233^*222211\ldots$ *or* $\ldots 2211233^*222212\ldots$ *up to transposition.*

Proof. By Corollary 2.7, $a = \ldots 221233^*22212\ldots$ or $\ldots 211233^*22221\ldots$. By Lemma 2.27 (xix) and Lemma 2.23 (i), (ii), the sole possible extensions a are

$$\ldots 221233^*222122\ldots \qquad \ldots 221233^*2221233\ldots$$

$$\ldots 211233^*222211\ldots \qquad \ldots 211233^*222212\ldots$$

By Lemma 2.27 (xx), (xxi), (xxii), (xxiii), (xxiv), (xxv), the sole possible continuations of a are

$$\ldots 3221233^*222122\ldots \qquad \ldots 2221233^*2221233\ldots$$

$$\ldots 3211233^*222211\ldots \qquad \ldots 2211233^*222212\ldots$$

By Lemma 2.23 (i), (ii) and Lemma 2.25 (vi), we obtain that a is one of the words

$$\ldots 233221233^*222122\ldots \qquad \ldots 2221233^*2221233\ldots$$

$$\ldots 33211233^*222211\ldots \qquad \ldots 2211233^*222212\ldots$$

However, Lemma 2.25 (vii) forbids the word $\ldots 233221233^*222122\ldots$, so that we end up with the following three possibilities

$$\ldots 2221233^*2221233\ldots$$

$$\ldots 33211233^*222211\ldots \qquad \ldots 2211233^*222212\ldots$$

for a. □

Lemma 2.28.

(xxvi) $\lambda_i(\ldots 2221233^*22212333\ldots) > 3.7097$ *for some* $i \in \{-7,0\}$

(xxvii) $\lambda_0(\ldots 33211233^*2222112\ldots) < 3.70968$

(xxviii) $\lambda_0(\ldots 2211233^*2222121\ldots) > 3.7097$

(xxix) $\lambda_i(\ldots 2211233^*2222122\ldots) > 3.7097$ *for some* $i \in \{-7,0\}$.

Corollary 2.9. *If* $3.709698 < \lambda_0(a) < 3.7097$ *and* $\lambda_i(a) < 3.7097$ *for* $|i| \leq 9$, *then* $a = \ldots 2221233^*22212332\ldots$ *or* $\ldots 33211233^*2222111\ldots$ *or* $\ldots 2211233^*22221233\ldots$ *up to transposition.*

Proof. By Corollary 2.8, $a = \ldots 2221233^*2221233\ldots$ or $\ldots 33211233^*222211\ldots$ or $\ldots 2211233^*222212\ldots$. By Lemma 2.23 (i), (ii) and Lemma 2.28, we see that the sole possible continuations of these words are $a = \ldots 2221233^*22212332\ldots$ or $\ldots 33211233^*2222111\ldots$ or $\ldots 2211233^*22221233\ldots$ □

Lemma 2.29.

(xxx) $\lambda_0(\ldots 12221233^*22212332\ldots) > \lambda_0(\ldots 22221233^*22212332\ldots) > 3.7097$

(xxxi) $\lambda_0(\ldots 12211233^*22221233\ldots) > \lambda_0(\ldots 22211233^*22221233\ldots) > 3.7097$

(xxxii) $\lambda_0(\ldots 2332211233^*222212333\ldots) < \lambda_0(\ldots 2332211233^*222212332\ldots) < 3.7096992$

(xxxiii) $\lambda_0(\ldots 2332221233^*222123321\ldots) > 3.7096999$

(xxxiv) $\lambda_0(\ldots 233211233^*22221111\ldots) < \lambda_0(\ldots 333211233^*22221111\ldots) < 3.709696$

(xxxv) $\lambda_0(\ldots 333211233^*22221112\ldots) > \lambda_0(\ldots 233211233^*22221112\ldots) > 3.7097$.

At this point, we are ready to prove the following key local uniqueness property:

Corollary 2.10. *If* $3.7096992 < \lambda_0(a) < 3.7096999$ *and* $\lambda_i(a) < 3.7096999$ *for* $|i| \leq 9$, *then*

$$a = \ldots 2332221233^*222123322 \ldots$$

up to transposition.

Proof. By Corollary 2.9, a is one of the words $\ldots 2221233^*22212332 \ldots$ or $\ldots 33211233^*2222111 \ldots$ or $\ldots 2211233^*22221233 \ldots$

By Lemma 2.29 (xxx), (xxxi), Lemma 2.23 (i), (ii), and Lemma 2.25 (vi), the sole possible continuations for these words are

$$\ldots 2332221233^*22212332 \ldots \qquad \ldots 233211233^*2222111 \ldots$$

$$\ldots 333211233^*2222111 \ldots \qquad \ldots 2332211233^*22221233 \ldots$$

However, Lemma 2.23 (i) and Lemma 2.29 (xxxii), (xxxiv), (xxxv) rules out all possibilities except for

$$a = \ldots 2332221233^*22212332 \ldots$$

Finally, Lemma 2.29 (xxxiii) and Lemma 2.24 (v), this word is forced to extend as

$$a = \ldots 2332221233^*222123322 \ldots$$

$\square$

2.5.2 *Replication mechanism*

Once again, we recall without proofs here two lemmas from [MM3 (2020)] and we derive as a corollary the relevant replication mechanism.

Lemma 2.30.

(xxxvi) $\lambda_i(\ldots 2_23_22_21233^*2_3123_22_2 \ldots) > 3.70969986$ *for some* $i \in \{0, 7\}$
(xxxvii) $\lambda_0(\ldots 123_22_31233^*2_3123_22_23 \ldots) > 3.70969986$
(xxxviii) $\lambda_0(\ldots 1_223_22_31233^*2_3123_22_2312 \ldots) > 3.70969986$
(xxxix) $\lambda_0(\ldots 32123_22_31233^*2_3123_22_312 \ldots) > 3.7096998599$.

Lemma 2.31. $\lambda_i(\ldots 12212332221233^*222123322212 \ldots)$ $>$ 3.70969985975033 *for some* $i \in \{-17, -15, 0, 13, 15\}$.

Corollary 2.11. *Let* $a = \ldots 2332221233^*222123322\ldots$ *where the asterisk indicates the position* $j \in \mathbb{Z}$. *If* $\lambda_i(a) < 3.70969985975033$ *for all* $|i-j| \leq 17$, *then*

$$a = \ldots 23322212332221233^*222123322212\ldots$$

and the vicinity of the position $j - 7$ *is* $\ldots 2332221233^*222123322\ldots.$

Proof. By Lemma 2.24 (v) and Lemma 2.30 (xxxvi), our word must extends as

$$a = \ldots 12332221233^*222123322\ldots$$

By Lemma 2.25 (vii) and Lemma 2.30 (xxxvii), our word is forced to continue as

$$a = \ldots 12332221233^*2221233222\ldots$$

By Lemma 2.23 (i), we have the following possibilities

$$\ldots 112332221233^*2221233222\ldots \quad \text{or} \quad \ldots 212332221233^*2221233222\ldots$$

for the word a.

By Lemma 2.26 (x), (xii), these two words can continue only as

$$\ldots 112332221233^*22212332221\ldots \quad \text{or} \quad \ldots 212332221233^*22212332221\ldots$$

By Lemma 2.26 (xiii), these words are obliged to extend as

$$\ldots 112332221233^*222123322212\ldots \text{ or } \ldots 212332221233^*222123322212\ldots$$

However, we can apply Lemma 2.30 (xxxviii) to rule out the first case, so that

$$a = \ldots 212332221233^*222123322212\ldots$$

By Lemma 2.26 (xvii) and Lemma 2.30 (xxxix), this word continues as

$$a = \ldots 2212332221233^*222123322212\ldots$$

By Lemma 2.31 and Lemma 2.27 (xxii), we have to extend as

$$a = \ldots 22212332221233^*222123322212\ldots$$

By Lemma 2.29 (xxx), we are forced to continue as

$$a = \ldots 322212332221233^*222123322212\ldots$$

Finally, Lemma 2.23 (i), (ii) and Lemma 2.25 (vi) reveal that

$$a = \ldots 23322212332221233^*222123322212\ldots$$

$\square$

2.5.3   End of the proof of Theorem 2.12

The local uniqueness property and replication mechanism allow to exhibit a gap of the Lagrange spectrum near 3.7:

Proposition 2.12. $L \cap (3.70969985968, 3.70969985975033) = \emptyset$.

Proof. Suppose that $\ell \in L \cap (3.70969985968, 3.70969985975033)$ and let $a \in \{1, 2, 3\}^{\mathbb{Z}}$ be a sequence such that $\ell = \limsup_{n \to \infty} \lambda_n(a)$. By repeatedly applying Corollaries 2.10 and 2.11, we would deduce that

$$\ell = \lambda_0(\overline{33^*22212}) = 3.709699859679 \cdots < 3.70969985968$$

a contradiction.   □

On the other hand, a straightforward computation shows that the gap above contains many elements of M:

Proposition 2.13. $C := \{\lambda_0(\overline{332221233}^*2221233221221212120) : \theta \in \{1, 2\}^{\mathbb{N}}\}$ *is contained in* $M \cap (3.70969985975024, 3.70969985975028)$.

The previous two propositions together with Jenkinson–Pollicott work [JePo2 (2018)] on the computation of the Hausdorff dimension of $C(2) := \{[0; \theta] : \theta \in \{1, 2\}^{\mathbb{N}}\}$ imply that the Hausdorff dimension of $(M \setminus L) \cap (3.7096, 3.7097)$ is

$$\geq 0.531280506277205141624468647368471785493059109018 3 \ldots$$

In particular, this proves Theorem 2.12.

As it turns out, the estimate above is fairly close to the bound announced in Remark 2.15. Nevertheless, before proving the claim in Remark 2.15, we need a better control of $M \setminus L$ near C.

2.5.4   Full description of $M \setminus L$ near 3.7

Let us now take the opportunity to give a complete description of $M \setminus L$ near the set C from Proposition 2.13: for this sake, we will determine the maximal gap (j_0, j_1) of the Lagrange spectrum containing the set C and we will exhibit a Cantor set Ω of continued fraction expansions such that $HD(\Omega) = HD((M \setminus L) \cap (j_0, j_1))$.

Consider the quantities

$$j_0 := \lambda_0(\overline{33^*22212}) = \frac{\sqrt{330629}}{155} = 3.70969985967967 \ldots$$

and

$$j_1 := \lambda_0(\overline{21}12212332221233^*2221233222123332\overline{112}) = 3.70969985975042\ldots$$

By Corollary 2.10, Lemma 2.30, and the proof of Corollary 2.11, we have that

Proposition 2.14. *If $j_0 < m(a) = \lambda_0(a) < 3.7096998599$, then (up to transposition)*

(1) either $a = \ldots 12212332221233^*222123322212\ldots$
(2) or $a = \ldots 23322212332221233^*222123322212\ldots$ and the vicinity of the position -7 is $\ldots 2332221233^*222123322\ldots$

Indeed, this happens because $m(a) < 3.7096998599$ allows to use all results from §2.5.1 and §2.5.2 *except* for Lemma 2.31.

Proposition 2.15. *If $m(a) < 3.71$ and a contains $12212332221233^*222123322212$, then $m(a) \geq j_1$.*

Proof. By Lemma 2.23 (i), (ii),

$$\lambda_0(a) = [3; 2, 2, 2, 1, 2, 3, 3, 2, 2, 2, 1, 2, \ldots]$$
$$+[0; 3, 2, 1, 2, 2, 2, 3, 3, 2, 1, 2, 2, 1, \ldots]$$
$$\geq [3; 2, 2, 2, 1, 2, 3, 3, 2, 2, 2, 1, 2, 3, 3, 3, 2, \ldots]$$
$$+[0; 3, 2, 1, 2, 2, 2, 3, 3, 2, 1, 2, 2, 1, \overline{1, 2}].$$

By Lemma 2.24 (v) and Lemma 2.25 (vi),

$$\lambda_0(a) \geq [3; 2, 2, 2, 1, 2, 3, 3, 2, 2, 2, 1, 2, 3, 3, 3, 2, \ldots]$$
$$+[0; 3, 2, 1, 2, 2, 2, 3, 3, 2, 1, 2, 2, 1, \overline{1, 2}]$$
$$\geq [3; 2, 2, 2, 1, 2, 3, 3, 2, 2, 2, 1, 2, 3, 3, 3, 2, 1, \ldots]$$
$$+[0; 3, 2, 1, 2, 2, 2, 3, 3, 2, 1, 2, 2, 1, \overline{1, 2}].$$

By Lemma 2.23 (i), we conclude that

$$\lambda_0(a) \geq [3; 2, 2, 2, 1, 2, 3, 3, 2, 2, 2, 1, 2, 3, 3, 3, 2, 1, \ldots]$$
$$+[0; 3, 2, 1, 2, 2, 2, 3, 3, 2, 1, 2, 2, 1, \overline{1, 2}]$$
$$\geq [3; 2, 2, 2, 1, 2, 3, 3, 2, 2, 2, 1, 2, 3, 3, 3, 2, 1, \overline{1, 2}]$$
$$+[0; 3, 2, 1, 2, 2, 2, 3, 3, 2, 1, 2, 2, 1, \overline{1, 2}],$$

that is, $\lambda_0(a) \geq j_1$. □

By putting together Propositions 2.14 and 2.15, we obtain that:

Proposition 2.16. *The interval $J = (j_0, j_1)$ is the maximal gap of L containing C.*

Proof. On one hand, the fact that J contains C is an immediate consequence of Proposition 2.13. On the other hand, if $j_0 < m(a) < j_1$ for a periodic sequence a, then, thanks to Proposition 2.15, we would be able to iteratively apply Proposition 2.14 to obtain that $m(a) = \lambda_0(\overline{33^*22212}) = j_0$, a contradiction. Since the Lagrange spectrum is the closure of Markov values associated to periodic sequences, we derive that $(j_0, j_1) \cap L = \emptyset$. In particular, (j_0, j_1) is a maximal gap of L because it is not hard to see that $j_0, j_1 \in L$. This completes the proof. $\square$

At this point, we have that

$$HD((M \setminus L) \cap (j_0, j_1)) = HD(\Omega) \qquad (2.16)$$

where Ω is the Gauss–Cantor set

$$\Omega := \{[0; \gamma] : \gamma \in \{1, 2, 3\}^{\mathbb{N}} \text{ doesn't contain subwords in } R\}$$

with R consisting of the words 13, 232, 323, 1223, 33322, 12332223, 212332222, 2123322211, 1112332222, 121233222, 3211233222212, 2211233222212, 321123322221112, 2221233222123 and their transposes. Indeed, in principle, we should prohibit the "big words"[12] appearing in items (i), (ii), (v), (vi), (vii), (x), (xii), (xiii), (xiv), (xv), (xvii), (xxii), (xxv), (xxvi), (xxviii), (xxix), (xxx), (xxxi), (xxxiii), (xxxv), (xxxvi), (xxxvii), (xxxviii), (xxxix) in §2.5.1 and §2.5.2 and their transposes, and the "self-replicating" word 2332221233222123322 in Corollary 2.11 and its transpose, but an elementary combinatorial analysis shows that to prohibit these words is equivalent to prohibit the words in R.

Remark 2.16. Ω is a topologically mixing dynamically defined Cantor set. Indeed, since the set of prohibited words R consist of words whose size is bounded by 15, we can consider a Markov partition determined by the set U of all words of size 14 not containing any factor in R and, for $\alpha, \beta \in U$, allow the transition $\alpha\beta$ if, and only if, $\alpha\beta$ does not contain any factor in R. This Markov partition is topologically mixing, since any word (or factor of a word) in U can be continued up to connect with the symmetric block

---

[12]In the sense that the appearance of these words implies that the value of λ_0 surpasses j_1.

$[1,2]$ — indeed it is enough to verify this for the words obtained from the elements of R by removing their last entry:

$\quad 1 \quad \to \quad 1\bar{1}, (123)3 \quad \to \quad (123)32\bar{1}, (223)3 \quad \to \quad (223)3\bar{2}, (123_{k+2})3 \quad \to$ $(123_{k+2})32\bar{1}$, $k \geq 0$,

$\quad (123)32 \to (123)32\bar{1}, (223)32 \to (223)32\bar{2}, (33)32 \to (33)32\bar{1}, 122 \to$ $122\bar{2}$,

$\quad (23)322 \to (23)322332\bar{1}, 2233 \to 2233223332\bar{1}, 3332 \to 3332\bar{1}, 1233222 \to$ $123322212\bar{1}$,

$\quad 3222332 \quad \to \quad 3222332\bar{2}, 21233222 \quad \to \quad 212332221 2\bar{1}, 22223321 \quad \to$ $2222332112\bar{1}$,

$\quad 212332221 \to 212332221 2\bar{1}, 112223321 \to 1122233211\bar{1}, 111233222 \to$ $111233222\bar{1}$,

$\quad 222233211 \quad \to \quad 2222332112\bar{1}, 12123322 \to 121233223332\bar{1}, 22233212 \to$ $222332123332\bar{1}$,

$\quad 321123322221 \to 321123322221 1\bar{2}, 212222332112 \to 212222332112\bar{1}$,

$\quad 221123322221 \to 221123322221\bar{1}, 212222332112 \to 212222332112\bar{1}$,

$\quad 321123322222111 \quad \to \quad 321123322222111\bar{1}, 211122223 32112 \quad \to$ $211122223321 12\bar{1}$,

$\quad 222123322212 \to 222123322212\bar{1}$ and $321222332122 \to 321222332122\bar{1}$ (where $(123), (223), (33)$, etc. indicate possible prefixes).

By a direct calculation, we get the following description of $M \setminus L$ near 3.7:

Theorem 2.13. $(M \setminus L) \cap (j_0, j_1)$ *is equal to the union of*

$$C''' = \{[0; \overline{3, 2, 1, 2_3, 3}] + [3; 2_3, 1, 2, 3_2, 2_3, 1, 2_2, 1, 2, 1, 2, 1, \theta, \overline{2, 1, 2_3, 3_2}] :$$
$$\theta \text{ is a finite word in}$$

$$1 \text{ and } 2 \text{ such that } [0; 2, 1, 2, 1, 2, 1, \theta] \geq [0; \theta^T, 1, 2, 1, 2, 1, 2]$$
$$\text{and no subword of } 121\theta 212_3 \text{ belongs to } R\}$$

and

$$\mathcal{D}''' = \{[0; \overline{3, 2, 1, 2_3, 3}] + [3; 2_3, 1, 2, 3_2, 2_3, 1, 2_2, 1, 2, 1, 2, 1, \gamma] :$$
$$\text{no subword of } 121\gamma \in \{1, 2\}^{\mathbb{N}} \text{ belongs to } R\}.$$

2.5.5 *Improved lower bound of the Hausdorff dimension of $M \setminus L$*

At this stage, we are ready to complete the claim in Remark 2.15. In this direction, we observe that Ω contains the Gauss-Cantor set $\tilde{K} = \{[0; \gamma] :$

$\gamma \in \{111, 112, 121, 122, 211, 212, 221, 222, 123321\}^{\mathbb{N}}\}$. In view of (2.16), our task is reduced to prove that

$$HD(\tilde{K}) > 0.53224.$$

Given a finite sequence $\alpha = (a_1, a_2, \dots, a_n) \in (\mathbb{N}^*)^n$, let $s(\alpha) := |I(\alpha)|$ be the length of the interval $I(\alpha) := \{x \in [0,1] \mid x = [0; a_1, a_2, \dots, a_n, \alpha_{n+1}], \alpha_{n+1} \geq 1\}$ whose endpoints are p_n/q_n and $\frac{p_n + p_{n-1}}{q_n + q_{n-1}}$, i.e., $s(\alpha) = \frac{1}{q_n(q_n + q_{n-1})}$ (compare with Chapter 1, §1.5). As it was shown in Lemma A.2 of [Mor1 (2018)], the elementary properties of continued fractions imply that

$$\frac{1}{2}s(\alpha)s(\beta) < s(\alpha\beta) < 2s(\alpha)s(\beta) \tag{2.17}$$

for all finite words α and β.

Given $n, k \in \mathbb{N}$, let $X_{n,k}$ be the set of words obtained from $\theta = \theta_1 \dots \theta_n$, $\theta_i \in \{1, 2\}^3 \ \forall 1 \leq i \leq n$, by inserting k copies of $\gamma = 123321$, and $X_{n,\infty} = \bigcup_{k \geq 0} X_{n,k}$. Next, we introduce the quantities $d_{n,k}$ and d_n defined by

$$\sum_{\tau \in X_{n,k}} s(\tau)^{d_{n,k}} = 1 \quad \text{and} \quad \sum_{\tau \in X_{n,\infty}} s(\tau)^{d_n} = 1,$$

and we set $\hat{d}_n := \sup_{k \geq 0} d_{n,k}$. Note that $d_n \geq d_{n,k}$ for all $n, k \geq 0$, so that $d_n \geq \hat{d}_n$ for all $n \in \mathbb{N}$. On the other hand, for any given $\varepsilon > 0$, $n, k \geq 0$, we have that

$$\sum_{\beta \in X_{n,k}} s(\beta)^{\hat{d}_n} \leq 1 \implies \sum_{\beta \in X_{n,k}} s(\beta)^{\hat{d}_n + \varepsilon} \leq \hat{c}^\varepsilon \left(\frac{1 + \sqrt{5}}{2}\right)^{-2\varepsilon(3n + 6k)},$$

where $\hat{c}$ is a positive constant.[13] Thus, for any $\varepsilon > 0$, there exists $n_\varepsilon \in \mathbb{N}$ such that

$$\sum_{\tau \in X_{n,\infty}} s(\tau)^{\hat{d}_n + \varepsilon} = \sum_{k \geq 0} \sum_{\beta \in X_{n,k}} s(\beta)^{\hat{d}_n + \varepsilon} \leq \frac{\hat{c}^\varepsilon \left(\frac{1 + \sqrt{5}}{2}\right)^{-6\varepsilon n}}{1 - \left(\frac{1 + \sqrt{5}}{2}\right)^{-12\varepsilon}} < 1$$

for all $n \geq n_\varepsilon$.

In summary, we have that, for each $\varepsilon > 0$, there exists $n_\varepsilon \in \mathbb{N}$ such that

$$\hat{d}_n \leq d_n \leq \hat{d}_n + \varepsilon \tag{2.18}$$

---

[13]Here, we are using the facts that $s((b_1, \dots, b_m)) \leq 1/q_m^2$ and the recurrence relations from Chapter 1, Proposition 1.1 imply that $q_m \geq \hat{c}^{-1}(\frac{1 + \sqrt{5}}{2})^m$ for some positive constant $\hat{c}$ and for all $n \in \mathbb{N}$.

for all $n \geq n_\varepsilon$. Now, let us observe that by modifying Proposition E.3 in Appendix E, one can see that $HD(\tilde{K}) \geq d_{n,k} - O(1/n)$, so that

$$HD(\tilde{K}) \geq \hat{d}_n - O(1/n). \tag{2.19}$$

Finally, if $\tau \in X_{n,k}$, say $\tau = \theta_0 \gamma \theta_1 \gamma \ldots \theta_{k-1} \gamma \theta_k$ with $\theta_i \in \{\emptyset\} \cup \{1,2\}^3$, $\theta = \theta_0 \ldots \theta_k \in \{1,2\}^{3n}$, then (2.17) ensures that

$$s(\tau) > \frac{1}{2^{2k}} s(\theta_1) \ldots s(\theta_k) s(\gamma)^k > \frac{1}{2^{3k}} s(\theta) s(\gamma)^k.$$

Therefore,

$$\sum_{\tau \in X_{n,\infty}} s(\tau)^d > \sum_{k \geq 0} \binom{n+k}{k} \sum_{\theta \in \{1,2\}^{3n}} s(\theta)^d \left(\frac{s(\gamma)}{8} \right)^{kd}$$

$$= \sum_{\theta \in \{1,2\}^{3n}} s(\theta)^d \cdot \sum_{k \geq 0} \binom{n+k}{k} t^k,$$

where $t := (\frac{s(\gamma)}{8})^d = 1/161320^d$ (as $s(\gamma) = \frac{1}{109 \cdot 185} = \frac{1}{20165}$). Since $\sum_{k \geq 0} \binom{n+k}{k} x^k = (1-x)^{-(n+1)}$ for $x \in (-1,1)$, the previous estimate (together with (2.17)) yields

$$\sum_{\tau \in X_{n,\infty}} s(\tau)^d > \sum_{\theta \in \{1,2\}^{3n}} s(\theta)^{d_0} s(\theta)^{d-d_0} (1-t)^{-n-1}$$

$$> c_0 \sum_{\theta \in \{1,2\}^{3n}} s(\theta)^{d_0} s(2_n)^{d-d_0} (1-t)^{-n-1}$$

$$> \tilde{c}_0 \sum_{\theta \in \{1,2\}^{3n}} s(\theta)^{d_0} (1+\sqrt{2})^{-2n(d-d_0)} (1-t)^{-n-1},$$

where $d > d_0 = 0.53128$ is arbitrary and c_0 and $\tilde{c}_0$ are positive universal constants. Because $d_0 < HD(C(2))$, the inequality above implies that if $d^* > d_0$ satisfies

$$\left(1 - \frac{1}{161320^{d^*}} \right)^{-1} (1+\sqrt{2})^{-2(d^*-d_0)} \geq 1,$$

then

$$\sum_{\tau \in X_{n,\infty}} s(\tau)^{d^*} > \tilde{c}_0 (1-t)^{-1} \sum_{\theta \in \{1,2\}^{3n}} s(\theta)^{d_0} > 1$$

for all n sufficiently large and, *a fortiori*, $d_n > d^*$ for all n sufficiently large. Since $f(x) = (1 - \frac{1}{161320^x})^{-1} (1+\sqrt{2})^{-2(x-d_0)}$ is decreasing for $x \in (0, +\infty)$ and $f(0.53224) > 1$ (but $f(0.53225) < 1$), we deduce that

$$d_n > 0.53224$$

for all n sufficiently large. In view of (2.18) and (2.19), this completes the proof of the desired estimate $HD(\tilde{K}) > 0.53224$.

2.6   The complement of the Lagrange spectrum in the Markov spectrum IV

We saw in the previous three sections that $M \setminus L$ has a rich fractal structure including some portions with Hausdorff dimension $> 1/2$. Nevertheless, Matheus–Moreira proved in [MM3 (2020)] that there is a bound on the complexity of $M \setminus L$:

Theorem 2.14. *The Hausdorff dimension of $M \setminus L$ is at most* 0.987.

In this section, we will sketch the proof of the following particular case of this theorem:

$$HD((M \setminus L) \cap [\sqrt{5}, \sqrt{12}]) < 0.987. \tag{2.20}$$

Note that this estimate is not conceptually far from Theorem 2.14: on one hand, it is non-trivial because Moreira's theorem (cf. Theorem 1.7 in Chapter 1) says that $HD(L \cap [\sqrt{5}, \sqrt{12}]) = HD(M \cap [\sqrt{5}, \sqrt{12}]) = 1$; on the other hand, the proof of this estimate already contains all ideas needed to establish Theorem 2.14.

After that, we will briefly outline the key steps taken in [MM3 (2020)] to extend the analysis of $(M \setminus L) \cap [\sqrt{5}, \sqrt{12}]$ in order to prove Theorem 2.14. Then, we will close this discussion with a topological consequence of Theorem 2.14.

2.6.1   *Upper bound on the Hausdorff dimension of $(M \setminus L) \cap [\sqrt{5}, \sqrt{12}]$*

Recall from Proposition 1.7 that Perron showed in 1921 that if $\theta \in (\mathbb{N}^*)^{\mathbb{Z}}$, then

$$m(\theta) \leq \sqrt{12} \iff \theta \in \{1, 2\}^{\mathbb{Z}}.$$

Moreover, Hall proved[14] in 1971 that

$$HD(M \cap [3, \sqrt{10}]) \leq 0.93.$$

Since $(M \setminus L) \cap [3, \sqrt{12}] \subset (M \cap [3, \sqrt{10}]) \cup ((M \setminus L) \cap [\sqrt{10}, \sqrt{12}])$, it follows from Exercise 2.1 (b) that

$$HD((M \setminus L) \cap [\sqrt{5}, \sqrt{12}]) \leq \max \left\{ 0.93, HD((M \setminus L) \cap [\sqrt{10}, \sqrt{12}]) \right\}. \tag{2.21}$$

---

[14]Actually, Hall gave an upper bound on $\dim(\{[0; \theta] : \theta \in \{1, 2\}^{\mathbb{N}}$ not containing $121\})$. Since $m(x) < \sqrt{10}$ if and only if $x \in \{1, 2\}^{\mathbb{Z}}$ does not contain 121, Hall's estimate can also be used to give an upper bound for $\dim(M \cap [3, \sqrt{10}])$: see Chapter 6 of Cusick–Flahive book [CF (1989)] for more details.

At this stage, the key idea is to put restrictions on the past or future shift dynamics of sequences $\theta \in \{1,2\}^{\mathbb{Z}}$ leading to Markov values in $(M \setminus L) \cap [\sqrt{10}, \sqrt{12}]$.

Very roughly speaking, the *shadowing lemma*[15] implies that if $m(\theta) \in (M \setminus L) \cap [\sqrt{10}, \sqrt{12}]$, then, up to transposition, the future dynamics of θ lives in the gaps of the *symmetric block* $B = \{11, 22\}^{\mathbb{Z}} \subset \{x \in \{1,2\}^{\mathbb{Z}} : x$ not containing $121\}$. We will make this statement more precise in what follows.

Here, we have a setting which can be repeatedly used in the analysis of other portions of $M \setminus L$: we have a symmetric block C, which here is $\{1,2\}^{\mathbb{Z}}$, together with its corresponding Gauss-Cantor set $K_C = C(2)$ and a smaller symmetric block B (together with its corresponding Gauss-Cantor set $K_B = K(\{11, 22\})$). The elements of $(M \setminus L) \cap [\sqrt{10}, \sqrt{12}]$ will be of the form $f(\theta) = \lambda_0(\theta)$ for some $\theta \in C$ such that $m(\theta) = f(\theta)$. Let $Y = \{\theta \in C; m(\theta) = f(\theta) \in (M \setminus L) \cap [\sqrt{10}, \sqrt{12}]\}$. In order to prove that $HD((M \setminus L) \cap [\sqrt{10}, \sqrt{12}]) \leq d$, it is enough (by taking countable subcoverings and applying Exercise 2.1 (b)) to prove that, for every $\theta = (\theta_j)_{j \in \mathbb{Z}} \in Y$, there is $N \in \mathbb{N}$ such that $HD(f(V_N(\theta) \cap Y)) \leq d$, where $V_N(\theta) = \{\tilde{\theta} = (\tilde{\theta}_j)_{j \in \mathbb{Z}}; \tilde{\theta}_j = \theta_j, -N \leq j \leq N\}$.

We say that a finite sequence $\omega = (\theta_j)_{-N \leq j \leq n}$ has an *allowed continuation* τ, where τ is a finite sequence whose terms belong to $\{1,2\}$ if there are sequences $\alpha, \beta \in \{1,2\}^{\mathbb{N}}$ such that $\beta^t \omega \tau \alpha \in Y$. Concretely, we *claim* that the statement in the previous paragraph related to the shadowing lemma means that, if $\theta \in Y$, then (perhaps replacing θ by its transpose θ^t) there is $N \in \mathbb{N}$ such that, for every $n \geq N$, the bifurcation tree (of allowed continuations) of such a θ is severely constrained:

(1) either $\theta_{-N} \ldots \theta_n$ has an unique 1-element continuation θ_{n+1},
(2) or its (3-elements) allowed continuations are 112 and 221.

In order to show this, observe that if $\theta \in Y$, then, since L is closed, there is $N \in \mathbb{N}$ such that the distance of $m(\theta)$ to L is larger than $\frac{1}{2^{N-3}}$. Also, if $\tilde{\theta} \in V_N(\theta) \cap Y$, then, by Remark 2.2, $|m(\tilde{\theta}) - m(\theta)| = |f(\tilde{\theta}) - f(\theta)| < \frac{1}{2^{N-2}}$ and $\lambda_j(\tilde{\theta}) \leq f(\tilde{\theta}), \forall j \in \mathbb{Z}$.

If the claimed statement above is not true for this N, there are $n \geq N$,

---

[15]This classical result from the theory of uniformly hyperbolic dynamical systems asserts that pseudo-orbits are tracked by genuine orbits: see Hasselblatt–Katok book [HK (1995)] and the introduction of the original article [MM3 (2020)] for more comments on this point.

$\tau = (1, a, b) \neq (1, 1, 2)$, $\tau' = (2, c, d) \neq (2, 2, 1)$, $\alpha, \alpha', \beta, \beta' \in \{1, 2\}^{\mathbb{N}}$ such that $\beta^t \omega \tau \alpha, \beta'^t \omega \tau' \alpha' \in Y$, where $\omega = (\theta_j)_{-N \leq j \leq n}$. Since the smallest continued fractions in $C(2)$ beginning by $[0; 1]$ begin indeed by $[0; 1, 1, 2]$ and the largest continued fractions in $C(2)$ beginning by $[0; 2]$ begin by $[0; 2, 2, 1]$, and there are elements of K_B beginning by $[0; 1, 1, 2]$ and by $[0; 2, 2, 1]$, it follows that there is $\tilde{\alpha} \in \{11, 22\}^{\mathbb{N}}$ such that $[0; \tilde{\alpha}] \in K_B \cap ([0; \tau\alpha], [0; \tau'\alpha'])$. By monotonicity of maps of the type $h(y) = [c_0, c_1, \ldots, c_k + y]$ (and by Remark 2.2), it follows that, for $-N \leq j \leq n+1$, $\lambda_j(\beta^t \omega \tilde{\alpha}) < m(\theta) + \frac{1}{2^{N-2}} + \frac{1}{2^{N-1}} < m(\theta) + \frac{1}{2^{N-3}}$. On the other hand, since $\tilde{\alpha} \in \{11, 22\}^{\mathbb{N}}$, we have, for $j > n + 1$, $\lambda_j(\beta^t \omega \tau \tilde{\alpha}) \leq [2; \overline{1, 1}] + [0; 2, \overline{2, 1}] < 3.0407 < \sqrt{10} \leq m(\theta)$.

Hence, if the claimed statement above on allowed continuations fails both for θ and for θ^t, we may find $n, n' \geq N$ and $\tilde{\alpha}, \hat{\alpha} \in \{11, 22\}^{\mathbb{N}}$ such that $\lambda_j(\hat{\alpha}^t \hat{\omega} \tilde{\alpha}) < m(\theta) + \frac{1}{2^{N-3}}$ for all $j \in \mathbb{Z}$, where $\hat{\omega} = (\theta_j)_{-n' \leq j \leq n}$. This leads to a contradiction since, if $\tilde{\alpha} = (\tilde{a}_1, \tilde{a}_2, \ldots)$ and $\hat{\alpha} = (\hat{a}_1, \hat{a}_2, \ldots)$, and $\breve{\alpha} := (\tilde{a}_1, \tilde{a}_2, \ldots, \tilde{a}_{2N}, \hat{a}_{2N}, \ldots, \hat{a}_2, \hat{a}_1) \in \{1, 2\}^{4N}$, then, if $\gamma = \overline{\hat{\omega} \breve{\alpha}}$ is the periodic sequence with period $\hat{\omega} \breve{\alpha}$, then $|m(\gamma) - m(\theta)| < \frac{1}{2^{N-3}}$ and $m(\gamma) \in L$, an absurd.

The result above on allowed continuations can be used to get that the set $\{[\theta_0; \theta_1, \theta_2, \ldots]\}$ related to those allowed continuations of $\theta_{-N} \ldots \theta_N$ is contained in a small Cantor set (a so-called "Cantor set of gaps") K_G and, *a fortiori*, we have $(M \setminus L) \cap [\sqrt{10}, \sqrt{12}] \subset C(2) + K_G$.

This non-trivial control of the bifurcation tree allows[16] to infer that

$$HD((M \setminus L) \cap [\sqrt{10}, \sqrt{12}]) \leq HD(C(2)) + HD(K_G) \leq HD(C(2)) + s_0 \tag{2.22}$$

where $s_0 \in (0, 1)$ is any number such that

$$|I(\theta_1, \ldots, \theta_n, 1, 1, 2)|^{s_0} + |I(\theta_1, \ldots, \theta_n, 2, 2, 1)|^{s_0} \leq |I(\theta_1, \ldots, \theta_n)|^{s_0} \tag{2.23}$$

for all $(\theta_0, \ldots, \theta_n) \in \{1, 2\}^n$. Here, $I(a_1, \ldots, a_k) := \{[0; a_1, \ldots, a_k, z] : z > 1\}$ is an interval with extremities $\frac{p_k}{q_k}$ and $\frac{p_k + p_{k-1}}{q_k + q_{k-1}}$, and length

$$\frac{1}{q_k(q_k + q_{k-1})},$$

where $\frac{p_m}{q_m} = [0; a_1, \ldots, a_m]$ satisfies the recurrence relation

$$q_k = a_k q_{k-1} + q_{k-2} \tag{2.24}$$

---

[16]Here, we are using two useful facts: the box dimension of a Gauss-Cantor set (such as $C(2)$) coincides with its Hausdorff dimension (cf. Chapter 4 of Palis–Takens book [PT (1993)]) and $HD(X \times Y) \leq \dim(X) + HD(Y)$ where $\dim(X)$ is the box dimension of X (cf. Chapter 7 of Falconer's book [F (1986)]).

(these facts follow from the basic properties of continued fractions discussed in Chapter 1, §1.5.)

At this point, we can give a good upper bound on $HD((M \setminus L) \cap [\sqrt{10}, \sqrt{12}])$ along the following lines. On one hand, it is known that $HD(C(2)) < 0.531281$: cf. Jenkinson–Pollicott article [JePo2 (2018)]. On the other hand, the recurrence relation (2.24) says that (2.23) can be rewritten as

$$\left(\frac{r+1}{(3r+5)(4r+7)}\right)^{s_0} + \left(\frac{r+1}{(3r+7)(5r+12)}\right)^{s_0} < 1$$

where $0 < r = q_{n-1}/q_n < 1$. Because $\frac{r+1}{(3r+5)(4r+7)} \leq \frac{1}{35}$, $\frac{r+1}{(3r+7)(5r+12)} \leq \frac{1}{81.98}$ for $r \in (0,1)$, and $\left(\frac{1}{35}\right)^{0.174813} + \left(\frac{1}{81.98}\right)^{0.174813} < 1$, we can take $s_0 = 0.174813$. By plugging these informations into (2.22), we derive that

$$HD((M \setminus L) \cap [\sqrt{10}, \sqrt{12}]) < 0.531281 + 0.174813 = 0.706094.$$

This inequality together with (2.21) completes our sketch of proof of (2.20).

2.6.2 *End of the proof of Theorem 2.14*

Since $(\sqrt{12}, \sqrt{13})$ is a gap of M (cf. Proposition 1.7 in Chapter 1) and $[\sqrt{21}, +\infty) \subset L \subset M$ (this follows from Freiman's theorem on Hall's ray, but was also proved by Schecker in [Sch (1977)]), we can complete the estimate on $HD(M \setminus L)$ by giving a bound on the Hausdorff dimension of $(M \setminus L) \cap [\sqrt{13}, \sqrt{21}]$. This is done in [MM3 (2020)] by dividing $[\sqrt{13}, \sqrt{21}]$ into some smaller intervals, and performing an argument analogous to the above, with different choices of B and C:

• In $[\sqrt{13}, 3.84]$: Take C as the set of sequences in $\{1, 2, 3\}^{\mathbb{Z}}$ not containing the factors 13 and 31 and $B = \{1, 2\}^{\mathbb{Z}}$. We get the estimate

$$HD((M \setminus L) \cap [\sqrt{13}, 3.84]) \leq HD(K_C) + HD(K_G) \leq HD(C(3)) + HD(K_G)$$

$$\leq 0.705661 + 0.281266 = 0.986927.$$

• In $[3.84, \sqrt{20}]$: Take $C = \{1, 2, 3\}^{\mathbb{Z}}$ and $B = \{1, 2, 2321, 1232\}^{\mathbb{Z}}$. We get the estimate

$$HD((M \setminus L) \cap [3.84, \sqrt{20}]) \leq HD(C(3)) + HD(K_G) \leq 0.705661 + 0.281266,$$

that is, $HD((M \setminus L) \cap [3.84, \sqrt{20}]) \leq 0.986927$.

• In $[\sqrt{20}, \sqrt{21}]$: Take

$$C = \{\gamma \in \{1, 2, 3, 4\}^{\mathbb{Z}} \text{ not containing factors } 14, 41, 24, 42\}$$

and $B \subset \{21312, 232, 3, 11313, 31311\}^{\mathbb{Z}}$ with the restrictions that 31311 follows only 3, and 11313 is followed only by 3. We get the estimate

$$HD((M \setminus L) \cap [3.84, \sqrt{20}]) \leq HD(K_C) + HD(K_G) \leq HD(C(4)) + HD(K_G)$$

$$\leq 0.788947 + 0.172825 = 0.961772.$$

Remark 2.17. In [MM3 (2020)] it is also given the *empiric* estimate $\dim(M \setminus L) < 0.888$, based on *heuristic* estimates using Jenkinson–Pollicott algorithm in [JePol (2001)] to compute the Hausdorff dimensions of Gauss-Cantor sets K_C and the following decomposition of $[\sqrt{13}, \sqrt{21}]$ into smaller intervals:

• In $(-\infty, 3.06]$: In this case, as it is explained in Table 1 of Chapter 5 of Cusick–Flahive's book [CF (1989)], a result due to Jackson implies that if the Markov value of a sequence $\underline{a} \in \Sigma$ is $m(\underline{a}) < 3.06$, then $\underline{a}$ doesn't contain $1, 2, 1$ nor $2, 1, 2$. Thus, $HD((M \setminus L) \cap (-\infty, 3.06]) \leq 2 \cdot HD(K_C)$ where

$$C = \{\gamma \in \{1, 2\}^{\mathbb{Z}} \text{ not containing factors } 121, 212\}.$$

A quick implementation of the Jenkinson–Pollicott algorithm seems to indicate that $HD(K_C) < 0.365$, and thus $HD((M \setminus L) \cap (-\infty, 3.06]) < 0.73$.

• In $[3.06, \sqrt{12}]$: The estimate we did in the case $m \leq \sqrt{12}$ implies
$$HD((M \setminus L) \cap [3.06, \sqrt{12}]) < 0.531281 + 0.174813 = 0.706094.$$

• In $[\sqrt{13}, 3.84]$: Take C as the set of sequences in $\{1, 2, 3\}^{\mathbb{Z}}$ not containing the factors 13 and 31 and $B = \{1, 2\}^{\mathbb{Z}}$. We get the empirical estimate

$$HD((M \setminus L) \cap [\sqrt{13}, 3.84]) \leq HD(K_C) + HD(K_G) \leq 0.574 + 0.281266 < 0.856.$$

• In $[3.84, 3.92]$: Take $C =$ words in $\{1, 2, 3\}^{\mathbb{Z}}$ not containing factors $131, 313, 231, 132$ and $B \subset \{1, 2, 2321, 1232, 33\}^{\mathbb{Z}}$ with the restrictions that 33 doesn't follow 1 or 2321, and 33 is not followed by 1 or 1232. We get the empirical estimate

$$HD((M \setminus L) \cap [3.84, 3.92]) \leq HD(K_C) + HD(K_G) \leq 0.612 + 0.25966 < 0.872.$$

• In $[3.92, 4.01]$: Take $C =$ words in $\{1, 2, 3\}^{\mathbb{Z}}$ not containing factors $131, 313, 2312, 2132$ and $B = \{1, 2, 211, 112, 232, 1133, 3311\}^{\mathbb{Z}}$ with the restrictions that 3311 comes only after 211 and 3311 has to be followed by 2, and 1133 has to appear after 2, and 1133 has to be followed by 112. We get the empirical estimate

$$HD((M \setminus L) \cap [3.92, 4.01]) \leq HD(K_C) + HD(K_G) \leq 0.65 + 0.177655 < 0.828.$$

- In $[4.01, \sqrt{20}]$: Take $C = \{1, 2, 3\}^{\mathbb{Z}}$ and $B = \{11, 2, 232, 213312, 33\}^{\mathbb{Z}}$. We get the estimate

$$HD((M \setminus L) \cap [4.01, \sqrt{20}]) \leq HD(C(3)) + HD(K_G)$$
$$\leq 0.705661 + 0.167655 = 0.873316.$$

- In $[\sqrt{20}, \sqrt{21}]$: Take

$$C = \{\gamma \in \{1, 2, 3, 4\}^{\mathbb{Z}} \text{ not containing factors } 14, 41, 24, 42\}$$

and $B \subset \{21312, 232, 3, 11313, 31311\}^{\mathbb{Z}}$ with the restrictions that 31311 follows only 3, and 11313 is followed only by 3.

We get the empirical estimate

$$HD((M \setminus L) \cap [3.84, \sqrt{20}]) \leq HD(K_C) + HD(K_G)$$
$$\leq 0.715 + 0.172825 = 0.887825 < 0.888.$$

2.6.3 *A topological consequence of Theorem 2.14*

Closing this section, let us observe that Theorem 2.14 has the following topological consequence about the interior of the Lagrange and Markov spectra:

Corollary 2.12. *The interior of the Lagrange spectrum coincides with the interior of the Markov spectrum.*

Proof. This happens because $L \subset M$ are closed subsets of $\mathbb{R}$ such that $M \setminus L$ has empty interior (thanks to the Hausdorff dimension estimate in Theorem 2.14). $\qquad\square$

2.7 The complement of the Lagrange spectrum in the Markov spectrum V

The descriptions of the portions of $M \setminus L$ near 3.11, 3.29 and 3.7 in Remark 2.13 and Theorems 2.11 and 2.13 reveal a curious fact: the intersections of $M \setminus L$ with the corresponding maximal gaps of the Lagrange spectrum are actually *closed* subsets of the real line. In other words, these three portions of $M \setminus L$ stay at *positive* distance from the Lagrange spectrum.

This scenario led Bousch to ask whether $M \setminus L$ is a closed subset of $\mathbb{R}$. In a first attempt to negatively answer this question, Lima–Matheus–Moreira–Vieira [LMMV1 (2019)] gave some evidence towards the possibility that $3 \in L \cap \overline{(M \setminus L)}$. More precisely, they proved that

$$m_k := m(\overline{2_{2k}1_22_{2k+1}1_22_{2k+2}1_2}2_{2k}1_22^*2_{2k}1_22_{2k+2}1_22_{2k}1_22_{2k+1}1_22_{2k+2}1_2\overline{2})$$

is a decreasing sequence of Markov values converging to 3 such that its first four terms are elements of $M \setminus L$ lying in distinct maximal gaps of L. Furthermore, they could establish the analog of the replication mechanism described in Section 2.5 for all $k \geq 1$, but they could *not* conclude that $m_k \in M \setminus L$ for $k \geq 5$ because unfortunately they were unable to derive the analog of the local uniqueness property described in Section 2.5 for $k \geq 5$.

As it turns out, a negative answer to Bousch's question was recently announced by Lima–Matheus–Moreira–Vieira in [LMMV2 (2019)], where they employed the ideas of local uniqueness property and replication mechanism described in Section 2.5 to show that the sequence of Markov values

$$\widetilde{m}_k := m(\overline{2_{2k-1}1}2_{2k}1\overline{2}_{2k+1}\overline{1}2_{2k-1}12_{2k}12_{2k+1}12_{2k-1}12_{2k}1\overline{1}1\overline{2})$$

is a decreasing sequence of elements of $M \setminus L$ lying in distinct maximal gaps of L such that $\lim\limits_{k \to \infty} \widetilde{m}_k = 1 + \frac{3}{\sqrt{2}} \in L \cap \overline{(M \setminus L)}$.

2.8 Rigorous drawings of the classical Lagrange and Markov spectra

Pictorially, our discussions of classical Lagrange and Markov spectra so far can be summarized as:

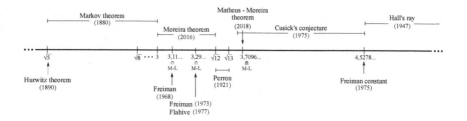

In a recent work, Delecroix–Matheus–Moreira [DMM (2019)] found a polynomial time algorithm allowing to rigorously draw better pictures of the classical spectra:

Theorem 2.15. *Let $R > 0$. Then there exists an algorithm that given Q provide finite sets $(1/Q)$-close[17] respectively to the Lagrange and Markov*

---

[17]We are using the so-called *Hausdorff topology* here, i.e., X and Y are ϵ-close (in Hausdorff distance) whenever $\forall x \in X, \exists y \in Y$ with $|x - y| < \epsilon$ and $\forall y \in Y, \exists x \in X$ with $|x - y| < \epsilon$.

spectrum in $[0, R]$. There exists a constant $0 < d_R < 1$ such that its running time is $O(Q^{3d_R})$ with the following upper bounds

(1) $d_R < 0.532$ for $R < \sqrt{13} \sim 3.606$,
(2) $d_R < 0.706$ for $R \leq 2\sqrt{5} \sim 4.472$,
(3) $d_R < 0.789$ for $R \leq \sqrt{21} \sim 4.583$.

Using this algorithm, Delecroix–Matheus–Moreira drew the following rigorous approximations for the portions $L_2 := \{\ell(\theta) : \theta \in \{1, 2\}^{\mathbb{Z}}\}$ and $L_3 := \{\ell(\theta) : \theta \in \{1, 2, 3\}^{\mathbb{Z}}\}$ of the classical Lagrange spectrum:

Let us now briefly describe the proof of Theorem 2.15. In order to determine the portions of the Lagrange and Markov spectra in a given interval $[0, R]$, we can restrict ourselves to the Lagrange and Markov values of sequences in $\Sigma_K := \{1, \ldots, K\}^{\mathbb{Z}}$, where $K := \lfloor R \rfloor$.

We will first describe a simplified version of the algorithm whose complexity is worst, but still polynomial in Q: we choose conveniently a positive integer n such that, for any $b_{-n}, \ldots, b_0, \ldots, b_n \in \{1, \ldots, K\}$, the cylinder set $\vec{b} = (b_{-n}, \ldots, b_0, \ldots, b_n) = \{u = (c_j)_{j\in\mathbb{Z}} \in \Sigma_K; c_j = b_j \text{ for } |j| \leq n\}$ satisfies

$$\sup_{\underline{a} \in \vec{b}} \lambda_0(\underline{a}) - \inf_{\underline{a} \in \vec{b}} \lambda_0(\underline{a}) < \frac{1}{Q}. \tag{2.25}$$

We consider a directed graph $\tilde{G}_{K,Q} = (\tilde{V}_{K,Q}, \tilde{E}_{K,Q})$ whose vertices are sequences $(a_{-n}, \ldots, a_0, \ldots, a_{n-1}) \in \{1, \ldots, K\}^{2n}$, and there is an edge frum $u = (a_{-n}, \ldots, a_0, \ldots, a_{n-1})$ and $v = (b_{-n}, \ldots, b_0, \ldots, b_{n-1})$ if and only if $b_j = a_{j+1}$ for $-n \leq j \leq n-2$. So $\tilde{G}_{K,Q}$ is a de Brujin graph [dBr (1946)] and its edges correspond to sequences $(b_{-n}, \ldots, b_0, \ldots, b_n) \in \{1, \ldots, K\}^{2n+1}$. To each such edge, we consider the corresponding cylinder $\vec{b}$, and associate its weight

$$w(\vec{b}) = \frac{\inf_{\underline{a} \in \vec{b}} \lambda_0(\underline{a}) + \sup_{\underline{a} \in \vec{b}} \lambda_0(\underline{a})}{2}.$$

We call an edge e in $G = \tilde{G}_{K,Q}$ a Lagrange edge if there exists a cycle γ in G that passes through e and so that the weight of the edge e is maximal among the weights of edges in γ. An edge is called a Markov edge if there exist two cycles γ^- and γ^+ and a path p from γ^- to γ^+ so that the edge e is maximal among the weights of edges in $\gamma^- \cup p \cup \gamma^+$.

Let $L_K = \{\ell(\theta) : \theta \in \Sigma_K\}$ and $M_K = \{m(\theta) : \theta \in \Sigma_K\}$. The algorithm is based on the following fact: for any K, Q, the set of weights of, respectively, Lagrange and Markov edges in $\tilde{G}_{K,Q}$ is $1/Q$-close to, respectively, L_K and M_K.

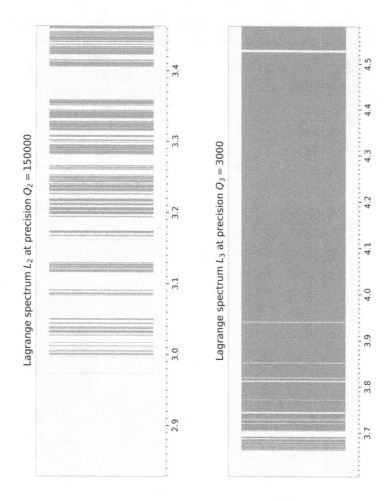

Fig. 2.2   Pictures of intervals whose unions have Hausdorff distance at most $1/Q_2$, resp. $1/Q_3$, from the portion L_2, resp. L_3, of the Lagrange spectrum.

Now we describe some features of the actual algorithm of Delecroix–Matheus–Moreira.

The first step of the algorithm consists in constructing a subshift of finite type $\Sigma_{K,Q}$ on the alphabet $\{1,\ldots,K\}$ depending on the desired quality of approximation Q. As it turns out, $\Sigma_{K,Q}$ is the set of infinite paths on a graph $G_{K,Q}$ whose edges correspond to certain cylinder sets

$\vec{b} = (b_{-m}, \ldots, b_0, \ldots, b_n)$ of the shift $\Sigma_{K,Q}$ satisfying

$$\sup_{\underline{a} \in \vec{b}} \lambda_0(\underline{a}) - \inf_{\underline{a} \in \vec{b}} \lambda_0(\underline{a}) < \frac{1}{Q}. \tag{2.26}$$

In the above equation, we identified a finite pointed word $\vec{b}$ (i.e., with a distinguished origin b_0) and there associated cylinder set $\{\underline{a} \in \Sigma_K : \forall k \in \{-m, \ldots, n\}, a_k = b_k\}$.

Next, we turn the graph $G_{K,Q}$ into a weighted graph by considering on the edge associated to $\vec{b}$ the weight $\dfrac{\inf_{\underline{a} \in \vec{b}} \lambda_0(\underline{a}) + \sup_{\underline{a} \in \vec{b}} \lambda_0(\underline{a})}{2}$. The weighted graph $G_{K,Q}$ provides an approximation of the shift Σ_K together with the height function λ_0. The condition (2.26) gives upper bound on the quality of approximation. The rest of the algorithm consists in studying directly on the graph $G_{K,Q}$ the discrete analogue of Lagrange and Markov spectrum.

The overall complexity of the algorithm is governed by the size of $G_{K,Q}$. In particular, we control it thanks to a simple relation (explained in Section 6 of [DMM (2019)]) between the size of $G_{K,Q}$ and the cardinality of the set

$$C_{K,Q} := \left\{ \vec{b} \in \{1, \ldots, K\}^{\mathbb{N}} : \operatorname{diam}_K(\vec{b}) \leq \frac{1}{Q} < \operatorname{diam}_K(\vec{b}') \right\},$$

where $\vec{b}'$ denotes the prefix of length $|\vec{b}| - 1$ of b, and $\operatorname{diam}_K(\vec{b})$ is the diameter of the interval with extremities

$$\frac{p_n(\vec{b}) + [0; \overline{1,K}] \, p_{n-1}(\vec{b})}{q_n(\vec{b}) + [0; \overline{1,K}] \, q_{n-1}(\vec{b})} \quad \text{and} \quad \frac{p_n(\vec{b}) + [0; \overline{K,1}] \, p_{n-1}(\vec{b})}{q_n(\vec{b}) + [0; \overline{K,1}] \, q_{n-1}(\vec{b})}.$$

Since it is not hard to compute the cardinality of $C_{K,Q}$ in terms of the Hausdorff dimension of the set of continued fraction expansions in $\{1, \ldots, K\}^{\mathbb{N}}$ (see again Section 6 of [DMM (2019)]), we can conclude that the running time of our algorithm is at most $O(Q^{2.367})$ in the less favourable situations.

We close this chapter with some remarks about the statement of Theorem 2.15.

Remark 2.18. Among our several motivations to get Theorem 2.15, we had in mind the possibility that an efficient algorithm producing high resolution drawings of classical Lagrange and Markov spectra could lead the way to solve Berstein's conjecture that $[4.1, 4.52] \subset L$ (cf. Remark 1.15 in Chapter 1). Unfortunately, the algorithm provided by Theorem 2.15 is not sufficiently powerful yet to put us in good position to attack Berstein's conjecture.

Remark 2.19. An alternative idea to approach the Lagrange spectrum would be to use Remark 2.1 to produce an algorithm based on the calculation of Markov values of periodic sequences in $\{1, \ldots, K\}^{\mathbb{Z}}$. As it is explained in Section 7 of [DMM (2019)], a naive implementation of this alternative algorithm leads to complexity bounds which are much worse than those in Theorem 2.15: for instance, this alternative algorithm obliges one to perform calculations with 4^{Q^4} periodic orbits in order to rigorously ensure that we got a $(1/Q)$-dense subset of L!

Remark 2.20. The Markov spectrum M is originally defined in terms of the values of real indefinite binary quadratic forms. An attempt to draw some portions of M using certain binary quadratic forms of bounded heights was performed by T. Morrison [Mo (2012)], but this text does not discuss the quality of the approximation of M obtained by this method.

Chapter 3

Continuity of Hausdorff dimension across classical and dynamical spectra

This chapter is dedicated to the study of the behaviour of the Hausdorff dimension of the intersections of classical and dynamical Lagrange and Markov spectra with half-infinite rays of the form $(-\infty, t)$, $t \in \mathbb{R}$. In particular, we will begin by giving a sketch of the proof of Moreira's theorem (stated as Theorem 1.7 in Chapter 1) asserting the continuity of $t \mapsto HD((-\infty, t) \cap L) = HD((-\infty, t) \cap M)$ (where L and M are the *classical* Lagrange and Markov spectra), and then we discuss the work of Cerqueira–Matheus–Moreira [CMM (2018)] extending Moreira's theorem to the *dynamical* context of horseshoes of area-preserving diffeomorphisms of surfaces. Finally, we will conclude with a quick application of Cerqueira–Matheus–Moreira work to the continuity of the Hausdorff dimension across the so-called Dirichlet spectrum.

3.1 Proof of Moreira's theorem

We recall that Moreira's theorem asserts that, for each $t \in \mathbb{R}$, the sets $L \cap (-\infty, t)$ and $M \cap (-\infty, t)$ have the same Hausdorff dimension, say $d(t) \in [0, 1]$, and, moreover,

- the function $t \mapsto d(t)$ is continuous,
- $d(3 + \varepsilon) > 0$ for all $\varepsilon > 0$ and
- $d(\sqrt{12}) = 1$ (even though $\sqrt{12} = 3.4641... < 4.5278... = c_F$),

cf. the statement of Theorem 1.7 in Chapter 1.

Before explaining the proof of this result, let us derive the following interesting corollary (also contained in Moreira's orginal article [Mor1 (2018)]):

Corollary 3.1 (Moreira). *The function $t \mapsto HD(L \cap (-\infty, t))$ is not α-Hölder continuous for any $\alpha > 0$.*

Proof. The map d sends $L \cap [3, 3+\varepsilon]$ to the non-trivial interval $[0, d(3+\varepsilon)]$ for any $\varepsilon > 0$. By item (c) of Exercise 2.1 in Chapter 2, if $t \mapsto d(t) = HD(L \cap (-\infty, t))$ were α-Hölder continuous for some $\alpha > 0$, then it would follow that

$$0 < \alpha = \alpha \cdot HD([0, d(3+\varepsilon)]) \leq HD(L \cap [3, 3+\varepsilon]) = d(3+\varepsilon)$$

for all $\varepsilon > 0$. On the other hand, the continuity of d (and item (b) of Exercise 2.1) also says that

$$\lim_{\varepsilon \to 0} d(3+\varepsilon) = d(3) = HD(L \cap (-\infty, 3)) = 0.$$

In summary, $0 < \alpha \leq \lim_{\varepsilon \to 0} d(3+\varepsilon) = 0$, a contradiction.   □

Let us now briefly outline the strategy of proof of Moreira's theorem. Very roughly speaking, the continuity of $d(t) = HD(L \cap (-\infty, t))$ is proved in four steps:

- if $0 < d(t) < 1$, then one shows that for all $\eta > 0$ there exists $\delta > 0$ such that $L \cap (-\infty, t - \delta)$ can be "*approximated from inside*" by $K + K' = f(K \times K')$ where K and K' are *Gauss-Cantor sets* with $HD(K) + HD(K') = HD(K \times K') > (1-\eta)d(t)$ (and $f(x, y) = x+y$);
- by *Moreira's dimension formula* (derived from profound works of Moreira and Yoccoz [MY (2001)] on the geometry of Cantor sets), we have that

$$HD(f(K \times K')) = HD(K \times K')$$

- thus, if $0 < d(t) < 1$, then for all $\eta > 0$ there exists $\delta > 0$ such that

$$d(t - \delta) \geq HD(f(K \times K')) = HD(K \times K') \geq (1-\eta)d(t);$$

  hence, $d(t)$ is *lower semicontinuous*;
- finally, an elementary compactness argument shows the *upper semicontinuity* of $d(t)$.

Remark 3.1. This strategy is *purely dynamical* because the particular forms of the height function f and the Gauss map G are *not* used. Instead, we just need the *transversality* of the gradient of f to the stable and unstable manifolds (vertical and horizontal axis) and the *non-essential affinity* of Gauss-Cantor sets. This fact was the starting point of Cerqueira–Matheus–Moreira work [CMM (2018)] and it is discussed in Section 3.2 below.

In the remainder of this section, we will implement (a version of) this strategy in order to sketch the proof of Moreira's theorem.

3.1.1 An interesting "symmetry" on the Hausdorff dimension of Gauss-Cantor sets

Let $K(B)$ be a Gauss-Cantor set associated to a finite, primitive alphabet B of finite words in $\bigcup_{n \geq 1} (\mathbb{N}^*)^n$: recall from §2.3.1.1 in Chapter 2 that this means that

$$K(B) = \{[0; \beta_1, \beta_2, \ldots] : \beta_i \in B \,\forall i \geq 1\}.$$

For later use, denote by $B^T = \{\beta^T : \beta \in B\}$ the *transpose* of B, where $\beta^T := (a_n, \ldots, a_1)$ stands as usual for the transpose of $\beta = (a_1, \ldots, a_n)$ and $K^+(B)$ is the G-invariant version of $K(B)$.

The following proposition (due to Euler) is proved in Appendix B:

Proposition 3.1 (Euler). *If $[0; \beta] = \frac{p_n}{q_n}$, then $[0; \beta^T] = \frac{r_n}{q_n}$.*

A striking consequence of this proposition is the following "symmetry" on the Hausdorff dimension of Gauss-Cantor sets:

Corollary 3.2. $HD(K(B)) = HD(K(B^T))$.

Sketch of proof. The lengths of the intervals $I(\beta) = \{[0; \beta, a_1, \ldots] : a_i \in \mathbb{N} \,\forall i\}$ in the construction of $K(B)$ depend only on the denominators of the partial quotients of $[0; \beta]$. Therefore, we have from Proposition 3.1 that $K(B)$ and $K(B^T)$ are Cantor sets constructed from intervals with same lengths, and, *a fortiori*, they have the Hausdorff dimension. $\square$

Remark 3.2. Dynamically, this corollary is explained by the existence of *area-preserving* natural extensions of Gauss map (see [Ar (1994)]) and the coincidence of stable and unstable dimensions of a horseshoe of an area-preserving surface diffeomorphism (see [McCMa (1983)]). We will come back to this point later in Section 3.2.

3.1.2 Non-essentially affine Cantor sets

We say that a dynamically defined Cantor set (cf. Chapter 2, §2.3.1)

$$K = \bigcap_{n \in \mathbb{N}} \psi^{-n}(I_1 \cup \cdots \cup I_r)$$

is *non-essentially affine* if there is *no* global conjugation $h \circ \psi \circ h^{-1}$ such that *all* branches

$$(h \circ \psi \circ h^{-1})|_{h(I_j)}, \quad j = 1, \ldots, r$$

are affine maps of the real line.

Equivalently, if $p \in K$ is a periodic point of ψ of period k and $h : I \to I$ is a diffeomorphism of the convex hull I of $I_1 \cup \cdots \cup I_r$ such that $h \circ \psi^k \circ h^{-1}$ is affine[1] on $h(J)$ where J is the connected component of the domain of ψ^k containing p, then K is non-essentially affine if and only if $(h \circ \psi \circ h^{-1})''(x) \neq 0$ for some $x \in h(K)$.

Proposition 3.2. *Gauss-Cantor sets are non-essentially affine.*

Proof. The basic idea is to explore the non-linearity of typical Möbius transformations. More concretely, let $B = \{\beta_1, \ldots, \beta_m\}$, $\beta_j \in (\mathbb{N}^*)^{r_j}$, $1 \leq j \leq m$. For each β_j, let

$$x_j := [0; \beta_j, \beta_j, \ldots] \in I_j = I(\beta_j) \subset \{[0; \beta_j, \alpha] : \alpha \geq 1\}$$

be the fixed point of the branch $\psi|_{I_j} = G^{r_j}$ of the expanding map ψ naturally defining the Gauss-Cantor set $K(B)$ (cf. Exercise 2.2 in Chapter 1).

By Corollary 1.2 in Chapter 1, we have that $\psi|_{I_j}(x) = \frac{q_{r_j-1}^{(j)} x - p_{r_j-1}^{(j)}}{p_{r_j}^{(j)} - q_{r_j}^{(j)} x}$, where $\frac{p_k^{(j)}}{q_k^{(j)}} = [0; b_1^{(j)}, \ldots, b_k^{(j)}]$ and $\beta_j = (b_1^{(j)}, \ldots, b_{r_j}^{(j)})$.

Note that the fixed point x_j of $\psi|_{I_j}$ is the positive solution of the second degree equation

$$q_{r_j}^{(j)} x^2 + (q_{r_j-1}^{(j)} - p_{r_j}^{(j)}) x - p_{r_j-1}^{(j)} = 0.$$

In particular, x_j is a *quadratic surd*.

For each $1 \leq j \leq k$, the Möbius transformation $\psi|_{I_j}$ has a hyperbolic fixed point x_j. It follows (from Poincaré linearization theorem) that there exists a Möbius transformation

$$\alpha_j(x) = \frac{a_j x + b_j}{c_j x + d_j}$$

linearizing $\psi|_{I_j}$, i.e., $\alpha_j(x_j) = x_j$, $\alpha'(x_j) = 1$ and $\alpha_j \circ (\psi|_{I_j}) \circ \alpha_j^{-1}$ is an affine map.

Since non-affine Möbius transformations have non-vanishing second derivative, the proof of the proposition will be complete once we show that $\alpha_1 \circ (\psi|_{I_2}) \circ \alpha_1^{-1}$ is not affine. So, let us suppose by contradiction that $\alpha_1 \circ (\psi|_{I_2}) \circ \alpha_1^{-1}$ is affine. In this case, ∞ is a common fixed point of the (affine) maps $\alpha_1 \circ (\psi|_{I_2}) \circ \alpha_1^{-1}$ and $\alpha_1 \circ (\psi|_{I_1}) \circ \alpha_1^{-1}$, and, *a fortiori*,

---

[1]Such a diffeomorphism h linearizing *one* branch of ψ always exists by Poincaré's linearization theorem (see, e.g., [HK (1995)]).

$\alpha_1^{-1}(\infty) = -d_1/c_1$ is a common fixed point of $\psi|_{I_1}$ and $\psi|_{I_2}$. Thus, the second degree equations

$$q_{r_1}^{(1)} x^2 + (q_{r_1-1}^{(1)} - p_{r_1}^{(1)}) x - p_{r_1-1}^{(j)} - 0 \quad \text{and} \quad q_{r_2}^{(2)} x^2 + (q_{r_2-1}^{(2)} \ p_{r_2}^{(2)}) x \ p_{r_2-1}^{(2)} = 0$$

would have a common root. This implies that these polynomials coincide (because they are polynomials in $\mathbb{Z}[x]$ which are irreducible[2]) and, hence, their other roots x_1, x_2 must coincide, a contradiction. $\qquad\square$

3.1.3 *Moreira's dimension formula*

The Hausdorff dimension of projections of products of non-essentially affine Cantor sets is given by the following formula:

Theorem 3.1 (Moreira). *Let K and K' be two C^2 dynamical Cantor sets. If K is non-essentially affine, then the projection $f(K \times K') = K + K'$ of $K \times K'$ under $f(x,y) = x + y$ has Hausdorff dimension*

$$HD(f(K + K')) = \min\{1, HD(K) + HD(K')\}.$$

This statement is a *particular* case of Moreira's dimension formula (which is sufficient for our current purposes because Gauss-Cantor sets are non-essentially affine): see [Mor2 (2016)] for the general statement (and also [HS (2012)] for more informations on this type of "dimension formula").

As it turns out, Moreira's dimension formula is proved using some techniques introduced in two works (from 2001 and 2010) by Moreira and Yoccoz [MY (2001)], [MY (2010)] such as fine analysis of *limit geometries* and *renormalization operators*, "recurrence on scales", "compact recurrent sets of relative configurations", and *Marstrand's theorem*.

The proof of Moreira's dimension formula is out of the scope of this book. Nevertheless, even though we will not give a proof of Moreira's dimension formula here, we will come back to the techniques of Moreira–Yoccoz in Chapter 5 during our discussions of intervals and rays in dynamical Lagrange and Markov spectra. For now, let us just observe that Moreira's dimension formula is coherent with Hall's Lemma 1.1 in Chapter 1: in fact, since $HD(C(4)) > 1/2$ (cf. Remark 2.9), Moreira's dimension formula predicts that $HD(C(4) + C(4)) = 1$ which is certainly compatible with Hall's lemma asserting that $C(4) + C(4)$ contains a non-trivial interval.

---

[2]Thanks to the fact that their roots x_1, x_2 are irrational.

3.1.4   First step towards Moreira's theorem: projections of Gauss-Cantor sets

Let $\Sigma(B) \subset (\mathbb{N}^*)^{\mathbb{Z}}$ be a complete shift of finite type. Denote by $\ell(\Sigma(B))$, resp. $m(\Sigma(B))$, the pieces of the Lagrange, resp. Markov, spectrum generated by $\Sigma(B)$, i.e.,

$$\ell(\Sigma(B)) = \{\ell(\underline{\theta}) : \underline{\theta} \in \Sigma(B)\}, \text{ resp. } m(\Sigma(B)) = \{m(\underline{\theta}) : \underline{\theta} \in \Sigma(B)\}$$

where $\ell(\underline{\theta}) = \limsup\limits_{n \to \infty} f(\sigma^n(\underline{\theta}))$, $m(\underline{\theta}) = \sup\limits_{n \in \mathbb{Z}} f(\sigma^n(\underline{\theta}))$, $f((\theta_i)_{i \in \mathbb{Z}}) = [\theta_0; \theta_1, \dots] + [0; \theta_{-1}, \dots]$ and $\sigma((\theta_i)_{i \in \mathbb{Z}}) = (\theta_{i+1})_{i \in \mathbb{Z}}$ is the shift map.

The following proposition relates the Hausdorff dimensions of the pieces of the Lagrange and Markov spectra associated to $\Sigma(B)$ and the projection $f(K(B) \times K(B^T))$:

Proposition 3.3. *One has* $HD(\ell(\Sigma(B))) = HD(m(\Sigma(B))) = \min\{1, 2 \cdot HD(K(B))\}$.

Sketch of proof. By definition,

$$\ell(\Sigma(B)) \subset m(\Sigma(B)) \subset \bigcup_{a=1}^{R}(a + K^+(B) + K^+(B^T))$$

where $R \in \mathbb{N}$ is the largest entry among all words of B.

By item (b) of Exercise 2.1 in Chapter 2, it follows that $HD(\ell(\Sigma(B))) \leq HD(m(\Sigma(B))) \leq HD(K(B)) + HD(K(B^T))$. Thus, the symmetry in Hausdorff dimension of Gauss-Cantor sets in Corollary 3.2 allows us to conclude that

$$HD(\ell(\Sigma(B))) \leq HD(m(\Sigma(B))) \leq \min\{1, 2 \cdot HD(K(B))\}.$$

By Moreira's dimension formula (cf. Theorem 3.1), our task is now reduced to show that for all $\varepsilon > 0$, there are "replicas" K and K' of Gauss-Cantor sets such that

$$HD(K), HD(K') > HD(K(B)) - \varepsilon \quad \text{and} \quad f(K \times K') = K + K' \subset \ell(\Sigma(B)).$$

In this direction, let us order B and B^T by declaring that $\gamma < \gamma'$ if and only if $[0; \gamma] < [0; \gamma']$.

Given $\varepsilon > 0$, we can replace if necessary B and/or B^T by $B^n = \{\gamma_1 \dots \gamma_n : \gamma_i \in B \; \forall i\}$ and/or $(B^T)^n$ for some large $n = n(\varepsilon) \in \mathbb{N}$ in such a way that

$$HD(K(B^*)), HD(K((B^T)^*)) > HD(K(B)) - \varepsilon$$

where $A^* := A \setminus \{\min A, \max A\}$. Indeed, this holds because the Hausdorff dimension of a Gauss-Cantor set $K(A)$ associated to an alphabet A with a large number of words does not decrease too much after removing only two words from A (cf. the Hausdorff dimension bounds in Appendix E).

We *expect* the values of ℓ on $((B^T)^*)^{\mathbb{Z}^-} \times (B^*)^{\mathbb{N}}$ to *decrease* because we removed the minimal and maximal elements of B and B^T (and, in general, $[a_0; a_1, a_2, \ldots] < [b_0; b_1, b_2, \ldots]$ if and only if $(-1)^k(a_k - b_k) < 0$ where k is the smallest integer with $a_k \neq b_k$).

In particular, this gives *some* control on the values of ℓ on $((B^T)^*)^{\mathbb{Z}^-} \times (B^*)^{\mathbb{N}}$, but this does *not* mean that $K(B^*) + K((B^T)^*) \subset \ell(\Sigma(B))$.

We overcome this problem by studying *replicas* of $K(B^*)$ and $K((B^T)^*)$. More precisely, let $\widetilde{\theta} = (\ldots, \widetilde{\gamma}_0, \widetilde{\gamma}_1, \ldots) \in \Sigma(B)$, $\widetilde{\gamma}_i \in B$ for all $i \in \mathbb{Z}$, such that

$$m(\widetilde{\theta}) = \max m(\Sigma(B))$$

is attained at a position in the block $\widetilde{\gamma}_0$.

By compactness, there exists $\eta > 0$ and $m \in \mathbb{N}$ such that any

$$\theta = (\ldots, \gamma_{-m-2}, \gamma_{-m-1}, \widetilde{\gamma}_{-m}, \ldots, \widetilde{\gamma}_0, \ldots, \widetilde{\gamma}_m, \gamma_{m+1}, \gamma_{m+2}, \ldots)$$

with $\gamma_i \in B^*$ for all $i > m$ and $\gamma_i \in (B^T)^*$ for all $i < -m$ satisfies:

- $m(\theta)$ is attained in a position in the *central block* $(\widetilde{\gamma}_{-m}, \ldots, \widetilde{\gamma}_0, \ldots, \widetilde{\gamma}_m)$;
- $f(\sigma^n(\theta)) < m(\theta) - \eta$ for any *non-central position* n.

By exploring these properties, it is possible to enlarge the central block to get a word called $\tau^{\#} = (a_{-N_1}, \ldots, a_0, \ldots, a_{N_2})$ in Moreira's paper [Mor1] (2018)] (here the position 0 of a_0 belongs to the central block, but does not necessarily coincide with its position 0) such that, for every $\gamma_i \in B^*, \hat{\gamma}_i \in (B^T)^*, i \geq 1$, for

$$\theta = (\ldots, \hat{\gamma}_{-2}, \hat{\gamma}_{-1}, a_{-N_1}, \ldots, a_0, \ldots, a_{N_2}, \gamma_1, \gamma_2, \ldots),$$

$m(\theta)$ is attained at the position 0, and so is equal to $[a_0; a_1, \ldots, a_{N_2}, \gamma_1, \gamma_2, \ldots] + [0; a_{-1}, \ldots, a_{-N_1}, \gamma_{-1}, \gamma_{-2}, \ldots]$ (we leave the construction of such an extended central block as an exercise to the reader).

This implies that the replicas

$$K = \{[a_0; a_1, \ldots, a_{N_2}, \gamma_1, \gamma_2, \ldots] : \gamma_i \in B^* \ \forall i > 0\}$$

and

$$K' = \{[0; a_{-1}, \ldots, a_{-N_1}, \gamma_{-1}, \gamma_{-2}, \ldots] : \gamma_i \in (B^T)^* \ \forall i < 0\}$$

of $K(B^*)$ and $K((B^T)^*)$ have the desired property that

$$K + K' = f(K \times K') \subset \ell(\Sigma(B)).$$

Indeed, given $x = [a_0; a_1, \ldots, a_{N_2}, \gamma_1, \gamma_2, \ldots] \in K$ and $y = [0; a_{-1}, \ldots, a_{-N_1}, \gamma_{-1}, \gamma_{-2}, \ldots] \in K'$, for

$$\beta = [0; b_1, b_2, \ldots]$$
$$= [0; \hat{\gamma}_{-1}, \tau^{\#}, \gamma_1, \hat{\gamma}_{-2}, \hat{\gamma}_{-1}, \tau^{\#}, \gamma_1, \gamma_2, \hat{\gamma}_{-3}, \hat{\gamma}_{-2}, \hat{\gamma}_{-1}, \tau^{\#}, \gamma_1, \gamma_2, \gamma_3, \ldots],$$

we have $\ell(\beta) = x + y \in \ell(\Sigma(B))$. Since we have

$$HD(K) = HD(K(B^*)) > HD(K) - \varepsilon,$$
$$HD(K') = HD(K((B^T)^*)) > HD(K(B^T)) - \varepsilon,$$

this completes our sketch of proof of the proposition. □

We use this technique of construction of elements of L (which already appears in the construction of Hall's ray in Chapter 1) in order to prove two results: one is a slightly more precise version of statement Theorem 1.8 in Chapter 1 on the topological structure of L and the other which shows that ℓ is highly non-injective at large values:

Theorem 3.2. *L' is a perfect set, i.e., $L'' = L'$. Moreover, every neighbourhood in L of every element of L' contains a Cantor set of positive Hausdorff dimension.*

Proof. Let $x \in L'$. Consider a sequence x_n converging to x, $x_n \in L$, $x_n \neq x$. Choose $\underline{\theta}^{(n)} \in \Sigma$ such that $x_n = \ell(\underline{\theta}^{(n)})$. Let $\underline{\theta}^{(n)} = (b_j^{(n)})_{j \in \mathbb{Z}}$ and assume $b_j^{(n)} \leq 4$, $\forall j$, $\forall n$ (which is possible since we may assume that the x_n are not in Hall's ray). We have $x_n = \limsup_{j \to \infty} (\alpha_j^{(n)} + \beta_j^{(n)})$. Given $\delta > 0$, $\exists n_0 \in \mathbb{N}$ large such that $n \geq n_0 \Rightarrow |\ell(\underline{\theta}^{(n)}) - x| < \delta$ and there are infinitely many $j \in \mathbb{N}$ such that $|\alpha_j^{(n)} + \beta_j^{(n)} - x| < \delta$. Let $N = \lceil \delta^{-1} \rceil$. Given such a pair (j, n) consider the finite sequence with $2N + 1$ terms $(b_{j-N}^{(n)}, b_{j-N+1}^{(n)}, \ldots, b_j^{(n)}, \ldots, b_{j+N}^{(n)}) =: S(j, n)$. There is a sequence S such that for infinitely many values of n, S appears infinitely many times as $S(j, n)$, $j \in \mathbb{N}$, i.e., there are $j_1(n) < j_2(n) < \ldots$ with $\lim_{i \to \infty} (j_{i+1}(n) - j_i(n)) = \infty$ and $S(j_i(n), n) = S$, $\forall i \geq 1$, for all n in some infinite set $A \subset \mathbb{N}$.

Consider the sequences $\beta(i, n)$ for $i \geq 1$, $n \in A$ given by

$$\beta(i, n) = (b_{j_i(n)+N+1}^{(n)}, b_{j_i(n)+N+2}^{(n)}, \ldots, b_{j_{i+1}(n)+N}^{(n)}).$$

There are (i_1, n_1) and (i_2, n_2) for which there is no sequence γ such that $\beta(i_1, n_1)$ and $\beta(i_2, n_2)$ are concatenations of copies of γ, otherwise x_n would be constant for $n \in A$. This implies that, taking $B = \{\beta(i_1, n_1)\beta(i_2, n_2), \beta(i_2, n_2)\beta(i_1, n_1)\}$, $K(B)$ is a Gauss-Cantor set, so, as in the proof of Proposition 3.3, $\ell(K(B))$ contains a Cantor set $\hat{K}$ of positive Hausdorff dimension with $d(x, \hat{K}) \leq 2\delta$. $\qquad\square$

Theorem 3.3. *One has* $\lim_{t \to \infty} HD(\ell^{-1}(t)) = 1$. *Thus,* $\lim_{t \to \infty} HD(\ell^{-1}(-\infty, t)) = \lim_{t \to \infty} \Delta^+(t) = 1$.

Proof. Given $m \geq 2$, let $C(m) = \{\alpha = [0; a_1, a_2, a_3, \ldots] \in [0, 1] | a_k \leq m, \forall k \geq 1\}$. As we discussed in Chapter 1, M. Hall proved that $C(4) + C(4) = [\sqrt{2} - 1, 4(\sqrt{2} - 1)]$. On the other hand, we have $\lim_{m \to \infty} HD(C(m)) = 1$. In fact, Jarník proved[3] in [J (1929)] that

$$HD(C(m)) > 1 - \frac{1}{m \cdot \log 2}, \forall m > 8.$$

Let now $t \geq 7$ be given. Let $m = \lfloor t \rfloor - 3$. There are an integer $n \in \{m + 2, m + 3\}$ and $\alpha = [0; a_1, a_2, a_3, \ldots], \beta = [0; b_1, b_2, b_3, \ldots] \in C(4)$ such that $t = n + \alpha + \beta$. For each $r \geq 1$, let

$$\tau^{(r)} = (m + 1, b_r, b_{r-1}, \ldots, b_2, b_1, n, a_1, a_2, \ldots, a_{r-1}, a_r, m + 1).$$

Consider now the map $h : C(m) \to [0, 1]$ given by

$$h([0; c_1, c_2, c_3, \ldots])$$
$$= [0; c_{1!}, \tau^{(1)}, c_{2!}, \tau^{(2)}, c_3, c_4, c_5, c_{3!}, \tau^{(3)}, \ldots, c_{r!}, \tilde{\tau}^{(r)}, c_{r!+1}, \ldots].$$

It is easy to see that $\ell(\tilde{h}(z)) = t$ for every $z \in C(m)$. On the other hand, given any $\rho > 0$, we have $|z - z'| = O(|h(z) - h(z')|^{1-\rho})$ for $|z - z'|$ small, so $HD(\ell^{-1}(t)) \geq HD(C(m))$. Since $\lim_{m \to \infty} HD(C(m)) = 1$, we are done. $\qquad\square$

3.1.5 *Second step towards Moreira's theorem: upper semi-continuity*

Let $\Sigma_t := \{\theta \in (\mathbb{N}^*)^{\mathbb{Z}} : m(\theta) \leq t\}$ for $3 \leq t < 5$.

Our long term goal is to compare Σ_t with its projection $K_t^+ := \{[0; \gamma] : \gamma \in \pi^+(\Sigma_t)\}$ on the unstable part (where $\pi^+ : (\mathbb{N}^*)^{\mathbb{Z}} \to (\mathbb{N}^*)^{\mathbb{N}}$ is the natural projection).

---

[3]See also Remark 2.9 in Chapter 2 for more precise estimates.

Given $\alpha = (a_1, \ldots, a_n)$, its *unstable scale* $r^+(\alpha)$ is

$$r^+(\alpha) = \lfloor \log 1/(\text{length of } I^+(\alpha)) \rfloor$$

where $I^+(\alpha)$ is the interval with extremities $[0; a_1, \ldots, a_n]$ and $[0; a_1, \ldots, a_n + 1]$.

Denote by

$$P_r^+ := \{\alpha = (a_1, \ldots, a_n) : r^+(\alpha) \geq r, r^+(a_1, \ldots, a_{n-1}) < r\}$$

and

$$C^+(t, r) := \{\alpha \in P_r^+ : I^+(\alpha) \cap K_t^+ \neq \emptyset\}.$$

Remark 3.3. By symmetry (i.e., replacing γ's by γ^T's), we can define K_t^-, $r^-(\alpha)$, etc.

For later use, we observe that the unstable scales have the following behaviour under concatenations of words:

Exercise 3.1

Show that $r^+(\alpha \beta k) \geq r^+(\alpha) + r^+(\beta)$ for all α, β finite words and for all $k \in \{1, 2, 3, 4\}$.

In particular, since the family of intervals

$$\{I^+(\alpha \beta k) : \alpha \in C^+(t, r), \beta \in C^+(t, s), 1 \leq k \leq 4\}$$

covers K_t^+, it follows from Exercise 3.1 that

$$\#C^+(t, r + s) \leq 4\#C^+(t, r)\#C^+(t, s)$$

for all $r, s \in \mathbb{N}$ and, hence, the sequence $(4\#C^+(t, r))_{r \in \mathbb{N}}$ is *submultiplicative*.

So, the *box-counting dimension* (cf. Remark 2.6) $\Delta^+(t)$ of K_t^+ is

$$\Delta^+(t) = \inf_{m \in \mathbb{N}} \frac{1}{m} \log(4\#C^+(t, m)) = \lim_{m \to \infty} \frac{1}{m} \log \#C^+(t, m).$$

An elementary compactness argument shows that the upper-semicontinuity of $\Delta^+(t)$:

Proposition 3.4. *The function $t \mapsto \Delta^+(t)$ is upper-semicontinuous.*

Proof. For the sake of contradiction, assume that there exist $\eta > 0$ and t_0 such that $\Delta^+(t) > \Delta^+(t_0) + \eta$ for all $t > t_0$.

By definition, this means that there exists $r_0 \in \mathbb{N}$ such that

$$\frac{1}{r} \log \#C^+(t, r) > \Delta^+(t_0) + \eta$$

for all $r \geq r_0$ and $t > t_0$.

On the other hand, $C^+(t, r) \subset C^+(s, r)$ for all $t \leq s$ and, by compactness, $C^+(t_0, r) = \bigcap_{t > t_0} C^+(t, r)$. Thus, if $r \to \infty$ and $t \to t_0$, the inequality of the previous paragraph would imply that

$$\Delta^+(t_0) > \Delta^+(t_0) + \eta,$$

a contradiction. $\qquad\square$

3.1.6 *Third step towards Moreira's theorem: lower semi-continuity*

The heart of the proof of Moreira's theorem is the following "lower semi-continuity" theorem allowing us to "approximate from inside" Σ_t by Gauss-Cantor sets.

Theorem 3.4. *Given $\eta > 0$ and $3 \leq t < 5$ with $d(t) := HD(L \cap (-\infty, t)) > 0$, we can find $\delta > 0$ and a Gauss-Cantor set $K(B)$ associated to $\Sigma(B) \subset \{1, 2, 3, 4\}^{\mathbb{Z}}$ such that*

$$\Sigma(B) \subset \Sigma_{t-\delta} \quad \text{and} \quad HD(K(B)) \geq (1 - \eta)\Delta^+(t).$$

In fact, this theorem allows us to derive the continuity statement in Moreira's Theorem 1.7 from Chapter 1:

Corollary 3.3. $\Delta^-(t) = \Delta^+(t)$ *is a continuous function of t and $d(t) = \min\{1, 2 \cdot \Delta^+(t)\}$.*

Proof. By Corollary 3.2 and Theorem 3.4, we have that

$$\Delta^-(t - \delta) \geq HD(K(B^T)) = HD(K(B)) \geq (1 - \eta)\Delta^+(t).$$

Also, a "symmetric" estimate holds after exchanging the roles of Δ^- and Δ^+. Hence, $\Delta^-(t) = \Delta^+(t)$. Moreover, the inequality above says that $\Delta^-(t) = \Delta^+(t)$ is a lower-semicontinuous function of t. Since we already know that $\Delta^+(t)$ is an upper-semicontinuous function of t thanks to Proposition 3.4, we conclude that $t \mapsto \Delta^-(t) = \Delta^+(t)$ is continuous. Finally, by Proposition 3.3, from $\Sigma(B) \subset \Sigma_{t-\delta}$, we also have that

$$d(t-\delta) \geq HD(\ell(\Sigma(B))) = \min\{1, 2 \cdot HD(K(B))\} \geq (1-\eta)\min\{1, 2\Delta^+(t)\}.$$

Since $d(t) \leq \min\{1, \Delta^+(t) + \Delta^-(t)\}$ (because $\Sigma_t \subset \pi^-(\Sigma_t) \times \pi^+(\Sigma_t)$), the proof is complete. $\qquad\square$

Let us now sketch the construction of the Gauss-Cantor sets $K(B)$ approaching Σ_t from inside (while referring to Appendix F below for a detailed proof of this theorem).

Sketch of proof of Theorem 3.4. Fix $r_0 \in \mathbb{N}$ large enough so that

$$\left| \frac{\log \#C^+(t,r)}{r} - \Delta^+(t) \right| < \frac{\eta}{80} \Delta^+(t)$$

for all $r \geq r_0$.

Set $B_0 := C^+(t, r_0)$, $k = 8(\#B_0)^2 \lceil 80/\eta \rceil$ and

$$\widetilde{B} := \{\beta = (\beta_1, \ldots, \beta_k) : \beta_j \in B_0 \text{ and } I^+(\beta) \cap K_t^+ \neq \emptyset\} \subset B_0^k.$$

It is not hard to show that $\widetilde{B}$ has a significant cardinality in the sense that

$$\#\widetilde{B} > 2(\#B_0)^{(1-\frac{\eta}{40})k}.$$

In particular, one can use this information to prove that $HD(K(\widetilde{B}))$ is not far from $\Delta^+(t)$, i.e.

$$HD(K(\widetilde{B})) \geq \left(1 - \frac{\eta}{20}\right) \Delta^+(t).$$

Unfortunately, since we have no control on the values of m on $\Sigma(\widetilde{B})$, there is no guarantee that $\Sigma(\widetilde{B}) \subset \Sigma_{t-\delta}$ for some $\delta > 0$.

We can overcome this issue with the aid of the notion of *left-good* and *right-good* positions. More concretely, we say that $1 \leq j \leq k$ is a right-good position of $\beta = (\beta_1, \ldots, \beta_k) \in \widetilde{B}$ whenever there are two elements $\beta^{(s)} = \beta_1 \ldots \beta_j \beta_{j+1}^{(s)} \ldots \beta_k^{(s)} \in \widetilde{B}$, $s \in \{1,2\}$ such that

$$[0; \beta_j^{(1)}] < [0; \beta_j] < [0; \beta_j^{(2)}].$$

Similarly, $1 \leq j \leq k$ is a left-good position $\beta = (\beta_1, \ldots, \beta_k) \in \widetilde{B}$ whenever there are two elements $\beta^{(s)} = \beta_1 \ldots \beta_j \beta_{j+1}^{(s)} \ldots \beta_k^{(s)} \in \widetilde{B}$, $s \in \{3,4\}$ such that

$$[0; (\beta_j^{(3)})^T] < [0; \beta_j^T] < [0; (\beta_j^{(2)})^T].$$

Furthermore, we say that $1 \leq j \leq k$ is a *good position* of $\beta = (\beta_1, \ldots, \beta_k) \in \widetilde{B}$ when it is both a left-good and a right-good position.

Since there are at most two choices of $\beta_j \in B_0$ when $\beta_1, \ldots, \beta_{j-1}$ are fixed and j is a right-good position, one has that the subset

$$\mathcal{E} := \{\beta \in \widetilde{B} : \beta \text{ has } 9k/10 \text{ good positions (at least)}\}$$

of *excellent* words in $\widetilde{B}$ has cardinality

$$\#\mathcal{E} > \frac{1}{2} \#\widetilde{B} > (\#B_0)^{(1-\frac{\eta}{40})k}.$$

We *expect* the values of m on $\Sigma(\mathcal{E})$ to *decrease* because excellent words have many good positions. Also, the Hausdorff dimension of $K(\mathcal{E})$ is not far from $\Delta^+(t)$ thanks to the estimate above on the cardinality of $\mathcal{E}$. However, there is no reason for $\Sigma(\mathcal{E}) \subset \Sigma_{t-\delta}$ for some $\delta > 0$ because an *arbitrary* concatenation of words in $\mathcal{E}$ might not belong to Σ_t.

At this point, the idea is to build a complete shift $\Sigma(B) \subset \Sigma_{t-\delta}$ from $\mathcal{E}$ with the following combinatorial argument. Since $\beta = (\beta_1, \ldots, \beta_k) \in \mathcal{E}$ has $9k/10$ good positions, we can find good positions $1 \leq i_1 \leq i_2 \leq \cdots \leq i_{\lceil 2k/5 \rceil} \leq k-1$ such that $i_s + 2 \leq i_{s+1}$ for all $1 \leq s \leq \lceil 2k/5 \rceil - 1$ and $i_s + 1$ are also good positions for all $1 \leq s \leq \lceil 2k/5 \rceil$. Because $k := 8(\#B_0)^2 \lceil 80/\eta \rceil$, the pigeonhole principle reveals that we can choose positions $j_1 \leq \cdots \leq j_{3(\#B_0)^2}$ and words $\widehat{\beta}_{j_1}, \widehat{\beta}_{j_1+1}, \ldots, \widehat{\beta}_{j_{3(\#B_0)^2}}, \widehat{\beta}_{j_{3(\#B_0)^2}+1} \in B_0$ such that $j_s + 2\lceil 80/\eta \rceil \leq j_{s+1}$ for all $s < 3(\#B_0)^2$ and the subset

$$X = \{(\beta_1, \ldots, \beta_k) \in \mathcal{E} : j_s, j_s + 1 \text{ are good positions and }$$
$$\beta_{j_s} = \widehat{\beta}_{j_s}, \beta_{j_s+1} = \widehat{\beta}_{j_s+1} \,\forall\, s \leq 3(\#B_0)^2\}$$

of excellent words with prescribed subwords $\widehat{\beta}_{j_s}, \widehat{\beta}_{j_s+1}$ at the good positions $j_s, j_s + 1$ has cardinality

$$\#X > (\#B_0)^{(1-\frac{\eta}{20})k}.$$

Next, we convert X into the alphabet B of an appropriate complete shift with the help of the projections $\pi_{a,b} : X \to B_0^{j_b - j_a}$, $\pi_{a,b}(\beta_1, \ldots, \beta_k) = (\beta_{j_a+1}, \beta_{j_a+2}, \ldots, \beta_{j_b})$. More precisely, an elementary counting argument shows that we can take $1 \leq a < b \leq 3(\#B_0)^2$ such that $\widehat{\beta}_{j_a} = \widehat{\beta}_{j_b}$, $\widehat{\beta}_{j_a+1} = \widehat{\beta}_{j_b+1}$, and the image $\pi_{a,b}(X)$ of some projection $\pi_{a,b}$ has a significant cardinality

$$\#\pi_{a,b}(X) > (\#B_0)^{(1-\frac{\eta}{4})(j_b - j_a)}.$$

From these properties, we get an alphabet $B = \pi_{a,b}(X)$ whose words concatenate in an appropriate way (because $\widehat{\beta}_{j_a} = \widehat{\beta}_{j_b}$, $\widehat{\beta}_{j_a+1} = \widehat{\beta}_{j_b+1}$), the Hausdorff dimension of $K(B)$ is $HD(K(B)) > (1 - \eta)\Delta^+(t)$ (because $\#B > (\#B_0)^{(1-\frac{\eta}{4})(j_b - j_a)}$ and $j_b - j_a > 2\lceil \frac{80}{\eta} \rceil$), and $\Sigma(B) \subset \Sigma_{t-\delta}$ for some $\delta > 0$ (because the features of good positions forces the values of m on $\Sigma(B)$ to decrease). This completes our sketch of proof. $\square$

3.1.7 *End of proof of Moreira's theorem*

By Corollary 3.3, the function

$$t \mapsto d(t) = HD(L \cap (-\infty, t))$$

is continuous. Moreover, an inspection of the proof of Corollary 3.3 shows that we have also proved the equality $HD(M \cap (-\infty, t)) = HD(L \cap (-\infty, t))$.

Therefore, it remains only to prove that $d(3 + \varepsilon) > 0$ for all $\varepsilon > 0$ and $d(\sqrt{12}) = 1$.

The fact that $d(3 + \varepsilon) > 0$ for any $\varepsilon > 0$ uses explicit sequences $\theta_m \in \{1, 2\}^{\mathbb{Z}}$ such that $\lim_{m \to \infty} m(\theta_m) = 3$ in order to exhibit non-trivial Cantor sets in $M \cap (-\infty, 3 + \varepsilon)$. More precisely, consider[4] the periodic sequences

$$\theta_m := \overline{2 \underbrace{1 \ldots 1}_{2m \text{ times}} 2}$$

where $\overline{a_1 \ldots a_k} := \ldots a_1 \ldots a_k \, a_1 \ldots a_k \ldots$. Since the sequence $\theta_\infty = \overline{1}, 2, 2, \overline{1}$ has the property that $m(\theta_\infty) = [2; \overline{1}] + [0; 2, \overline{1}] = 3$, and $|[a_0; a_1, \ldots, a_n, b_1, \ldots] - [a_0; a_1, \ldots, a_n, c_1, \ldots]| < \frac{1}{2^{n-1}}$ in general (cf. Remark 2.2 in Chapter 2), we have that the alphabet B_m consisting of the two words $2 \underbrace{1 \ldots 1}_{2m \text{ times}} 2$ and $2 \underbrace{1 \ldots 1}_{2m+2 \text{ times}} 2$ satisfies

$$\Sigma(B_m) \subset \Sigma_{3 + \frac{1}{2^m}}.$$

Thus, $d(3 + \frac{1}{2^m}) = HD(M \cap (-\infty, 3 + \frac{1}{2^m})) \geq HD(\Sigma(B_m)) = 2 \cdot HD(K(B_m)) > 0$ for all $m \in \mathbb{N}$.

Finally, the fact that $d(\sqrt{12}) = 1$ follows from Corollary 3.3 and Remark 2.9. Indeed, recall that Perron showed that $m(\theta) \leq \sqrt{12}$ if and only if $\theta \in \{1, 2\}^{\mathbb{Z}}$ (cf. Proposition 1.7 in Chapter 1). Thus, $K^+_{\sqrt{12}} = C(2)$. By Corollary 3.3, it follows that

$$d(\sqrt{12}) = \min\{1, 2 \cdot \Delta^+(\sqrt{12})\} = \min\{1, 2 \cdot HD(C(2))\}.$$

Since Remark 2.9 tells us that $HD(C(2)) > 1/2$, we conclude that $d(\sqrt{12}) = 1$. This completes our discussion of the proof of Moreira's Theorem 1.7 in Chapter 1.

3.2 Continuity of the dimension across dynamical spectra of thin horseshoes

In their article [CMM (2018)], Cerqueira–Matheus–Moreira extended the continuity portion of Moreira's theorem to the context of dynamical Lagrange and Markov spectra associated to typical thin horseshoes of area-preserving diffeomorphisms of surfaces. Before stating their result, let us introduce the relevant notions of dynamical Lagrange and Markov spectra and horseshoes.

---

[4] This choice of θ_m is motivated by the discussion in Chapter 1 of Cusick-Flahive book [CF (1989)].

3.2.1 *Dynamical Lagrange and Markov spectra*

Recall that Perron's characterization of the classical spectra in Chapter 1, §1.6 says that

$$L = \left\{ \limsup_{n \to \infty} f(\sigma^n(\underline{\theta})) : \underline{\theta} \in \Sigma \right\} \quad \text{and} \quad M = \left\{ \sup_{n \in \mathbb{Z}} f(\sigma^n(\underline{\theta})) : \underline{\theta} \in \Sigma \right\},$$

where $\Sigma = (\mathbb{N}^*)^{\mathbb{Z}}$, $\sigma \colon \Sigma \to \Sigma$ is the shift map and $f \colon \Sigma \to \mathbb{R}$ is the height function $f((a_n)_{n \in \mathbb{Z}}) = \alpha_0 + \beta_0 = [a_0, a_1, \dots] + [0, a_{-1}, a_{-2}, \dots]$.

The *dynamical* Lagrange and Markov spectra are obtained after replacing σ by a *general* dynamical system and f by a general height function.

Definition 3.1. Let $\varphi : M \to M$ be a homeomorphism of a topological space M. Given a compact φ-invariant subset Λ of M and a continuous function $f : M \to \mathbb{R}$, the *dynamical Lagrange and Markov spectra* associated to (f, Λ) are

$$L(f, \Lambda) = \left\{ \limsup_{n \to \infty} f(\varphi^n(x)) : x \in \Lambda \right\} \quad \text{and}$$

$$M(f, \Lambda) = \left\{ \sup_{n \in \mathbb{Z}} f(\varphi^n(x)) : x \in \Lambda \right\}.$$

Remark 3.4. An analogous definition can be made when the discrete time dynamical system $\varphi : M \to M$ is replaced by a continuous time dynamical system $(\varphi^t)_{t \in \mathbb{R}}$.

Remark 3.5. The quantities $\limsup_{n \to \infty} f(\varphi^n(x))$ and $\sup_{n \in \mathbb{Z}} f(\varphi^n(x))$ depend only on the orbit of x by φ. Therefore, the dynamical Lagrange and Markov spectra $L(f, \Lambda)$ and $M(f, \Lambda)$ can be viewed as images of the *orbit space* of the map $\varphi|_\Lambda$.

Exercise 3.2
Use the compactness of Λ to prove that $L(f, \Lambda) \subset M(f, \Lambda) \subset f(\Lambda)$.

3.2.2 *Horseshoes of surface diffeomorphisms*

Let $\varphi \colon M^2 \to M^2$ be a diffeomorphism of a surface M^2. We say that a compact φ-invariant subset $\Lambda \subset M^2$ is *hyperbolic of saddle-type* if there are some constants $C > 0$, $0 < \lambda < 1$ and a splitting $T_x M = E^s(x) \oplus E^u(x)$ for each $x \in H$ such that:

- *the splitting is equivariant*: $d\varphi(x) \cdot E^s(x) = E^s(\varphi(x))$ and $d\varphi(E^u(x)) = E^u(\varphi(x))$;

- *the subbundles of the splitting are asymptotically contracted[5]:* $\|d\varphi^n(x) \cdot v^s\| \leq C\lambda^n \|v^s\|$ *and* $\|d\varphi^{-n}(x) \cdot v^u\| \leq C\lambda^n \|v^u\|$ *for all* $n \geq 0$, $v^s \in E^s(x)$ *and* $v^u \in E^u(x)$;
- *for all* $x \in \Lambda$, *the subbundles* $E^s(x)$ *and* $E^u(x)$ *are one-dimensional.*

In this context, a *horseshoe* Λ is an infinite, totally disconnected, hyperbolic set of saddle-type of the form $\Lambda = \overline{\{\varphi^n(x) : n \in \mathbb{Z}\}} = \bigcap\limits_{n \in \mathbb{Z}} \varphi^n(U)$ for some $x \in \Lambda$ and some neighborhood U of Λ. This notion was discovered by Smale in 1967 and it became since then a paradigm for the theory of *uniformly hyperbolic systems*: the reader is invited to consult the books of Shub [Sh (1987)], Hasselblatt–Katok [HK (1995)], and Palis–Takens [PT (1993)] for (some pictures justifying the nomenclature "horseshoe" and) nice introductions to this beautiful portion of the theory of Dynamical Systems.

Remark 3.6. Similarly to the case of classical Lagrange and Markov spectra, the dynamical Lagrange and Markov spectra associated to horseshoes are closed subsets of $\mathbb{R}$: this fact was observed in the similar context of geodesic flows on negatively curved manifolds in the articles of Maucourant [Mac (2003)] and Parkkonen–Paulin [PP0 (2009)], and we offer a proof of this result (based on the exposition in Vieira's PhD thesis [V (2020)]) in the setting of horseshoes in Appendix G.

3.2.3 *Horseshoes and dynamically defined Cantor sets*

Let Λ be a horseshoe of a smooth diffeomorphism φ of a surface. As it is explained in Palis–Takens book [PT (1993)], the stable and unstable manifolds[6] of periodic points in a horseshoe Λ of φ can be used to produce a geometrical Markov partition $\{R_a\}_{a \in \mathcal{A}}$ with sufficiently small diameter consisting of rectangles $R_a \simeq I_a^s \times I_a^u$ delimited by compact pieces I_a^s, resp. I_a^u, of stable, resp. unstable, manifolds of certain points of Λ. In this setting, the subset $\mathcal{T} \subset \mathcal{A}^2$ of admissible transitions is the subset of pairs $(a_0, a_1) \in \mathcal{A}^2$ such that $\varphi(R_{a_0}) \cap R_{a_1} \neq \emptyset$. In this way, the dynamics of φ on Λ is topologically conjugated to a Markov shift $\Sigma_\mathcal{T} := \{(a_n)_{n \in \mathbb{Z}} : (a_i, a_{i+1}) \in \mathcal{T} \ \forall i\} \subset \mathcal{A}^{\mathbb{Z}}$ of finite type associated to $\mathcal{T}$, namely, there is a homeomorphism $\Pi \colon \Sigma_\mathcal{T} \to \Lambda$ such that $\varphi|_\Lambda \circ \Pi = \Pi \circ \sigma|_{\Sigma_\mathcal{T}}$ (where σ is the shift map).

---

[5]Here, $\|.\|$ stands for a norm associated to some choice of Riemannian metric on M.
[6]Recall that the stable manifold of $x \in \Lambda$ is $W^s(x) := \{y \in M : \lim\limits_{n \to +\infty} \text{dist}(\varphi^n(y), \varphi^n(x)) = 0\}$ (and its unstable manifold is obtained after replacing φ by φ^{-1} in the definition of stable manifold).

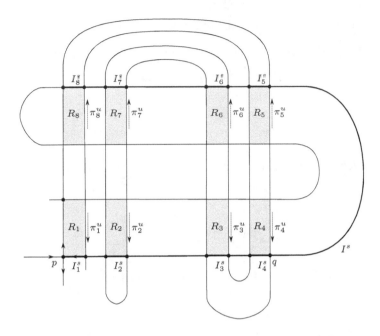

Fig. 3.1   Geometry of the horseshoe Λ.

Next, we recall that the laminations associated to the stable and unstable manifolds of points in Λ can be extended to locally invariant $C^{1+\varepsilon}$-foliations in a neighborhood of Λ for some $\varepsilon > 0$. Therefore, we can use these foliations to define projections $\pi_a^u : R_a \to I_a^s \times \{i_a^u\}$ and $\pi_a^s : R_a \to \{i_a^s\} \times I_a^u$ of the rectangles into the connected components $I_a^s \times \{i_a^u\}$ and $\{i_a^s\} \times I_a^u$ of the stable and unstable boundaries of R_a where $i_a^u \in \partial I_a^u$ and $i_a^s \in \partial I_a^s$ are fixed arbitrarily. Using these projections, we have the stable and unstable Cantor sets

$$K^s = \bigcup_{a \in \mathcal{A}} \pi_a^u(\Lambda \cap R_a) \quad \text{and} \quad K^u = \bigcup_{a \in \mathcal{A}} \pi_a^s(\Lambda \cap R_a)$$

associated to Λ.

The stable and unstable Cantor sets K^s and K^u are $C^{1+\varepsilon}$-dynamically defined Cantor sets in the sense of Chapter 2, §2.3.1 because the $C^{1+\varepsilon}$-maps

$$g_s(\pi_{a_1}^u(y)) = \pi_{a_0}^u(\varphi^{-1}(y))$$

for $y \in R_{a_1} \cap \varphi(R_{a_0})$ and

$$g_u(\pi_{a_0}^s(z)) = \pi_{a_1}^s(\varphi(z))$$

for $z \in R_{a_0} \cap \varphi^{-1}(R_{a_1})$ are expanding of type $\Sigma_{\mathcal{T}}$ defining K^s and K^u in the sense that

- the domains of g_s and g_u are disjoint unions $\bigsqcup\limits_{(a_0,a_1)\in\mathcal{T}} I^s(a_1,a_0)$ and $\bigsqcup\limits_{(a_0,a_1)\in\mathcal{T}} I^u(a_0,a_1)$ where $I^s(a_1,a_0)$, resp. $I^u(a_0,a_1)$, are compact subintervals of $I^s_{a_1}$, resp. $I^u_{a_0}$;
- for each $(a_0,a_1) \in \mathcal{T}$, the restrictions $g_s|_{I^s(a_0,a_1)}$ and $g_u|_{I^u(a_0,a_1)}$ are $C^{1+\varepsilon}$ diffeomorphisms onto $I^s_{a_0}$ and $I^u_{a_0}$ with $|Dg_s(t)| > 1$, resp. $|Dg_u(t)| > 1$, for all $t \in I^s(a_0,a_1)$, resp. $I^u(a_0,a_1)$ (for appropriate choices of the parametrization of I^s_a and I^u_a);
- K^s, resp. K^u, are the maximal invariant sets associated to g_s, resp. g_u, that is,

$$K^s = \bigcap_{n\in\mathbb{N}} g_s^{-n}\left(\bigcup_{(a_0,a_1)\in\mathcal{T}} I^s(a_1,a_0)\right) \text{ and}$$

$$K^u = \bigcap_{n\in\mathbb{N}} g_u^{-n}\left(\bigcup_{(a_0,a_1)\in\mathcal{T}} I^u(a_0,a_1)\right).$$

The stable and unstable Cantor sets K^s and K^u are closely related to the geometry of the horseshoe Λ: for instance, it is well-known that

$$HD(\Lambda) = HD(K^s) + HD(K^u) := d_s + d_u$$

and, furthermore, $d_s = d_u$ when φ is preserves a smooth area form ω.

Remark 3.7. The identity $d_s = d_u$ when φ preserves area is the generalization of Euler's remark leading to the symmetry on Hausdorff of Gauss-Cantor sets: compare with Remark 3.2 above.

The classical Markov and Lagrange spectra (or, more precisely, arbitrarily large compact parts of them) can be viewed as dynamical Markov and Lagrange spectra associated to conservative horseshoes, as in the discussion below, based in [Ar (1994)]:

Consider for instance $T_1 : (0,1) \times (0,1) \to [0,1) \times (0,1)$ given by

$$T_1(x,y) = \left(\left\{\frac{1}{x}\right\}, \frac{1}{y + \lfloor 1/x \rfloor}\right).$$

The maximal invariant set by T_1 in $(1/5,1) \times (0,1)$ is $C(4) \times C(4)$, a horseshoe (more generally, for any $m \geq 2$, the maximal invariant set by T_1 in $(\frac{1}{m+1},1) \times (0,1)$ is the horseshoe $C(m) \times C(m)$). Notice that

$$T_1([0;a_0,a_1,a_2,\dots],[0;b_1,b_2,b_3,\dots])$$
$$= ([0;a_1,a_2,a_3,\dots],[0;a_0,b_1,b_2,\dots]),$$

and that T_1 contracts in the vertical direction and expands in the horizontal direction.

For the real map $f(x, y) = y + \frac{1}{x}$, it follows from Perron's characterization that the corresponding dynamical Markov and Lagrange spectra have the same intersections with $(-\infty, 5]$ as the classical Markov and Lagrange spectra.

Pierre Arnoux observed that T_1 preserves a smooth measure in a neighbourhood of the horseshoe $C(4) \times C(4)$ (and indeed of all horseshoes $C(m) \times C(m), m \geq 2$).

Indeed, if $\Sigma = \{(x, y) \in \mathbb{R}^2 | 0 < x < 1, 0 < y < 1/(1+x)\}$ and $T : \Sigma \to \overline{\Sigma}$ is given by

$$T(x, y) = \left(\left\{ \frac{1}{x} \right\}, x - x^2 y \right),$$

then T preserves the Lebesgue measure in the plane.

If $h : \overline{\Sigma} \to [0, 1]^2$ is given by $h(x, y) = (x, y/(1 - xy))$ then h is a conjugation between T and T_1, and thus T_1 preserves the smooth measure $h_*(\text{Leb})$, which projects in the first coordinate to the Gauss measure $\frac{dx}{1+x}$, a smooth measure in $[0, 1]$ invariant by the Gauss map $g(x) = \{\frac{1}{x}\}$. This explicit measure (which is ergodic and mixing with respect to g) can be used for instance to prove the statements of Remark 1.4.

3.2.4   *Cerqueira–Matheus–Moreira extension of Moreira's continuity theorem*

In their article [CMM (2018)], Cerqueira–Matheus–Moreira managed to extend Moreira's ideas in [Mor1 (2018)] in order to establish the continuity of the Hausdorff dimension across the dynamical Lagrange and Markov spectra of typical thin horseshoes of area-preserving surface diffeomorphisms.

More precisely, let φ_0 be a smooth diffeomorphism of a surface M^2 preserving an area-form ω. Suppose that φ_0 possesses a thin horseshoe Λ_0 in the sense that its Hausdorff dimension is $HD(\Lambda_0) < 1$. Denote by $\mathcal{U}$ a small C^∞ neighborhood of φ_0 in the space $\text{Diff}_\omega^\infty(M)$ of smooth area-preserving diffeomorphisms of M such that Λ_0 admits a *continuation*[7] Λ for every $\varphi \in \mathcal{U}$.

Theorem 3.5. *If $\mathcal{U} \subset \text{Diff}_\omega^\infty(M)$ is sufficiently small, then there exists a Baire residual subset $\mathcal{U}^{**} \subset \mathcal{U}$ with the following property. For every $\varphi \in$*

---

[7]I.e., if U_0 is a neighborhood of Λ_0 such that $\Lambda_0 = \bigcap_{n \in \mathbb{Z}} \varphi_0^n(U_0)$, then $\mathcal{U}$ is taken small enough so that $\Lambda = \bigcap_{n \in \mathbb{Z}} \varphi^n(U_0)$ still is a horseshoe for any $\varphi \in \mathcal{U}$.

$\mathcal{U}^{**}$ *and* $r \geq 2$, *there exists a* C^r-*open and dense subset* $\mathcal{R}_{\varphi,\Lambda} \subset C^r(M,\mathbb{R})$ *such that the functions*

$$t \mapsto d_s(t) := HD(K_t^s) \quad \text{and} \quad t \mapsto d_u(t) := HD(K_t^u)$$

are continuous and

$$d_s(t) + d_u(t) = 2d_u(t) = HD(L_{\varphi,f} \cap (-\infty, t)) = HD(M_{\varphi,f} \cap (-\infty, t))$$

whenever $f \in \mathcal{R}_{\varphi,\Lambda}$.

Remark 3.8. The proof of Theorem 3.5 in [CMM (2018)] actually shows that $d_s(t)$, resp. $d_u(t)$, coincide with the box counting dimension of K_t^s, resp. K_t^u.

Remark 3.9. The first part of Theorem 3.5 (i.e., the continuity of $d_s(t)$ and $d_u(t)$) still holds in the broader context of diffeomorphisms which might not preserve area: cf. Remarks 1.4 and 2.10 in [CMM (2018)]. On the other hand, the second part of Theorem 3.5 relies in a crucial way on the invariance of an area-form ω.

In the sequel, we will sketch the proof of Theorem 3.5 by following the same type of strategy used in the proof of Moreira's theorem.

3.2.5 *Upper semi-continuity of dimension across dynamical spectra*

Our long-term goal is to study the Hausdorff dimension of the sets $M_{\varphi,f} \cap (-\infty, t)$ and $L_{\varphi,f} \cap (-\infty, t)$ as $t \in \mathbb{R}$ varies.

For this reason, for each $t \in \mathbb{R}$, we will consider

$$\Lambda_t := \bigcap_{n \in \mathbb{Z}} \varphi^{-n}(\{y \in \Lambda : f(y) \leq t\}) = \left\{ x \in \Lambda : m_{\varphi,f}(x) = \sup_{n \in \mathbb{Z}} f(\varphi^n(x)) \leq t \right\}$$

and its projections

$$K_t^s = \bigcup_{a \in \mathcal{A}} \pi_a^u(\Lambda_t \cap R_a) \text{ and } K_t^u = \bigcup_{a \in \mathcal{A}} \pi_a^s(\Lambda_t \cap R_a)$$

on the stable and unstable Cantor sets of Λ.

Recall that the horseshoe Λ has a combinatorial counterpart in terms of the Markov shift $\Sigma = \Sigma_{\mathcal{T}} \subset \mathcal{A}^{\mathbb{Z}}$. In particular, we have a homeomorphism $\Pi : \Lambda \to \Sigma$ conjugating φ to the shift map σ allowing to transfer the function f from Λ to a function (still denoted f) on Σ. In this setting, we have $\Pi(\Lambda_t) = \Sigma_t$ where

$$\Sigma_t := \left\{ \theta \in \Sigma : \sup_{n \in \mathbb{Z}} f(\sigma^n(\theta)) \leq t \right\}.$$

Using these notations, we can introduce the dynamical analogs of the objects introduced in §3.1.5 above. More precisely, given an admissible finite sequence $\alpha = (a_1, \ldots, a_n) \in \mathcal{A}^n$ (i.e., $(a_i, a_{i+1}) \in \mathcal{T}$ for all $i = 1, \ldots, n-1$), we define

$$I^u(\alpha) = \{x \in K^u : g_u^i(x) \in I^u(a_i, a_{i+1}) \,\forall\, i = 1, \ldots, n-1\}.$$

Similarly, given an admissible finite sequence $\alpha = (a_1, \ldots, a_n) \in \mathcal{A}^n$, we define

$$I^s(\alpha^T) =: \{y \in K^s : g_s^i(y) \in I^s(a_i, a_{i-1}) \,\forall\, i = 2, \ldots, n\}.$$

Here, $\alpha^T = (a_n, \ldots, a_1)$ denotes the *transpose* of α.

We say that the *unstable size* $s^{(u)}(\alpha)$ of α is the length $|I^u(\alpha)|$ of the interval $I^u(\alpha)$ and the *unstable scale* of α is $r^{(u)}(\alpha) = \lfloor \log(1/s^{(u)}(\alpha)) \rfloor$. Similarly, the *stable size* $s^{(s)}(\alpha)$ is $s^{(s)}(\alpha) = |I^s(\alpha^T)|$ and the *stable scale* of α is $r^{(s)}(\alpha) = \lfloor \log(1/s^{(s)}(\alpha)) \rfloor$.

Remark 3.10. In the context of $C^{1+\varepsilon}$-dynamically defined Cantor sets, we can relate the unstable and stable sizes of α to its length as a word in the alphabet $\mathcal{A}$ via the so-called *bounded distortion property* saying that there exists a constant $c_1 = c_1(\varphi, \Lambda) > 0$ such that:

$$e^{-c_1} \leq \frac{|I^u(\alpha\beta)|}{|I^u(\alpha)| \cdot |I^u(\beta)|} \leq e^{c_1}, \quad e^{-c_1} \leq \frac{|I^s((\alpha\beta)^T)|}{|I^s((\alpha)^T)| \cdot |I^s((\beta)^T)|} \leq e^{c_1}.$$

We refer the reader to [PT (1993), p. 59] for more details.

Given $r \in \mathbb{N}$, we define

$$P_r^{(u)} = \{\alpha = (a_1, \ldots, a_n) \in \mathcal{A}^n \text{ admissible } :$$
$$r^{(u)}(\alpha) \geq r \text{ and } r^{(u)}(a_1 \ldots a_{n-1}) < r\},$$

resp.

$$P_r^{(s)} = \{\alpha = (a_1, \ldots, a_n) \in \mathcal{A}^n \text{ admissible } :$$
$$r^{(s)}(\alpha) \geq r \text{ and } r^{(s)}(a_1 \ldots a_{n-1}) < r\},$$

and we consider the sets

$$\mathcal{C}_{\mathbf{u}}(t,r) = \{\alpha \in P_r^{(u)} : I^u(\alpha) \cap K_t^u \neq \emptyset\}, \quad \text{resp.}$$
$$\mathcal{C}_{\mathbf{s}}(t,r) = \{\alpha \in P_r^{(s)} : I^s(\alpha^T) \cap K_t^s \neq \emptyset\}$$

whose cardinalities are denoted by $N_{\mathbf{u}}(t,r) := \#\mathcal{C}_{\mathbf{u}}(t,r)$ and $N_{\mathbf{s}}(t,r) := \#\mathcal{C}_{\mathbf{s}}(t,r)$.

At this point, we can use the bounded distortion property in Remark 3.10 to derive the analogue of Exercise 3.1 from §3.1.5 in our current setting. More concretely, the bounded distortion property provides a constant $c_1 = c_1(\varphi, \Lambda)$ such that

$$|I^u(\alpha\beta\gamma)| \le e^{2c_1}|I^u(\alpha)| \cdot |I^u(\beta)| \cdot |I^u(\gamma)|$$

for all α, β, γ finite words such that the concatenation $\alpha\beta\gamma$ is admissible. Next, we observe that, if $\gamma = \gamma_1 \dots \gamma_c$ is a finite word in the letters $\gamma_i \in \mathcal{A}$, $1 \le i \le c$, then

$$|I^u(\gamma)| \le \frac{1}{\mu^c} \max_{a \in \mathcal{A}} |I^u_a|$$

where $\mu = \mu_{\mathbf{u}} := \min |Dg_u| > 1$. Now, we note that, for each $c \in \mathbb{N}$, one can cover K^u_t with $\le \#\mathcal{A}^c \cdot N_{\mathbf{u}}(t, n) \cdot N_{\mathbf{u}}(t, m)$ intervals $I^u(\alpha\beta\gamma)$ with $\alpha \in \mathcal{C}_{\mathbf{u}}(t, n)$, $\beta \in \mathcal{C}_{\mathbf{u}}(t, m)$, $\gamma \in \mathcal{A}^c$ and $\alpha\beta\gamma$ admissible. Therefore, by taking

$$\tilde{c}_1 = \tilde{c}_1(\varphi, \Lambda) = \lceil \frac{\log(e^{2c_1} \max_{a \in \mathcal{A}} |I^u_a|)}{\log \mu} \rceil \in \mathbb{N},$$

it follows that we can cover K^u_t with $\le \#\mathcal{A}^{\tilde{c}_1} \cdot N_{\mathbf{u}}(t, n) \cdot N_{\mathbf{u}}(t, m)$ intervals $I^u(\alpha\beta\gamma)$ whose scales satisfy

$$r^{(\mathbf{u})}(\alpha\beta\gamma) \ge r^{(\mathbf{u})}(\alpha) + r^{(\mathbf{u})}(\beta) \ge n + m$$

whenever $\alpha \in \mathcal{C}_{\mathbf{u}}(t, n)$, $\beta \in \mathcal{C}_{\mathbf{u}}(t, m)$, $\gamma \in \mathcal{A}^{\tilde{c}_1}$ and $\alpha\beta\gamma$ is admissible. From this inequality, we deduce that

$$N_{\mathbf{u}}(t, n + m) \le \#\mathcal{A}^{\tilde{c}_1} \cdot N_{\mathbf{u}}(t, n) \cdot N_{\mathbf{u}}(t, m)$$

for all $n, m \in \mathbb{N}$, and, *a fortiori*, $(\#\mathcal{A}^{\tilde{c}_1} \cdot N_{\mathbf{u}}(t, m))_{m \in \mathbb{N}}$ is a sub-multiplicative sequence. Moreover, the same fact is true when $N_{\mathbf{u}}$ is replaced by $N_{\mathbf{s}}$, so that we get well-defined quantities

$$D_u(t) := \inf_{m \in \mathbb{N}} \frac{1}{m} \log(\#\mathcal{A}^{\tilde{c}_1} \cdot N_{\mathbf{u}}(t, m)) \text{ and } D_s(t) := \inf_{m \in \mathbb{N}} \frac{1}{m} \log(\#\mathcal{A}^{c_3} \cdot N_{\mathbf{u}}(t, m))$$

coinciding with the box counting dimensions of K^u_t, resp. K^s_t.

The same compactness argument from the proof of Proposition 3.4 in §3.1.5 yields:

Proposition 3.5. *The functions $t \mapsto D_u(t)$ and $t \mapsto D_s(t)$ are upper semi-continuous.*

Proof. By symmetry between φ and φ^{-1}, it suffices to prove that, for each $t_0 \in \mathbb{R}$, the values $D_u(t)$ converge to $D_u(t_0)$ as $t > t_0$ approaches t_0. By contradiction, suppose that this is not the case. Then, there exists $\eta > 0$ such that $D_u(t) > D_u(t_0) + \eta$ for all $t > t_0$. By definition, this means that

$$\frac{1}{m} \log(\#\mathcal{A}^{c_3} \cdot N_{\mathbf{u}}(t, m)) > D_u(t_0) + \eta$$

for all $t > t_0$ and $m \in \mathbb{N}$.

On the other hand, by compactness, for each $m \in \mathbb{N}$, one has

$$\mathcal{C}_{\mathbf{u}}(t_0, m) = \bigcap_{t > t_0} \mathcal{C}_{\mathbf{u}}(t, m).$$

In particular, for each $m \in \mathbb{N}$, there exists $t(m) > t_0$ such that $N_{\mathbf{u}}(t(m), m) = N_{\mathbf{u}}(t_0, m)$.

Therefore, by putting these facts together, we would deduce that, for each $m \in \mathbb{N}$,

$$\frac{1}{m} \log(\#\mathcal{A}^{c_3} \cdot N_{\mathbf{u}}(t_0, m)) = \frac{1}{m} \log(\#\mathcal{A}^{c_3} \cdot N_{\mathbf{u}}(t(m), m)) > D_u(t_0) + \eta.$$

Hence, by letting $m \to \infty$, we would conclude that $D_u(t_0) > D_u(t_0) + \eta$, a contradiction. $\qquad\square$

3.2.6   *Genericity of height functions transverse to invariant manifolds*

Fix $r \geq 2$. We define

$$\mathcal{R}_{\varphi, \Lambda} = \{f \in C^r(M, \mathbb{R}) :$$

$$\nabla f(x) \text{ is not perpendicular to } E_x^s \text{ or } E_x^u \text{ for all } x \in \Lambda\}.$$

In other terms, $\mathcal{R}_{\varphi, \Lambda}$ is the class of C^r-functions $f : M \to \mathbb{R}$ that are locally monotone along stable and unstable directions.

It is well-known fact (cf. [PT (1993), pp. 162–165]) that the stable and unstable spaces E_x^s and E_x^u are Lipschitz functions of $x \in \Lambda$. As it is shown in Proposition 2.7 in [CMM (2018)], it is not difficult to explore this fact to show that a typical height function belongs to $\mathcal{R}_{\varphi, \Lambda}$ whenever $HD(\Lambda) < 1$:

Proposition 3.6. *Fix $r \geq 2$. If the horseshoe Λ has Hausdorff dimension $HD(\Lambda) < 1$, then $\mathcal{R}_{\varphi, \Lambda}$ is C^r-open and dense in $C^r(M, \mathbb{R})$.*

Proof. By definition, $\mathcal{R}_{\varphi,\Lambda}$ is C^r-open. Hence, our task is reduced to show that $\mathcal{R}_{\varphi,\Lambda}$ is C^r-dense. For this sake, we fix a smooth system of coordinates z on the initial Markov partition $\{R_a\}_{a\in\mathcal{A}}$ so that, given $f \in C^r(M,\mathbb{R})$ and $\varepsilon > 0$, the function $g(z) = f(z) + \langle v, z \rangle$ for $z \in \bigcup_{a\in\mathcal{A}} R_a$ can be extended (via a partition of unity) to a ε-C^r-perturbation of f whenever $v \in \mathbb{R}^2$ has norm $\|v\| \leq \varepsilon$.

Suppose that $f \notin \mathcal{R}_{\varphi,\Lambda}$. Given $0 < \varepsilon < 1$, we will construct $v \in \mathbb{R}^2$ such that $\|v\| \leq \varepsilon$ and $g(z) = f(z) + \langle v, z \rangle \in \mathcal{R}_{\varphi,\Lambda}$.

For each $\delta > 0$, let $\mathcal{C}(\delta)$ be the set of admissible finite words of the form $(a_{-m}, \ldots, a_0, \ldots, a_n)$, $m, n \in \mathbb{N}$, such that the rectangle $R(a_{-m}, \ldots, a_0, \ldots, a_n) = \bigcap_{j=-m}^{n} \varphi^{-j}(R_{a_j})$ has diameter $\leq \delta$ but the rectangles $R(a_{-m}, \ldots, a_0, \ldots, a_{n-1})$ and $R(a_{-m+1}, \ldots, a_0, \ldots, a_n)$ have diameters $> \delta$.

Since Λ is a horseshoe associated to a C^2-diffeomorphism φ, its Hausdorff dimension coincide with its box counting dimension (see, e.g., Palis–Takens book [PT (1993)]). Therefore, if we set $HD(\Lambda) < d := \sqrt{HD(\Lambda)} < 1$, then there exists $\delta_0 > 0$ such that $\#\mathcal{C}(\delta) \leq 1/\delta^d$ for all $0 < \delta < \delta_0$.

By the Lipschitz properties of $\nabla f(z)$, E_z^u and E_z^s, there exists a constant $c_4 = c_4(f, \Lambda) > 0$ such that, for all $z, w \in \Lambda$, one has

$$|\nabla f(z) - \nabla f(w)|, \ |v^s(z) - v^s(w)|, \ |v^u(z) - v^u(w)| \leq c_4|z - w|,$$

where v^s and v^u (resp.) are unitary vectors in E^s and E^u (resp.).

In this setting, for each rectangle $R(\alpha)$, $\alpha \in \mathcal{C}(\delta)$, such that $\nabla f(z_\alpha)$ is perpendicular to $v^*(z_\alpha)$, $* = s$ or u, for some $z_\alpha \in R(\alpha)$, we write

$$\langle \nabla f(z) + v, v^*(z) \rangle = \langle \nabla f(z) - \nabla f(z_\alpha), v^*(z) \rangle + \langle \nabla f(z_\alpha), v^*(z) - v^*(z_\alpha) \rangle$$
$$+ \langle \nabla f(z_\alpha), v^*(z_\alpha) \rangle + \langle v, v^*(z) - v^*(z_\alpha) \rangle + \langle v, v^*(z_\alpha) \rangle.$$

We control this quantity as follows. Since $R(\alpha)$ has diameter $\leq \delta$, we see from our previous discussion that

$$|\langle \nabla f(z) - \nabla f(z_\alpha), v^*(z) \rangle| \leq c_4\delta, \qquad |\langle \nabla f(z_\alpha), v^*(z) - v^*(z_\alpha) \rangle| \leq \rho c_5 \delta$$

$$\langle \nabla f(z_\alpha), v^*(z_\alpha) \rangle = 0, \qquad |\langle v, v^*(z) - v^*(z_\alpha) \rangle| \leq \|v\| c_5 \delta.$$

In particular, it follows that

$$S_\alpha := \{v \in \mathbb{R}^2 : \|v\| = \varepsilon, \langle \nabla f(z) + v, v^*(z) \rangle = 0 \text{ for some } z \in R(\alpha)\}$$
$$\subset \{v \in \mathbb{R}^2 : \|v\| = \varepsilon, |\langle v, v^*(z_\alpha) \rangle| \leq (\rho + 2)c_4\delta\}$$

for $0 < \varepsilon < 1$.

Since the function $g(z) = f(z) + \langle v, z \rangle$ satisfies $\nabla g(z) = \nabla f(z) + v$, it suffices to exhibit $v \in \mathbb{R}^2$ such that $\|v\| = \varepsilon$ and $v \notin \bigcup\limits_{\alpha \in \mathcal{C}(\delta)} S_\alpha$. As it turns out, this is not hard to do: by our previous discussion, for all $0 < \delta < \delta_0$, the relative Lebesgue measure of $\bigcup\limits_{\alpha \in \mathcal{C}(\delta)} S_\alpha$ is

$$\leq (\rho + 2)c_4 \delta^{1-d}/\varepsilon$$

because, for each $\alpha \in \mathcal{C}(\delta)$, the relative Lebesgue measure of S_α is $\leq (\rho + 2)c_4\delta/\varepsilon$, and the cardinality of $\mathcal{C}(\delta)$ is $\leq \delta^{-d}$. $\square$

3.2.7   *Approximation of Λ_t by subhorseshoes*

The first application of the transversality property satisfied by a height function $f \in \mathcal{R}_{\varphi,\Lambda}$ in Cerqueira–Matheus–Moreira work [CMM (2018)] consists in the proof of the following generalization of Theorem 3.4:

Proposition 3.7. *Let Λ be a horseshoe of an area-preserving C^2-diffeomorphism φ of a surface M. Let $f \in \mathcal{R}_{\varphi,\Lambda}$ with $\mathcal{R}_{\varphi,\Lambda}$ as defined in §3.2.6, and let us fix $t \in \mathbb{R}$ with $D_u(t) > 0$, resp. $D_s(t) > 0$. Then, for each $0 < \eta < 1$, there exists $\delta > 0$ and a complete subshift $\Sigma(\mathcal{B}_u) \subset \Sigma \subset \mathcal{A}^{\mathbb{Z}}$, resp. $\Sigma(\mathcal{B}_s) \subset \Sigma \subset \mathcal{A}^{\mathbb{Z}}$, associated to a finite set $\mathcal{B}_u = \{\beta_1^{(u)}, \ldots, \beta_m^{(u)}\}$, resp. $\mathcal{B}_s = \{\beta_1^{(s)}, \ldots, \beta_n^{(s)}\}$, of finite sequences $\beta_i^{(u)} = (a_1^{(i);u}, \ldots, a_{m_i}^{(i);u}) \in \mathcal{A}^{m_i}$, resp. $\beta_i^{(s)} = (a_1^{(i);s}, \ldots, a_{n_i}^{(i);s}) \in \mathcal{A}^{n_i}$ such that*

$$\Sigma(\mathcal{B}_u) \subset \Sigma_{t-\delta}, \quad resp. \ \Sigma(\mathcal{B}_s) \subset \Sigma_{t-\delta},$$

and

$$HD(K^u(\Sigma(\mathcal{B}_u))) > (1-\eta)D_u(t), \quad HD(K^s(\Sigma(\mathcal{B}_u^T))) > (1-\eta)D_u(t)$$

resp.

$$HD(K^s(\Sigma(\mathcal{B}_s))) > (1-\eta)D_s(t), \quad HD(K^u(\Sigma(\mathcal{B}_s^T))) > (1-\eta)D_s(t)$$

where $K^u(\Sigma(*))$, resp. $K^s(\Sigma(*))$, is the subset of K^u, resp. K^s, consisting of points whose trajectory under g_u, resp. g_s, follows an itinerary obtained from the concatenation of words in the alphabet $*$, and $*^T$ is the alphabet whose words are the transposes of the words of the alphabet $*$.

In particular, $D_s(t) = D_u(t) = d_u(t) = d_s(t)$ for all $t \in \mathbb{R}$.

The proof of this result follows closely the arguments in the sketch of proof of Theorem 3.4 presented above with additional complications related to the fact that the comparisons between continued fraction values need to be replaced by some computations using the transversality properties of $f \in \mathcal{R}_{\varphi,\Lambda}$. In particular, we refer the reader to the original article [CMM (2018)] for the details and we conclude our discussion of Proposition 3.7 with the following observation.

Remark 3.11. As it was pointed out in [CMM (2018)], even if φ is *not* necessarily conservative, for each $\eta > 0$, we can still find $\delta > 0$ and a complete subshift $\Sigma(\mathcal{B}_u) \subset \Sigma \subset \mathcal{A}^{\mathbb{Z}}$, resp. $\Sigma(\mathcal{B}_s) \subset \Sigma \subset \mathcal{A}^{\mathbb{Z}}$ such that $\dim(K^u(\Sigma(\mathcal{B}_u))) > (1 - \eta)D_u(t)$ and $\dim(K^s(\Sigma(\mathcal{B}_s))) > (1 - \eta)D_s(t)$. In particular, we can use this result together with Proposition 3.5 to get the continuity statement in Theorem 3.5. On the other hand, the proof in [CMM (2018)] of the facts that $\dim(K^s(\Sigma(\mathcal{B}_u^T))) > (1 - \eta)D_u(t)$, $\dim(K^u(\Sigma(\mathcal{B}_s^T))) > (1 - \eta)D_s(t)$ and $D_s(t) = D_u(t) = d_u(t) = d_s(t)$ requires φ to be an area-preserving diffeomorphism.

3.2.8 *Approximation of dynamical Lagrange spectra by images of "subhorseshoes"*

The second application of the transversality property satisfied by a height function $f \in \mathcal{R}_{\varphi,\Lambda}$ in Cerqueira–Matheus–Moreira work [CMM (2018)] consists in the approximation of dynamical Lagrange spectra by the values of height function on certain "subhorseshoes":

Proposition 3.8. *Let $f \in \mathcal{R}_{\varphi,\Lambda}$, $\Sigma(\mathcal{B}) \subset \Sigma \subset \mathcal{A}^{\mathbb{Z}}$ be a complete subshift associated to a finite alphabet $\mathcal{B}$ of finite words on $\mathcal{A}$, and $\Lambda(\Sigma(\mathcal{B})) \subset \Lambda$ the subhorseshoe of Λ associated to $\mathcal{B}$.*

Then, for each $\varepsilon > 0$, there exists a subhorseshoe $\Lambda(\mathcal{B}, \varepsilon) \subset \Lambda(\Sigma(\mathcal{B}))$, a rectangle $R(\mathcal{B}, \varepsilon)$ of some Markov partition of $\Lambda(\mathcal{B}, \varepsilon)$, a C^1-diffeomorphism $\mathfrak{A}$ defined in a neighborhood of $R(\mathcal{B}, \varepsilon)$ respecting the local stable and unstable foliations, and an integer $j(\mathcal{B}, \varepsilon) \in \mathbb{Z}$ such that

$$dim(\Lambda(\mathcal{B}, \varepsilon)) \geq (1 - \varepsilon)dim(\Lambda(\Sigma(\mathcal{B})))$$

and

$$f(\varphi^{j(\mathcal{B},\varepsilon)}(\mathfrak{A}(\Lambda(\mathcal{B}, \varepsilon) \cap R(\mathcal{B}, \varepsilon)))) \subset \left\{ \limsup_{n \to \infty} f(\varphi^n(x)) : x \in \Lambda(\Sigma(\mathcal{B})) \right\}.$$

Proof. Since $f \in \mathcal{R}_{\varphi,\Lambda}$ is locally monotone on stable and unstable manifolds, we see that $f \in \mathcal{R}_{\varphi,\Lambda}$ attains its maximal value on $\Lambda(\Sigma(\mathcal{B}))$ at finitely many points $x_1, \ldots, x_k \in \Lambda(\Sigma(\mathcal{B}))$.

For each $n \in \mathbb{N}$, denote by $\mathcal{B}^n := \{(\beta_{-n}, \ldots, \beta_{-1}; \beta_0, \ldots, \beta_n) : \beta_i \in \mathcal{B}, i = 1, \ldots n\}$.

Given $\varepsilon > 0$, we consider the symbolic sequences $(\beta_n^{(j)})_{n \in \mathbb{Z}} \in \mathcal{B}^{\mathbb{Z}}$ associated to x_j, $j = 1, \ldots, k$ and we take[8] $m = m(\varepsilon, k) \in \mathbb{N}$ large enough such that $m > k^2$ and the subhorseshoe $\Lambda(\mathcal{B}, \varepsilon) := \Lambda(\Sigma(\mathcal{B}^*))$ of $\Lambda(\Sigma(\mathcal{B}))$ associated to the complete subshift generated by the alphabet

$$\mathcal{B}^* := \mathcal{B}^m - \{\gamma^{(j,m)} := (\beta_{-m}^{(j)}, \ldots, \beta_{-1}^{(j)}; \beta_0^{(j)}, \beta_1^{(j)}, \ldots, \beta_m^{(j)}) : j = 1, \ldots, k\}$$

has Hausdorff dimension $\dim(\Lambda(\mathcal{B}, \varepsilon)) \geq (1 - \varepsilon)\dim(\Lambda(\Sigma(\mathcal{B})))$.

Denote by R_a the rectangles of the Markov partition of $\Lambda(\Sigma(\mathcal{B}^m))$ $(= \Lambda(\Sigma(\mathcal{B})))$ induced by $\mathcal{B}^m$. By construction, we can find an open set $U \subset M$ such that $\overline{U}$ is disjoint from the Markov rectangles R_j associated to $\gamma_m^{(j)} \in \mathcal{B}^m$, $j = 1, \ldots, k$, $U \cap \Lambda(\Sigma(\mathcal{B})) = \Lambda(\Sigma(\mathcal{B})) - \bigcup_{j=1}^{k} R_j$, and $\Lambda(\mathcal{B}, \varepsilon) = \bigcap_{n \in \mathbb{Z}} \varphi^n(U)$. In particular, there exists $\eta > 0$ such that $\max f(\overline{U}) < \min f(\cup_{j=1}^{k} R_j) - \eta$. Hence, if we take an appropriate periodic point in $\Lambda(\mathcal{B}, \varepsilon)$ with a symbolic sequence $\overline{d} = (\ldots, d_{\widehat{m}}, d_{-\widehat{m}}, \ldots, d_{-1}; d_0, \ldots, d_{\widehat{m}}, d_{-\widehat{m}}, \ldots)$, then the sequences of the form

$$\underline{\theta} = (\ldots, \beta_{-m-2}, \beta_{-m-1}, d_{-\widehat{m}}, \ldots, d_{\widehat{m}}, \gamma^{(1,m)}, d_{-\widehat{m}}, \ldots, d_{\widehat{m}}, \beta_{m+1}, \beta_{m+2}, \ldots)$$

with $\beta_j \in \mathcal{B}^*$, $|j| > m$, have the property that $m_{\varphi,f}(\underline{\theta}) := \sup_{n \in \mathbb{Z}} f(\sigma^n(\underline{\theta}))$ is attained for values of $|n| \leq n_0$ corresponding to the piece $\gamma^{(1,m)}$ and, moreover, $f(\sigma^n(\underline{\theta})) < m_{\varphi,f}(\underline{\theta}) - \eta$ whenever $|n| > n_0$ does not correspond to $\gamma^{(1,m)}$. Furthermore, $m_{\varphi,f}(\underline{\theta}') < m_{\varphi,f}(\underline{\theta}) - \eta$ for all $\underline{\theta}' \in \Sigma(\mathcal{B}^*)$ and $\underline{\theta}$ as above.

Next, we denote by $R_1(\mathcal{B}, \varepsilon) := R_{\overline{d}}$ the Markov rectangle associated to the periodic orbit $\overline{d}$ and we define two maps θ^* and $\underline{\theta}$ from $\Lambda(\mathcal{B}, \varepsilon) \cap R_1(\mathcal{B}, \varepsilon)$ to Λ in the following way. Given $x \in \Lambda(\mathcal{B}, \varepsilon) \cap R_1(\mathcal{B}, \varepsilon)$, let us denote by the corresponding sequence in $\Sigma(\mathcal{B})$ by $(\ldots, \gamma_{-1}, d_{-\widehat{m}}, \ldots; d_0, \ldots, d_{\widehat{m}}, \gamma_1, \ldots)$ where $\gamma_j \in \mathcal{B}^*$ for all $j \in \mathbb{Z}$. We set

$$\tau^{(j)} := (\gamma_{-|j|}, \ldots, \gamma_{-1}, d_{-\widehat{m}}, \ldots, d_{\widehat{m}}, \gamma^{(1,m)}, d_{-\widehat{m}}, \ldots, d_{\widehat{m}}, \gamma_1, \ldots, \gamma_{|j|})$$

for each $j \in \mathbb{Z}$, and we introduce the point $\theta^*(x)$, resp. $\underline{\theta}(x)$, of $\Lambda(\Sigma(\mathcal{B}))$ associated to the sequence

$$(\ldots, \tau^{(-2)}, \tau^{(-1)}; \tau^{(1)}, \tau^{(2)}, \ldots),$$

resp.

$$(\ldots, \gamma_{-2}, \gamma_{-1}, d_{-\widehat{m}}, \ldots, d_{\widehat{m}}, \gamma^{(1,m)}, d_{-\widehat{m}}, \ldots, d_{\widehat{m}}, \gamma_1, \gamma_2, \ldots).$$

---

[8]Compare with the arguments in Appendix E.

Observe[9] that the map $\underline{\theta} : \Lambda(\mathcal{B}, \varepsilon) \cap R_1(\mathcal{B}, \varepsilon) \to \Lambda(\Sigma(B))$ extends to a C^1 diffeomorphism $\mathfrak{A}$ on a neighborhood of $R_1(\mathcal{B}, \varepsilon)$ respecting the stable and unstable laminations.

Given $x \in \Lambda(\mathcal{B}, \varepsilon) \cap R_1(\mathcal{B}, \varepsilon)$, we affirm that

$$\limsup_{n \to \infty} f(\sigma^n(\theta^*(x))) := \ell_{\varphi, f}(\theta^*(x)) = f(\sigma^n(\underline{\theta}(x)))$$

for some $|n| \leq n_0$.

Indeed, we note that $\underline{\theta}(x)$ was constructed in such a way that $m_{\varphi, f}(\underline{\theta}(x)) = f(\sigma^n(\underline{\theta}(x)))$ for some $|n| \leq n_0$. Thus, we have a sequence $(n_k)_{k \in \mathbb{Z}}$ with n_k corresponding to a position in the piece $\tau^{(k)}$ of $\theta^*(x)$ so that $\sigma^{n_k}(\theta^*(x))$ converges to $\sigma^n(\underline{\theta}(x))$ and, *a fortiori*,

$$\ell_{\varphi, f}(\theta^*(x)) \geq m_{\varphi, f}(\underline{\theta}(x)) = f(\sigma^n(\underline{\theta}(x))). \tag{3.1}$$

Also, if $(m_k)_{k \in \mathbb{N}}$ and $(r_k)_{k \in \mathbb{N}}$ are sequences where m_k are positions in pieces $\tau^{(r_k)}$ of $\theta^*(x)$ with the property that

$$\ell_{\varphi, f}(\theta^*(x)) = \lim_{k \to \infty} f(\sigma^{m_k}(\theta^*(x))),$$

then one of the following two possibilities occur: the sequence $|m_k - n_{r_k}|$ has a bounded subsequence or the sequence $|m_k - n_{r_k}|$ is unbounded. In the former case, we can find $b \in \mathbb{Z}$ such that $\sigma^{m_k}(\theta^*(x))$ has a subsequence converging to $\sigma^b(\underline{\theta}(x))$ and, hence,

$$\ell_{\varphi, f}(\theta^*(x)) = \lim_{k \to \infty} f(\sigma^{m_k}(\theta^*(x))) = f(\sigma^b(\underline{\theta}(x))) \leq m_{\varphi, f}(\underline{\theta}(x)) = f(\sigma^n(\underline{\theta}(x))).$$

In the latter case, there exists a subsequence of $\sigma^{m_k}(\theta^*(x))$ converging to an element $\theta' \in \Sigma(\mathcal{B}^*)$, but this is a contradiction with (3.1) because

$$\ell_{\varphi, f}(\theta^*(x)) = \lim_{k \to \infty} f(\sigma^{m_k}(\theta^*(x))) = f(\theta') \leq m_{\varphi, f}(\theta') < m_{\varphi, f}(\underline{\theta}(x)) - \eta.$$

This proves our claim.

It follows from our claim that

$$\Lambda(\mathcal{B}, \varepsilon) \cap R_1(\mathcal{B}, \varepsilon) = \bigcup_{j=-n_0}^{n_0} \Lambda(\mathcal{B}, \varepsilon, n)$$

where $\Lambda(\mathcal{B}, \varepsilon, n) = \{x \in \Lambda(\mathcal{B}, \varepsilon) \cap R_1(\mathcal{B}, \varepsilon) : \ell_{\varphi, f}(\theta^*(x)) = f(\sigma^n(\underline{\theta}(x)))\}$. Therefore, one of the closed subsets $\Lambda(\mathcal{B}, \varepsilon, j(\mathcal{B}, \varepsilon))$ has non-empty interior in $\Lambda(\mathcal{B}, \varepsilon)$ (for some $|j(\mathcal{B}, \varepsilon)| \leq n_0$) and, *a fortiori*, we can choose a rectangle of $R(\mathcal{B}, \varepsilon)$ of some Markov partition of $\Lambda(\mathcal{B}, \varepsilon)$ such that

$$\Lambda(\mathcal{B}, \varepsilon) \cap R(\mathcal{B}, \varepsilon) \subset \Lambda(\mathcal{B}, \varepsilon, j(\mathcal{B}, \varepsilon)).$$

This completes the proof of the proposition. $\square$

---

[9]Because these laminations extend into C^1 local foliations, cf. §3.2.3.

3.2.9 Moreira's dimension formula and lower semicontinuity of $D_u(t)$ and $D_s(t)$

The third application of the transversality property of height functions in $\mathcal{R}_{\varphi,\Lambda}$ consists in getting Moreira's dimension formula [Mor2 (2016)] in our context:

Theorem 3.6. *There exists a Baire residual subset $\mathcal{U}^{**} \subset \mathcal{U}$ such that for any $g \in \mathcal{R}_{\varphi,\Lambda}$ and for every subhorseshoe $\widetilde{\Lambda} \subset \Lambda$ one has*

$$dim(g(\widetilde{\Lambda})) = dim(\widetilde{\Lambda}).$$

Remark 3.12. The Baire residual subset $\mathcal{U}^{**}$ is relatively explicit: it consists of the diffeomorphisms $\varphi \in \mathcal{U}$ such that all ratios of the logarithms of the multipliers of all pairs of distinct periodic orbits in Λ are irrational and all Birkhoff invariants of all periodic orbits in Λ are non-zero. Indeed:

- by Corollary 1 of [Mor2 (2016)], the conclusion of Theorem 3.6 whenever the ratios of the logarithms of the multipliers of periodic orbits are irrational and a certain property $(H\alpha)$ holds for some $\alpha > 0$;
- as it is explained in Sections 4 and 9 of Moreira–Yoccoz article [MY (2010)], the property $(H\alpha)$ is satisfied whenever all Birkhoff invariants of all periodic points are non-zero.

In a certain sense, these conditions on (φ, Λ) generalize the notion of "non-essentially affine Cantor sets" discussed in the beginning of this chapter.

At this stage, we are ready to conclude the lower semicontinuity of the Hausdorff dimension across generic Lagrange and Markov dynamical spectra:

Proposition 3.9. *There exists a Baire residual subset $\mathcal{U}^{**} \subset \mathcal{U}$ such that, for any $\varphi \in \mathcal{U}^{**}$ and $f \in \mathcal{R}_{\varphi,\Lambda}$, the functions $t \mapsto D_u(t)$ and $t \mapsto D_s(t)$ are lower semicontinuous and*

$$D_s(t) + D_u(t) = 2D_u(t) = dim(L_{\varphi,f} \cap (-\infty, t)) = dim(M_{\varphi,f} \cap (-\infty, t)).$$

Proof. Let $t \in \mathbb{R}$ with $D_u(t) > 0$ and $\eta > 0$. By Proposition 3.7, we can find $\delta > 0$ and a complete subshift $\Sigma(\mathcal{B}) \subset \Sigma_{t-\delta}$ on a finite alphabet $\mathcal{B}$ on finite words of $\mathcal{A}$ such that

$$(1 - \eta)(D_s(t) + D_u(t)) = 2(1 - \eta)D_u(t) \leq \dim(\Lambda(\Sigma(\mathcal{B}))).$$

By Proposition 3.8 and Theorem 3.6, we see that

$$\dim(\Lambda(\Sigma(\mathcal{B}))) \leq \dim(\ell_{\varphi,f}(\Lambda(\Sigma(\mathcal{B})))).$$

It follows that

$$2(1-\eta)D_u(t) \leq \dim(\Lambda(\Sigma(\mathcal{B}))) \leq \dim(\ell_{\varphi,f}(\Lambda(\Sigma(\mathcal{B}))))$$
$$\leq \dim(L_{\varphi,f} \cap (-\infty, t-\delta)) \leq \dim(M_{\varphi,f} \cap (-\infty, t-\delta))$$
$$\leq \dim(f(\Lambda_{t-\delta})) \leq \dim(\Lambda_{t-\delta}) \leq 2D_u(t-\delta).$$

Since $\eta > 0$ is arbitrary, this proves the proposition. $\square$

3.2.10 *End of the proof of Theorem 3.5*

Let $\varphi \in \mathcal{U}^{**}$ and $f \in \mathcal{R}_{\varphi,\Lambda}$: note that $\mathcal{U}^{**}$ is residual by Theorem 3.6 and $\mathcal{R}_{\varphi,\Lambda}$ is C^r-open and dense by Proposition 3.6.

By Propositions 3.5 and 3.9, the function

$$t \mapsto D_s(t) = D_u(t) = \frac{1}{2}\dim(L_{\varphi,f} \cap (-\infty, t)) = \frac{1}{2}\dim(M_{\varphi,f} \cap (-\infty, t))$$

is continuous. Since Proposition 3.7 says that $d_s(t) = D_s(t) = D_u(t) = d_u(t)$ for all $t \in \mathbb{R}$, the proof of Theorem 3.5 is complete.

3.3   Application of Cerqueira–Matheus–Moreira theorem to Dirichlet's spectrum

The statement of Dirichlet's theorem in the beginning of Chapter 1 actually tells us that if $\alpha \notin \mathbb{Q}$, then, for every $Q \in \mathbb{N}^*$, there exists $1 \leq q \leq Q$ and $p \in \mathbb{Z}$ such that

$$\left|\alpha - \frac{p}{q}\right| < \frac{1}{qQ}.$$

In particular, an alternative way (proposed by Hardy and Littlewood) to try to improve Dirichlet's theorem for certain irrational numbers $\alpha \notin \mathbb{Q}$ would be to consider the best constant $c(\alpha)$ for the inequality $|\alpha - \frac{p}{q}| < \frac{1}{qQ}$ in the place of the best constant of Diophantine approximation $\ell(\alpha)$ for the inequality $|\alpha - \frac{p}{q}| < \frac{1}{q^2}$.

In other terms, we can seek for the best constant $c(\alpha)$ such that the inequality

$$\left|\alpha - \frac{p}{q}\right| < \frac{1}{c(\alpha)qQ}$$

has infinitely many solutions with $p \in \mathbb{Z}$, $q, Q \in \mathbb{N}$, and $1 \leq q \leq Q$, i.e.,

$$c(\alpha) := \limsup_{Q \to \infty} \left(\min_{\substack{(p,q,Q) \in \mathbb{Z}^3, \\ 1 \leq q \leq Q}} Q|q\alpha - p| \right)^{-1}.$$

The analogue of Perron's characterization of $\ell(\alpha)$ (cf. Proposition 1.5) was found by Davenport–Schmidt [DS (1970)] in 1970: they showed that $c(\alpha) = 1 + \frac{1}{\mathcal{D}(\alpha)}$, where

$$\mathcal{D}(\alpha) := \limsup_{n\to\infty}[a_{n+1}; a_{n+2}, \dots] \cdot [a_n; a_{n-1}, \dots, a_1] = \limsup_{n\to\infty} \frac{\alpha_{n+1}}{\beta_{n+1}}$$

for $\alpha = [a_0; a_1, a_2, \dots]$. Similarly to the case of the Lagrange spectrum, we can collect all finite values of $\mathcal{D}(\alpha)$ in order to obtain an interesting set

$$D = \{\mathcal{D}(\alpha) < \infty : \alpha \in \mathbb{R} \setminus \mathbb{Q}\}$$

called *Dirichlet spectrum*. In terms of the shift dynamics $\sigma : \Sigma \to \Sigma$ on the shift space $\Sigma = (\mathbb{N}^*)^{\mathbb{Z}}$, the Dirichlet spectrum is the dynamical Lagrange spectrum

$$\mathcal{D} = \left\{ \limsup_{n\to\infty} g(\sigma^n(\theta)) < \infty : \theta \in \Sigma \right\}$$

with respect to the shift σ and the height function $g((\theta_n)_{n\in\mathbb{Z}}) := [\theta_0; \theta_1, \dots]/[0; \theta_{-1}, \dots]$.

The structure of the Dirichlet spectrum was studied by several authors including Davenport–Schmidt [DS (1970)], Diviš–Novák [DN (1971)], Ivanov [IvA (1978)], [IvB (1980)], Kopetzky [Kop (1989)], etc., and it shares many common features with the classical Lagrange spectrum: for instance,

- the smallest element of $\mathcal{D}$ is $\frac{3+\sqrt{5}}{2} = 2.618\dots$,
- its beginning $\mathcal{D} \cap (-\infty, \sqrt{5}+2)$ is

$$\left\{ \frac{3+\sqrt{5}}{2} < \frac{\sqrt{F_3}+\sqrt{F_1}}{\sqrt{F_3}-\sqrt{F_1}} < \cdots \right.$$
$$\left. < \frac{\sqrt{F_{k+2}}+\sqrt{F_k}}{\sqrt{F_{k+2}}-\sqrt{F_k}} < \cdots : (F_k)_{k\geq 1} \text{ is Fibonacci sequence} \right\},$$

- for all $\varepsilon > 0$, the set $\mathcal{D} \cap (\sqrt{5}+2, \sqrt{5}+2+\varepsilon)$ is uncountable,
- the interval $(3+\sqrt{3}, \frac{5+\sqrt{21}}{2})$ is a maximal[10] gap of $\mathcal{D}$,
- the portion $\mathcal{D} \cap (-\infty, \frac{5+3\sqrt{5}}{2})$ has zero Lebesgue measure, and it contains a Hall's ray $[\mu_0, \infty)$ starting at a constant $\frac{5+3\sqrt{5}}{2} \leq \mu_0 \leq 10+6\sqrt{2}$.

---

[10] Actually, Kopetzky [Kop (1989)] showed that this interval is a gap such that its right endpoint belongs to $\mathcal{D}$ and its left endpoint is accumulated by $\mathcal{D}([0; \overline{(2,1)_n, 2, 1, 1}])$, $n \in \mathbb{N}$, and this essentially proves the the maximality of this gap (because D is closed, cf. Remark 3.6).

As it turns out, we can use Cerqueira–Matheus–Moreira theorem to show the analogue of Moreira's theorem for Dirichlet spectrum:

Proposition 3.10. *The function* $\mathbb{R} \ni t \mapsto HD(\mathcal{D} \cap (-\infty, t)) \in [0, 1]$ *is continuous.*

Proof. We saw above that the Dirichlet spectrum is the dynamical Lagrange spectrum with respect to the shift map $\sigma : \Sigma \to \Sigma$ on $\Sigma = (\mathbb{N}^*)^{\mathbb{Z}}$ and the height function $g((\theta_n)_{n \in \mathbb{Z}}) = [\theta_0; \theta_1, \ldots]/[0; \theta_{-1}, \ldots]$. It is not hard to see that the geometric counterparts $\widetilde{G} : \Lambda \to \Lambda$ and $\widetilde{g}$ obtained from σ and g via the natural map $\Sigma \to \mathbb{R}^2$ sending $(\theta_n)_{n \in \mathbb{Z}}$ to $([\theta_0; \theta_1, \ldots], [0; \theta_{-1}, \ldots])$ satisfy the assumptions of Cerqueira–Matheus–Moreira theorem. Indeed, the local stable and unstable manifolds of $\widetilde{G}$ are pieces of vertical and horizontal lines, so that they are transverse to the gradient of the height function $\widetilde{g}(x, y) = x/y$, i.e., $\widetilde{g} \in \mathcal{R}_{\widetilde{G}, \Lambda}$ in the notation of §3.2.6. Furthermore, $(\widetilde{G}, \Lambda)$ verifies the explicit genericity condition $\mathcal{U}^{**}$ in Remark 3.12: in fact, this basically due to the fact that $\widetilde{G}$ is the natural extension of the Gauss map and we saw in Proposition 3.2 that Gauss-Cantor sets are non-essentially affine. Therefore, we can apply Theorem 3.5 to $\widetilde{G}$ and $\widetilde{g}$ in order to deduce the desired continuity result. $\qquad\qquad \square$

Chapter 4

Beginning of dynamical Lagrange and Markov spectra

In this chapter, we will discuss some results about the beginning of certain dynamical Lagrange and Markov spectra. In fact, we will see that the smallest element of the dynamical Lagrange and Markov spectra with respect to horseshoes of surface diffeomorphisms and typical height functions is attained by a periodic orbit: this result originally proved by Moreira in [Mor3 (2017)] is a positive answer to a question of Yoccoz asking for an extension of Hurwitz's theorem for the classical Lagrange and Markov spectra. After that, we will discuss some results by Ferenczi [Ferenczi (2012)] and Boshernitzan–Delecroix [BoDe (2017)] about an extension of Hurwitz's theorem in the context of the dynamical Lagrange spectrum associated to interval exchange transformations. Furthermore, we will also mention some extensions due to Lehner[Lehner (1985)], Haas–Series [HaSe (1986)], Hersonsky–Paulin [HePa (2002)], Hubert–Lelièvre–Marchese–Ulcigrai [HLMU (2015)], and Andersen–Duke [AD (2009)] of Hurwitz's theorem to geodesic flows on negatively curved manifolds.

4.1 Minima of typical dynamical spectra of horseshoes of surface diffeomorphisms

The context of this section is the same of Chapter 3, §3.2, that is, we have a diffeomorphism $\varphi : M^2 \to M^2$ of a surface M^2, a horseshoe $\Lambda \subset M^2$ of φ, a continuous height function $f : M^2 \to \mathbb{R}$, and we want to investigate the corresponding dynamical Lagrange and Markov spectra

$$L(f, \Lambda) = \left\{ \limsup_{n \to \infty} f(\varphi^n(x)) : x \in \Lambda \right\} \text{ and } M(f, \Lambda) = \left\{ \sup_{n \in \mathbb{Z}} f(\varphi^n(x)) : x \in \Lambda \right\}.$$

The main result of this section is the following theorem (originally proved in [Mor3 (2017)]):

Theorem 4.1. *Let Λ be a horseshoe associated to a C^2-diffeomorphism φ. Then there is a dense set $H \subset C^\infty(M, \mathbb{R})$, which is is C^0-open, such that for all $f \in H$, we have*

$$\min L(f, \Lambda) = \min M(f, \Lambda) = f(p),$$

where $p = p(f) \in \Lambda$ is a periodic point of φ. Moreover, $f(p)$ is an isolated point both in $L(f, \Lambda)$ and in $M(f, \Lambda)$.

Remark 4.1. This theorem answers positively a question posed by Yoccoz to Moreira in 1998. It can be thought as an extension of Hurwitz's theorem saying that $\sqrt{5}$, which is the common minimum of the classical Lagrange and Markov spectra, is the image by $f((\theta_n)_{n \in \mathbb{Z}}) = [\theta_0; \theta_1, \dots] + [0; \theta_{-1}, \dots]$ of the fixed point $(\dots, 1, 1, 1, \dots)$ of the shift map σ.

The remainder of this section evolves around the proof of Theorem 4.1. Before going into the details, let us recall some basic definitions and facts from the theory of uniformly hyperbolic dynamical systems (most of which were previously mentioned in Chapter 3, §3.2) for the sake of convenience of the reader.

4.1.1 *Some preliminaries from dynamical systems*

The reader is invited to consult the books of Palis–Takens [PT (1993)] and Shub [Sh (1987)] for more details about the notions and results concerning horseshoes and subshifts of finite types mentioned below.

Recall that the stable and unstable laminations $\mathcal{F}^s(\Lambda)$ and $\mathcal{F}^u(\Lambda)$ of a horseshoe of a C^2-diffeomorphism of a surface are $C^{1+\varepsilon}$ laminations (for some $\varepsilon > 0$) which can be used to fix a geometrical Markov partition $\{R_a\}_{a \in \mathbb{A}}$ with sufficiently small diameter consisting of rectangles $R_a \simeq I_a^s \times I_a^u$ delimited by compact pieces I_a^s and I_a^u of stable and unstable manifolds of certain periodic points of Λ.

The set $\mathbb{B} \subset \mathbb{A}^2$ of admissible transitions of such a Markov partition consist of pairs (a_0, a_1) such that $\varphi(R_{a_0}) \cap R_{a_1} \neq \emptyset$. We can encode $\mathbb{B} \subset \mathbb{A}^2$ using a transition matrix $B = (b_{kl})_{(k,l) \in \mathbb{A}^2}$ whose entries are $b_{a_i a_j} = 1$ if $\varphi(R_{a_i}) \cap R_{a_j} \neq \emptyset$ and $b_{a_i a_j} = 0$ otherwise.

Next, we recall that the stable and unstable manifolds of Λ can be extended to locally invariant $C^{1+\varepsilon}$-foliations in a neighborhood of Λ for some $\varepsilon > 0$, so that we can define projections $\pi_a^u : R_a \to I_a^s \times \{i_a^u\}$ and $\pi_a^s : R_a \to \{i_a^s\} \times I_a^u$, where $i_a^u \in \partial I_a^u$ and $i_a^s \in \partial I_a^s$ are fixed arbitrarily.

Using these projections, we have the stable and unstable Cantor sets

$$K^s = \bigcup_{a \in \mathbb{A}} \pi_a^u(\Lambda \cap R_a) \quad \text{and} \quad K^u = \bigcup_{a \in \mathbb{A}} \pi_a^s(\Lambda \cap R_a)$$

associated to Λ which are $C^{1+\varepsilon}$-dynamically defined because

$$g_s(\pi_{a_1}^u(y)) = \pi_{a_0}^u(\varphi^{-1}(y))$$

for $y \in R_{a_1} \cap \varphi(R_{a_0})$ and

$$g_u(\pi_{a_0}^s(z)) = \pi_{a_1}^s(\varphi(z))$$

for $z \in R_{a_0} \cap \varphi^{-1}(R_{a_1})$ are expanding maps defined on disjoint unions $\bigsqcup_{(a_0,a_1) \in \mathbb{B}} I^s(a_1, a_0)$ and $\bigsqcup_{(a_0,a_1) \in \mathbb{B}} I^u(a_0, a_1)$ of certain compact subintervals of $I_{a_1}^s$ and $I_{a_0}^u$ such that the restrictions $g_s|_{I^s(a_0,a_1)}$ and $g_u|_{I^u(a_0,a_1)}$ are $C^{1+\varepsilon}$ diffeomorphisms onto $I_{a_0}^s$ and $I_{a_0}^u$ with

$$K^s = \bigcap_{n \in \mathbb{N}} g_s^{-n}\left(\bigcup_{(a_0,a_1) \in \mathbb{B}} I^s(a_1, a_0) \right) \text{ and } K^u = \bigcap_{n \in \mathbb{N}} g_u^{-n}\left(\bigcup_{(a_0,a_1) \in \mathbb{B}} I^u(a_0, a_1) \right).$$

Also, the dynamics of $\varphi|_\Lambda$ is coded by a subshift of finite type. More concretely, let $\Sigma_B = \{\underline{a} \in \mathbb{A}^{\mathbb{Z}} : b_{a_n a_{n+1}} = 1\}$. This is a closed subset of the shift space $\mathbb{A}^{\mathbb{Z}}$ which is invariant under the shift map $\sigma((a_n)_{n \in \mathbb{Z}}) = (a_{n+1})_{n \in \mathbb{Z}}$. In this context, there is a homeomorphism $\Pi \colon \Sigma_B \to \Lambda$ such that $\varphi|_\Lambda \circ \Pi = \Pi \circ \sigma|_{\Sigma_B}$. In the literature, Π is called coding map and we say that $\theta \in \Sigma_B$ is the *kneading sequence* of $p = \Pi(\theta) \in \Lambda$.

For later reference, we recall two facts. First, if $N_n(x, y, B)$ denotes the number of admissible strings for $\mathbb{B}$ of length $n + 1$, beginning at $x \in \mathbb{A}$ and ending with $y \in \mathbb{A}$, then $N_n(x, y, B) = b_{xy}^n$ and the transitivity of $\varphi|_\Lambda$ implies that there is $N_0 \in \mathbb{N}^*$ such that $N_{N_0}(x, y, B) > 0$ for all $x, y \in \mathbb{A}$. Secondly, any subshift of finite type has a sort of local product structure in the sense that if

$$W_{1/3}^s(\underline{a}) = \{\underline{b} \in \Sigma_B : \forall n \geq 0, \ d(\sigma^n(\underline{a}), \sigma^n(\underline{b})) \leq 1/3\}$$
$$= \{\underline{b} \in \Sigma_B : \forall n \geq 0, \ a_n = b_n\},$$
$$W_{1/3}^u(\underline{a}) = \{\underline{b} \in \Sigma_B : \forall n \leq 0, \ d(\sigma^n(\underline{a}), \sigma^n(\underline{b})) \leq 1/3\}$$
$$= \{\underline{b} \in \Sigma_B : \forall n \leq 0, \ a_n = b_n\},$$

are the local stable and unstable sets (cf. [Sh (1987), Chap. 10]), where $d(\underline{a}, \underline{b}) = \sum_{n=-\infty}^{\infty} 2^{-(2|n|+1)} \delta_n(\underline{a}, \underline{b})$ and $\delta_n(\underline{a}, \underline{b})$ is 0 when $a_n = b_n$ and 1 otherwise, then $a_0 = b_0$ and $W_{1/3}^u(\underline{a}) \cap W_{1/3}^u(\underline{b})$ is a unique point, denoted by the bracket $[\underline{a}, \underline{b}] = (\cdots, b_{-n}, \cdots, b_{-1}, b_0, a_1, \cdots, a_n, \cdots)$ whenever $\underline{a}, \underline{b} \in \Sigma_B$ and $d(\underline{a}, \underline{b}) < 1/2$. Moreover, the homeomorphism Π respects the local product structure induced by the bracket, that is, $\Pi[\underline{a}, \underline{b}] = [\Pi(\underline{a}), \Pi(\underline{b})]$.

4.1.2   Typical injectivity near the bottom of the dynamical spectra

We will study the bottom part of the dynamical spectra via the auxiliary sets

$$\Lambda_t := \bigcap_{n \in \mathbb{Z}} \varphi^{-n}(\{y \in \Lambda : f(y) \le t\}) = \{x \in \Lambda : m_{\varphi,f}(x) = \sup_{n \in \mathbb{Z}} f(\varphi^n(x)) \le t\}$$

and

$$K_t^s = \bigcup_{a \in \mathbb{A}} \pi_a^u(\Lambda_t \cap R_a) \text{ and } K_t^u = \bigcup_{a \in \mathbb{A}} \pi_a^s(\Lambda_t \cap R_a)$$

for $t \in \mathbb{R}$.

By Cerqueira–Matheus–Moreira theorem (see Theorem 3.5 and also Remark 3.9 in Chapter 3), the box dimensions $D_s(t)$ of K_t^s and $D_u(t)$ of K_t^u depend continuously on t (even if φ does not preserve area). In particular, if $t_0 = \min M(f, \Lambda)$, then $D_s(t_0) = D_u(t_0) = 0$, since, for any $t < t_0$, the sets Λ_t, K_t^s and K_t^u are empty and, *a fortiori*, $D_s(t) = D_u(t) = 0$. Since the stable and unstable foliations of Λ are of class C^1, the box dimension of Λ_{t_0} is at most[1] the sum of the box dimensions of $K_{t_0}^s$ and $K_{t_0}^u$, and so is equal to 0. It follows that, for any $\epsilon > 0$, there is a locally maximal subhorseshoe (of finite type) $\tilde{\Lambda} \subset \Lambda$ with $\Lambda_{t_0} \subset \tilde{\Lambda}$ and $HD(\tilde{\Lambda}) < \epsilon$ (we may fix a large positive integer m and take $\tilde{\Lambda}$ as the set of points of Λ in whose kneading sequences all factors of size m are factors of the kneading sequence of some element of Λ_{t_0} — the number of such factors grow subexponentially in m since the box dimension of Λ_{t_0} is 0, so the Hausdorff dimension of $\tilde{\Lambda}$ is small when m is large by the estimates on fractal dimensions of [PT (1993), chap. 4]).

Proposition 4.1. *Let $k > 1$ be an integer. If $HD(\tilde{\Lambda}) < 1/2k$, then, for any $r \in \mathbb{N} \cup \{\infty\}$, there is a dense set (in the C^r topology) of C^r real functions f such that, for some $c > 0$, $|f(p) - f(q)| \ge c \cdot |p - q|^{k/(k-1)}, \forall p, q \in \tilde{\Lambda}$. In particular $f|_{\tilde{\Lambda}}$ is injective and its inverse is $(1 - 1/k)$-Hölder.*

Proof. Given a smooth function f, there are arbitrarily small perturbations of it whose derivative does not vanish at the stable and unstable directions in points of $\tilde{\Lambda}$ (cf. Proposition 3.6 in Chapter 3), so we will assume that f satisfies this property.

---

[1]Cf. Chapter 7 of Falconer's book [F (1986)] for the proof of the relevant box dimension estimate.

Given a Markov partition $\{R_a\}_{a \in \mathbb{A}}$ with sufficiently small diameter as before, we may perturb f by adding, for each a, independently, a small constant t_a to f in a small neighbourhood of $\tilde{\Lambda} \cap R_a$ (notice that the compact sets $(\tilde{\Lambda} \cap R_a)$ are mutually disjoint). Since, for $a \neq b$, the images $f(\tilde{\Lambda} \cap R_a)$ and $f(\tilde{\Lambda} \cap R_b)$ have box dimensions smaller than $1/2k$, their arithmetic difference $f(\tilde{\Lambda} \cap R_a) - f(\tilde{\Lambda} \cap R_b) = \{x - y, x \in f(\tilde{\Lambda} \cap R_a), y \in f(\tilde{\Lambda} \cap R_b)\} = \{t \in \mathbb{R} | f(\tilde{\Lambda} \cap R_a) \cap (f(\tilde{\Lambda} \cap R_b) + t) \neq \emptyset$ has box dimension smaller than $1/k < 1$, and so, for almost all t_a, t_b, the perturbed images $f(\tilde{\Lambda} \cap R_a) + t_a$ and $f(\tilde{\Lambda} \cap R_b) + t_b$ are disjoint.

Consider now parametrizations of small neighbourhoods of the pieces R_a according to which the stable leaves of $\tilde{\Lambda}$ are C^1 close to be horizontal and the unstable leaves of $\tilde{\Lambda}$ are C^1 close to be vertical, and such that, in the coordinates given by these parametrizations, $f(x, y)$ is C^1 close to an affine map $f(x, y) = ax + by + c$, with a and b far from 0 (these parametrizations exist since the pieces R_a are chosen very small, so f is close to affine in R_a). Then we may consider, in each coordinate system as above, perturbations of f of the type $f_\lambda(x, y) = f(x, \lambda y)$, where λ is a parameter close to 1.

Given $\rho > 0$ small, a ρ-decomposition of $\tilde{\Lambda} \cap R_a$ is a decomposition of it in a union of rectangles $I_j^s \times I_j^u$ intersected with $\tilde{\Lambda}$ such that both intervals I_j^s and I_j^u have length of the order of ρ. Let r be a large positive integer, and consider 2^{-kr} and $2^{-(k-1)r}$-decompositions of $\tilde{\Lambda} \cap R_a$. Given two rectangles of the 2^{-kr}-decomposition which belong to different rectangles of the $2^{-(k-1)r}$-decomposition (and so have distance at least of the order of $2^{-(k-1)r}$), the measure of the interval of values of λ such that the images by f_λ of the two rectangles of the 2^{-kr}-decomposition have distance smaller than 2^{-kr} is at most of the order of $2^{-kr}/2^{-(k-1)r} = 2^{-r}$. Since the box dimension of $\tilde{\Lambda}$ is $d < 1/2k$, the number of pairs of rectangles in the 2^{-kr}-decomposition is of the order of $(2^{-kr})^{-2d} = 2^{2dkr}$, and so the the measure of the set of values of λ such that the images of some pair as before of two rectangles of the 2^{-kr}-decomposition have non-empty intersection is at most of the order of $2^{2dkr} \cdot 2^{-r} = 2^{-(1-2dk)r} \ll 1$ (notice that $2dk < 1$). The sum of these measures for all $r \geq r_0$ is $O(2^{-(1-2dk)r_0}) \ll 1$, and so there is λ close to 1 such that, if ϵ is small and $p, q \in \tilde{\Lambda} \cap R_a$ are such that $|p - q| \geq \epsilon$ then $|f(p) - f(q)|$ is at least of the order of $\epsilon^{k/(k-1)}$ (consider r in the above discussion such that $2^{-(k-1)r}$ is of the order of ϵ). This implies the result: for some $c > 0$, $|f(p) - f(q)| \geq c \cdot |p - q|^{k/(k-1)}$, $\forall p, q \in \tilde{\Lambda}$. It follows that $f|_{\tilde{\Lambda}}$ is injective and its inverse function g is $(1 - 1/k)$-Hölder: indeed, it satisfies $|g(x) - g(y)| \leq (c^{-1}|x - y|)^{(1-1/k)}$, for any $x, y \in f(\tilde{\Lambda})$. $\qquad\square$

4.1.3   *Minimality of orbits at the bottom of typical dynamical spectra*

The typical injectivity of height functions near the bottom of the dynamical spectra forces the minimality of φ on the lowest lying orbit closure.

Proposition 4.2. *Assume that $f|_{\tilde{\Lambda}}$ is injective. Then $\min L(f, \Lambda) = \min M(f, \Lambda) = f(p)$ for only one value of $p \in \tilde{\Lambda}$ such that the restriction of φ to the closure of the orbit of p is minimal.*

Proof. Let $p \in \Lambda$ be such that $f(p) = \min M(f, \Lambda)$, which is unique since $f|_{\tilde{\Lambda}}$ is injective. We have $f(\varphi^j(p)) < f(p)$ for all integer j such that $\varphi^j(p) \neq p$. If some subsequence $\varphi^{n_k}(p)$ converges to a point q such that p does not belong to the closure of the orbit of q, $f(p)$ does not belong to the image by f of the closure of the orbit of q and so the Markov value of the orbit of q is strictly smaller than $f(p)$, a contradiction. This implies that the restriction of φ to the closure of the orbit of p is minimal and, in particular, $f(p)$ is the Lagrange value of its orbit, so we also have $f(p) = \min L(f, \Lambda)$ (thanks to the fact that $L(f, \Lambda) \subset M(f, \Lambda)$, cf. Appendix G). $\qquad\square$

Given Λ a horseshoe of a C^2-diffeomorphism φ of a surface, we define $X \subset C^0(M, \mathbb{R})$ as the set of real functions f for which

$$\min L(f, \Lambda) = \min M(f, \Lambda) = f(p),$$

where $p = p(f) \in \Lambda$ is a periodic point of φ and there is $\varepsilon > 0$ such that, for every $q \in \Lambda$ which does not belong to the orbit of p, $\sup_{n \in \mathbb{Z}} f(\varphi^n(q)) > f(p) + \varepsilon$.

Proposition 4.3. *X is open in $C^0(M, \mathbb{R})$.*

Proof. Suppose that $f \in X$. If $\varepsilon > 0$ is as in the definition of X, let $g \in C^0(M, \mathbb{R})$ such that $|g(x) - f(x)| < \varepsilon/3 \ \forall x \in M$. Then we have $\sup_{n \in \mathbb{Z}} g(\varphi^n(p)) = g(\tilde{p}) < f(\tilde{p}) + \varepsilon/3 \leq f(p) + \varepsilon/3$, for some point $\tilde{p}$ in the (finite) orbit of p. Moreover, for every $q \in \Lambda$ which does not belong to the orbit of p, since $g(\varphi^n(q)) > f(\varphi^n(q)) - \varepsilon/3$, we have $\sup_{n \in \mathbb{Z}} g(\varphi^n(q)) \geq \sup_{n \in \mathbb{Z}} f(\varphi^n(q)) - \varepsilon/3 > f(p) + \varepsilon - \varepsilon/3 = f(p) + 2\varepsilon/3 > g(\tilde{p}) + \varepsilon/3$. Thus, $\min L(g, \Lambda) = \min M(g, \Lambda) = g(\tilde{p})$ and $g \in X$. $\qquad\square$

4.1.4 *Proof of Theorem 4.1*

Let $t_0 = \min M(f, \Lambda)$. Fix a large positive integer K and consider a locally maximal subhorseshoe $\tilde{\Lambda} \subset \Lambda$ with $\Lambda_{t_0} \subset \tilde{\Lambda}$ and $HD(\tilde{\Lambda}) < 1/2K$. We take symbolic representations of points of Λ associated to a Markov partition of Λ. We will assume that f satisfies the conclusions of Proposition 4.1 (replacing k by K). We will prove that, under these conditions, we have $f \in X$, which concludes the proof because if $f \in X$, then clearly $\min M(f, \Lambda) = \min L(f, \Lambda)$ is isolated in $M(f, \Lambda)$, and thus also in $L(f, \Lambda)$ (and X is C^0-open, by Proposition 4.3).

Since the restriction of f to $\tilde{\Lambda}$ is injective, by Proposition 4.2 there is a unique $p \in \Lambda$ such that $f(p) = t_0$, and the restriction of φ to the closure of the orbit of p is minimal. Let $\theta = (\ldots, a_{-2}, a_{-1}, a_0, a_1, a_2, \ldots)$ be the kneading sequence of p. Assume by contradiction that p is not a periodic point.

We will consider the following dynamically defined Cantor sets, which we may assume, using parametrizations, to be contained in the real line: $K^s = W^s_{loc}(p) \cap \tilde{\Lambda}$, the set of points of $\tilde{\Lambda}$ whose kneading sequences are of the type $(\ldots, b_{-2}, b_{-1}, a_0, a_1, a_2, \ldots)$, for some $b_{-1}, b_{-2}, \ldots$ (the point corresponding to this sequence will be denoted by $\pi^s(b_{-1}, b_{-2}, \ldots)$) and $K^u = W^u_{loc}(p) \cap \tilde{\Lambda}$, the set of points of $\tilde{\Lambda}$ whose kneading sequences are of the type $(\ldots, a_{-2}, a_{-1}, a_0, b_1, b_2, \ldots)$, for some $b_1, b_2, \ldots$ (the point corresponding to this sequence will be denoted by $\pi^u(b_1, b_2, \ldots)$). Given a finite sequence $(c_{-1}, c_{-2}, \ldots, c_{-r})$ such that $(c_{-r}, \ldots, c_{-2}, c_{-1}, a_0)$ is admissible, we define the interval $I^s(c_{-1}, c_{-2}, \ldots, c_{-r})$ to be the convex hull of

$$\{(\ldots, b_{-2}, b_{-1}, a_0, a_1, a_2, \ldots) \in K^s | b_{-j} = c_{-j}, 1 \le j \le r\}.$$

Analogously, given a finite sequence $(d_1, d_2, \ldots, d_s)$ such that $(a_0, d_1, d_2, \ldots, d_s)$ is admissible, we define the interval $I^u(d_1, d_2, \ldots, d_s)$ to be the convex hull of

$$\{(\ldots, a_{-2}, a_{-1}, a_0, b_1, b_2, \ldots) \in K^u | b_j = d_j, 1 \le j \le s\}.$$

If a and b are the values of the derivative of f at p applied to the unit tangent vectors of $W^s_{loc}(p)$ and $W^u_{loc}(p)$, respectively, we have that a and b are non-zero and, considering local isometric parametrizations of $W^s_{loc}(p)$ and $W^u_{loc}(p)$ which send p to 0, we have that, locally, for $x \in W^s_{loc}(p) \supset K^s$ and $y \in W^u_{loc}(p) \supset K^u$, $f(x, y) = ax + by + O(x^2 + y^2)$, in coordinates given by extended local stable and unstable foliations of the horseshoe $\tilde{\Lambda}$ (here, the point p has coordinates $(x, y) = (0, 0)$; we will use this local form in small neighbourhoods of p, i.e., for $|x|$ and y small).

Let $k \neq 0$ such that $a_k = a_0$. We define

$$d_k = \max\{d(\pi^s(a_{k-1}, a_{k-2}, \dots), p), d(\pi^u(a_{k+1}, a_{k+2}, \dots), p)\}.$$

There are $0 < \lambda_1 < \lambda_2 < 1$ such that the norm of the derivative of φ restricted to a stable direction and the inverse of the norm of the derivative of φ restricted to a unstable direction always belong to (λ_1, λ_2). We say that $k > 0$ is a *weak record* if $d_k < d_j$ for all j with $1 \leq j < k$ such that $a_j = a_0$. We will construct the sequence $0 < k_1 < k_2 < \dots$ of the *records*. Let k_1 be the smallest $k > 0$ with $a_k = a_0$ and, given a record k_n, k_{n+1} will be the smallest weak record $k > k_n$ with $d_k < \min\{|a/b|, |b/a|\} \cdot \lambda_1^3 \cdot d_{k_n}$. We say that $k > 0$ such that $a_k = a_0$ is *left-good* if $f(\pi^s(a_{k-1}, a_{k-2}, \dots)) < f(p)$ and that k is *right-good* if $f(\pi^u(a_{k+1}, a_{k+2}, \dots)) < f(p)$. We say that k is *left-happy* if k is left-good but not right-good or if k is left-good and right-good and $d(\pi^s(a_{k-1}, a_{k-2}, \dots), p) \geq d(\pi^u(a_{k+1}, a_{k+2}, \dots), p)$. We say that k is *right-happy* if k is right-good but not left-good or if k is left-good and right-good and $d(\pi^s(a_{k-1}, a_{k-2}, \dots), p) < d(\pi^u(a_{k+1}, a_{k+2}, \dots), p)$. Notice that k cannot be simultaneously left-happy and right-happy. We say that an index $k > 0$ is *cool* if it is left-happy or right-happy and $a_{k+j} = a_j$ for every j with $|j| \leq K$. By the minimality, we have $\lim d_{k_n} = 0$, and, since $f(\varphi^j(p)) < f(p)$ for all $j \neq 0$, for all large values of n, k_n is cool.

If k is a cool index then we define its *basic cell* as follows: if k is left-happy, we take r_k to be the non-negative integer r such that $a_{k-j} = a_{-j}$ for $0 \leq j \leq r$ and $a_{k-r-1} \neq a_{-r-1}$, and s_k to be the smallest positive integer s such that

$$\min\{|a/b|, |b/a|\} \cdot \lambda_1^3 |I^u(a_{k+1}, a_{k+2}, \dots, a_{k+s})| \leq |I^s(a_{k-1}, a_{k-2}, \dots, a_{k-r_j})|.$$

Analogously, if k is right-happy, we take s_k to be the positive integer s such that $a_{k+j} = a_j$ for $0 \leq j \leq s$ and $a_{k+s+1} \neq a_{s+1}$, and r_k to be the smallest positive integer r such that $\min\{|a/b|, |b/a|\} \cdot \lambda_1^3 |I^s(a_{k-1}, a_{k-2}, \dots, a_{k-r})| \leq |I^u(a_{k+1}, a_{k+2}, \dots, a_{k+s_k})|$. Notice that, in any case,

$$(a_{k-r_k}, \dots, a_{k-1}, a_k, a_{k+1}, \dots, a_{k+s_k}) = (a_{-r_k}, \dots, a_{-1}, a_0, a_1, \dots, a_{s_k}).$$

The basic cell of k is the finite sequence $(a_{k-r_k}, \dots, a_{k-1}, a_k, a_{k+1}, \dots, a_{k+s_k})$ indexed by the interval $[-r_k, s_k]$ of integers (so that the index 0 in this interval corresponds to a_k). Notice that, by minimality, $\lim r_{k_n} = \lim s_{k_n} = +\infty$. Moreover, if k_n and k_{n+1} are left-happy (resp. right-happy), then $r_{k_{n+1}} > r_{k_n}$ (resp. $s_{k_{n+1}} > s_{k_n}$) for every n large. We define the *extended cell* of k as the finite sequence $(a_{k-\tilde{r}_k}, \dots, a_{k-1}, a_k, a_{k+1}, \dots, a_{k+\tilde{s}_k})$ indexed by the interval $[-\tilde{r}_k, \tilde{s}_k]$ of integers, where $\tilde{r}_k := \lfloor (1 + 2\hat{c}/K) r_k \rfloor$ and $\tilde{s}_k := \lfloor (1 + 2\hat{c}/K) s_k \rfloor$, and $\hat{c} = \log \lambda_1 / \log \lambda_2 > 1$.

A crucial remark is that, since f satisfies the conclusions of Proposition 4.1 and $f(\varphi^j(p)) \prec f(p)$ for all $j \neq 0$, one has.

• There is a positive integer t_0 such that for every $m > 0$ which is not cool (in particular if $a_m \neq a_0$), and any point $q \in \tilde{\Lambda}$ whose kneading sequence $(\ldots, b_{-1}, b_0, b_1, \ldots)$ satisfies $b_j = a_{m+j}$ for $-t_0 \leq j \leq t_0$, we have $f(q) < f(p)$. Indeed, there is a constant $\tilde{r}$ such that, if $m > 0$ is not cool, then $(a_{m-\tilde{r}}, \ldots, a_{m-1}, a_m, a_{m+1}, \ldots, a_{m+\tilde{r}}) \neq (a_{-\tilde{r}}, \ldots, a_{-1}, a_0, a_1, \ldots, a_{\tilde{r}})$. We have $f(\varphi^m(p)) < f(p)$, and, if t_0 is much larger than $\tilde{r}$, if q is a point in $\tilde{\Lambda}$ whose kneading sequence $(\ldots, b_{-1}, b_0, b_1, \ldots)$ satisfies $b_j = a_{m+j}$ for $-t_0 \leq j \leq t_0$, we have $f(q)$ much closer to $f(\varphi^m(p))$ than to $f(p)$, and so $f(q) < f(p)$ (recall that f is injective in $\tilde{\Lambda}$).

• For any cool index $k > 0$, if $(a_{k-\tilde{r}_k}, \ldots, a_{k-1}, a_k, a_{k+1}, \ldots, a_{k+\tilde{s}_k})$ is the extended cell of k, and $q \in \tilde{\Lambda}$ is any point whose kneading sequence $(\ldots, b_{-1}, b_0, b_1, \ldots)$ satisfies $b_j = a_{k+j}$ for $-\tilde{r}_k \leq j \leq \tilde{s}_k$, we have $f(q) < f(p)$. Indeed, since $|f(u) - f(v)| \geq c \cdot |u - v|^{K/(K-1)}, \forall u, v \in \tilde{\Lambda}$, by definition of the extended cells, $f(p)$ does not belong to the convex hull of the image by f of the image by Π of the cylinder $\{(\ldots, b_{-1}, b_0, b_1, \ldots) | b_j = a_{k+j}, -\tilde{r}_k \leq j \leq \tilde{s}_k\}$, which contains the point $f(\varphi^k(p)) < f(p)$ (indeed, $|\varphi^k(p) - q| = O(|\varphi^k(p) - p|^{1+\frac{2}{K}}) = o(|p - q|^{K/(K-1)}))$.

We now show the following

Claim: There is a positive integer m such that we never have $(a_{-mt}, \ldots, a_{-1}, a_0, a_1, \ldots, a_{mt-1}) = \gamma^{2m} = \gamma\gamma \ldots \gamma$ ($2m$ times), for any finite sequence γ, where $t = |\gamma|$.

Indeed, take a positive integer k_0 such that $\lambda_2^{k_0} < \lambda_1$, a large positive integer m_0 (with $m_0 \geq \max\{t_0, K\}$) and $m = m_0 k_0$. Suppose by contradiction that $(a_{-mt}, \ldots, a_{-1}, a_0, a_1, \ldots, a_{mt-1}) = \gamma^{2m}$ for some $\gamma = (c_1, c_2, \ldots, c_t)$. We may assume that γ is not of the form α^n for a smaller sequence α (otherwise we may replace γ by α). The Markov value $\tilde{t}_0$ of $\Pi(\overline{\gamma}) = \Pi(\ldots \gamma\gamma\gamma \ldots)$, which is larger than t_0, is attained at c_1. Indeed, if $2 \leq j \leq t$, either $j - 1$ is not cool or the extended cell of $j - 1$ (centered in c_j) is contained in γ^5, so $f(\varphi^{j-1}(\Pi(\overline{\gamma}))) < f(p) = t_0$. Take the maximum values of $m_1, m_2 \geq m$ for which $(a_{-m_1 t}, \ldots, a_{-1}, a_0, a_1, \ldots, a_{m_2 t - 1}) = \gamma^{m_1 + m_2}$. Since $f(\varphi^{-(m_1-m_0)t}(p))$, $f(\varphi^{-(m_1-m_0-1)t}(p))$, $f(\varphi^{(m_2-m_0-1)t}(p))$ and $f(\varphi^{(m_2-m_0)t}(p))$ are smaller than $f(p) = t_0$, it follows that, for every $j \geq 1$, $f(\pi^u(\gamma^j a_{m_2 t}, a_{m_2 t+1}, a_{m_2 t+2}, \ldots)) < f(\pi^u(\gamma^{j+1} a_{m_2 t}, a_{m_2 t+1}, a_{m_2 t+2}, \ldots))$ (the suffix $(a_{m_2 t}, a_{m_2 t+1}, a_{m_2 t+2}, \ldots)$ helps diminishing the value of f) and

$f(\pi^s((\gamma^t)^j a_{-m_1 t-1}, a_{-m_1 t-2}, \dots)) < f(\pi^s((\gamma^t)^{j+1} a_{-m_1 t-1}, a_{-m_1 t-2}, \dots))$,
where $\gamma^t = (c_t, \dots, c_2, c_1)$ (the prefix $(a_{-m_1 t-1}, a_{-m_1 t-2}, \dots)$ also helps
diminishing the value of f). Thus, by comparison, and using the
previous remark on extended cells, if we delete from each factor of
$(\dots, a_{-1}, a_0, a_1, \dots)$ equal to $(a_{-(m_1+1)t}, \dots, a_{-1}, a_0, a_1, \dots, a_{(m_2+1)t-1})$
the factor $(a_{-t}, \dots, a_{-1}, a_0, a_1, \dots, a_{t-1}) = \gamma^2$, we will reduce the Markov
value of the sequence (indeed, the suffix $(a_{m_2 t}, a_{m_2 t+1}, a_{m_2 t+2}, \dots)$ and the
prefix $(a_{-m_1 t-1}, a_{-m_1 t-2}, \dots)$ are closer and thus help even more diminishing the value of f), a contradiction (notice that the number of consecutive
copies of γ in a factor of $(\dots, a_{-1}, a_0, a_1, \dots)$ is bounded, since $\tilde{t}_0 > t_0$).
This concludes the proof of the Claim.

Notice that we can assume that K is much larger than m^2 ($K > 5\hat{c}m^2$
is enough for our purposes), by reducing $\tilde{\Lambda}$, if necessary.

In order to conclude the proof we will have two cases:

(i) There are arbitrarily large values of n for which k_n is left-happy.

In this case we will show that, for such a large value of n, the peri-
odic point q of $\tilde{\Lambda}$ whose kneading sequence $(\dots, b_{-1}, b_0, b_1, \dots)$ has period
$(a_0, a_1, \dots, a_{k_n-1})$ (and so satisfies $b_m = a_m \pmod{k_n}$, for all m; notice that,
since $\tilde{\Lambda}$ is locally maximal and $p \in \tilde{\Lambda}$, we have $q \in \tilde{\Lambda}$ for n large) has Markov
value smaller than t_0, a contradiction. In order to do this, notice that if
$r_{k_n} > 2m^2 k_n$, then we get a contradiction by the previous Claim. If there
is $0 < k < k_n$ which is cool and satisfies $r_k > 2m^2 k$ or $s_k > 2m^2(k_n - k)$,
we also get a contradiction by the previous Claim. For the terms whose
indices are multiple of k_n (which correspond to the point q), if $\hat{s}$ is the
positive integer such that $a_{k_n+j} = a_j$ for $0 \le j < \hat{s}$ and $a_{k_n+\hat{s}} \ne a_{\hat{s}}$,
then $b_j = a_j$ for all $-r_{k_n} \le j < k_n + \hat{s}$, and so $d(\pi^u(b_1, b_2, \dots), p) =$
$o(d(\pi^u(a_{k_n+1}, a_{k_n+2}, \dots), p)) = o(d(\pi^s(a_{k_n-1}, a_{k_n-2}, \dots), p))$, which implies $f(q) < f(p) = t_0$ (since $f(x, y) = ax + by + O(x^2 + y^2)$). Otherwise,
for the terms whose indices are not multiple of k_n, any term equal to a_0 in
the periodic sequence with period $(a_0, a_1, \dots, a_{k_n-1})$ has a neighbourhood
in this periodic sequence which coincides with the extended cell of a corresponding element of the original sequence $(\dots, a_{-1}, a_0, a_1, \dots)$, since this
extended cell is contained in $(a_j)_{-r_{k_n} \le j < k_n+\hat{s}}$ and so the Markov value of
the point corresponding to this periodic sequence is smaller than t_0.

(ii) For all n large, k_n is right-happy.

In this case we will show that, for n large, the periodic point
q' of $\tilde{\Lambda}$ whose kneading sequence $(\dots, b'_{-1}, b'_0, b'_1, \dots)$ has period
$(a_{k_n}, a_{k_n+1}, \dots, a_{k_{n+1}-1})$ (and so satisfies $b_m = a_m \pmod{k_{n+1}-k_n)+k_n}$, for

all m; notice that, since $\tilde{\Lambda}$ is locally maximal and $p \in \tilde{\Lambda}$, we have $q' \in \tilde{\Lambda}$ for n large) has Markov value smaller than t_0, a contradiction. In order to do this, notice that if $r_k > 2m^2(k - k_n)$ for some k which is cool and satisfies $k_n < k \leq k_{n+1}$ or $s_k > 2m^2(k_{n+1} - k)$ for some k which is cool and satisfies $k_n \leq k < k_{n+1}$, then we get a contradiction by the previous Claim. Otherwise, except perhaps for the terms whose indices are multiple of $k_{n+1} - k_n$, any term equal to a_0 in the periodic sequence with period $(a_{k_n}, a_{k_n+1}, \ldots, a_{k_{n+1}-1})$ has a neighbourhood in this periodic sequence which coincides with the extended cell of a corresponding element of the original sequence $(\ldots, a_{-1}, a_0, a_1, \ldots)$, and so the Markov value of the point corresponding to this periodic sequence is smaller than t_0. For the terms whose indices are multiple of $k_{n+1} - k_n$ (which correspond to the point q'), if $\hat{r}$ is the positive integer such that $a_{k_{n+1}-j} = a_{-j}$ for $0 \leq j < \hat{r}$ and $a_{k_{n+1}-\hat{r}} \neq a_{-\hat{r}}$, then $b_j = a_{k_{n+1}+j}$ for all $-\min\{\hat{r}, k_{n+1} - k_n + r_{k_n}\} < j \leq 0$ and $b_j = a_{k_n+j}$ for $-r_{k_n} \leq j \leq k_{n+1} - k_n + s_{k_n}$, and, since $k_{n+1} - k_n \geq \frac{s_{k_n}}{2m^2} > \frac{2\hat{c}s_{k_n}}{K}$, we have $b_j = a_{k_n+j}$ for $0 \leq j \leq \tilde{s}_{k_n}$ (notice that $b_j = a_j$ for $0 \leq j \leq s_{k_n}$, but $b_{s_{k_n}+1} \neq a_{s_{k_n}+1}$), and so, by the definition of record, $d(\pi^s(a_{k_{n+1}-1}, a_{k_{n+1}-2}, \ldots), p) \leq (1 + o(1))|b/a|\lambda_1^3 d(\pi^u(a_{k_n+1}, a_{k_n+2}, \ldots), p)$ and thus $d(\pi^s(b_{-1}, b_{-2}, \ldots), p) \leq (1 + o(1))|b/a|\lambda_1 d(\pi^u(b_1, b_2, \ldots), p)$, which implies $f(q) < f(p) = t_0$ (since $f(x, y) = ax + by + O(x^2 + y^2)$).

Now we concluded that p and $\theta = (\ldots, a_{-2}, a_{-1}, a_0, a_1, a_2, \ldots)$ are periodic. Let $\alpha = (a_0, a_1, \ldots, a_{s-1})$ be a minimal period of θ. If $f \notin X$ then, for each positive integer n, there is a point $q_n \in \Lambda$ which does not belong to the orbit of p such that $\sup_{r \in \mathbb{Z}} f(\varphi^r(q_n)) \leq f(p) + 1/n$. Given a kneading sequence $(\ldots, c_{-2}, c_{-1}, c_0, c_1, c_2, \ldots)$ of some point of Λ, we say that $k \in \mathbb{Z}$ is a *regular* position of it if there is an integer j such that $c_{k+i} = a_{j+i \pmod{s}}$ for $1 - s \leq i \leq 1$, and that k is a *strange* position otherwise. Since q_n does not belong to the orbit of p, there is $k_n \in \mathbb{Z}$ which is a strange position of the kneading sequence of q_n. Let $\tilde{q}_n := \varphi^{k_n}(q_n)$. Then 0 is a strange position of its kneading sequence. Take a subsequence of $(\tilde{q}_n)$ converging to a point $\tilde{q} \in \Lambda$. Then 0 is a strange position of the kneading sequence of $\tilde{q}$ (and thus $\tilde{q}$ does not belong to the orbit of p) and $\sup_{r \in \mathbb{Z}} f(\varphi^r(\tilde{q})) \leq f(p)$. This implies (since f is injective in $\tilde{\Lambda}$ and $f(p)$ is the smallest element of $L(f, \Lambda)$) that $f(\varphi^r(\tilde{q})) < f(p), \forall r \in \mathbb{Z}$ and $\limsup_{r \to +\infty} f(\varphi^r(\tilde{q})) = \limsup_{r \to -\infty} f(\varphi^r(\tilde{q})) = f(p)$.

Let $\tilde{\theta} = (\ldots, b_{-2}, b_{-1}, b_0, b_1, b_2, \ldots)$ be the kneading sequence of $\tilde{q}$, and let m and m_0 be as in the Claim. We should have factors of $\tilde{\theta}$

with (arbitrarily large) positive indices equal to α^{2m}, and, analogously, we should have factors of $\tilde{\theta}$ with negative indices equal to α^{2m}. This implies that there is a factor of $\tilde{\theta}$ of the form $\alpha^{2m}\beta\alpha^{2m}$, where β is a finite sequence which does not belong or end by α (and thus is not of the form α^r for any positive integer r), and which me may assume not to contain any factor of the form α^{2m} (otherwise we find a smaller factor with these properties). We claim that if $z \in \Lambda$ is a point whose kneading sequence is periodic with period $\alpha^{2m}\beta$ then $f(\varphi^r(z)) < f(p), \forall r \in \mathbb{Z}$, and so, since z is periodic, $\sup_{r\in\mathbb{Z}} f(\varphi^r(z)) < f(p)$, a contradiction. In order to do this, we use an argument somewhat analogous to the proof of the Claim: let j such that $(b_j, b_{j+1}, \ldots, b_{j+M}) = \alpha^{2m}\beta\alpha^{2m}$, where $M = 4ms + b - 1$, with $b = |\beta|$. Since $f(\varphi^{j+(2m-m_0)s-i}(\tilde{q})) < f(p)$ for $1 \leq i \leq 2s$ and $f(\varphi^{j+(2m+m_0)s+b+i}(\tilde{q})) < f(p)$ for $1 \leq i \leq 2s$, we have that $f(\pi^u(\alpha^j\beta\overline{\alpha})) < f(p)$ (so the suffix β helps diminishing the value of f) and $f(\pi^s((\alpha^t)^j\beta^t\overline{\alpha^t})) < f(p)$ for every $j \geq 0$ (so the prefix β helps diminishing the value of f). So, for the positions corresponding to a_0 in α^{2m}, we use the fact that β helps diminishing the value of f both as a suffix and as a prefix in order to show that the corresponding values of f are smaller than $f(p)$. For the other positions, including the positions inside β, we use the extended cell argument in order to show that the corresponding values of f are also smaller than $f(p)$. This concludes the proof of Theorem 4.1

4.2   Bottom of the Lagrange spectra associated to interval exchange transformations

The best constant $\ell(\alpha)$ of Diophantine approximation describe recurrence rates of the rotation R_α of angle α on the circle. More precisely, let $R_\alpha(x) = x + \alpha$ be the rotation of angle α on the circle $\mathbb{R}/\mathbb{Z}$. Since R_α is an isometry, the *recurrence rate*

$$\liminf_{n\to\infty} n \cdot \text{dist}(R_\alpha^n(x), x)$$

of the R_α-orbit of any point $x \in \mathbb{R}/\mathbb{Z}$ is

$$r(R_\alpha) := \liminf_{n\to\infty} n \cdot \text{dis}(R_\alpha^n(0), 0) = \liminf_{p,q\to\infty} q|q\alpha - p| = 1/\ell(\alpha).$$

This point of view can be naturally extended to study recurrence rates and Lagrange spectra for *interval exchange transformations*.

In fact, it is well-known that a rotation R_α of the circle can be thought as an interval exchange map $R_\alpha : (0,1) \to (0,1)$ of two intervals: $R_\alpha(x) = x + \alpha$ for $x \in (0, 1-\alpha)$ and $R_\alpha(x) = x - (1-\alpha)$ for $x \in (1-\alpha, 1)$. In general, an *interval exchange transformation* (i.e.t.) of d subintervals of a bounded open interval $I \subset \mathbb{R}$ is a bijection $T : D_T \to D_{T^{-1}}$ where $I \backslash D_T$ and $I \backslash D_{T^{-1}}$ are finite sets with the same cardinality $d-1$ and the restriction of T to each connected component of D_T is a translation onto a connected component of $D_{T^{-1}}$. By definition, a d-i.e.t. T is completely determined by a *length data* $\lambda = (\lambda_1, \ldots, \lambda_d) \in \mathbb{R}_+^d$ describing the sizes of the connected components of D_T (listed from left to right) and *combinatorial data* $\pi \in \mathrm{Sym}_d$ describing how the connected components of D_T are permuted by T to produce the connected components of $D_{T^{-1}}$.

If an interval exchange transformation T satisfies the so-called *Keane condition* that there are no $x \in I \backslash D_{T^{-1}}$, $y \in I \backslash D_T$ and $n \geq 0$ such that $T^n(x) = y$, then T is minimal in the sense that all half-orbit is dense. Also, a rotation $R_\alpha : (0,1) \to (0,1)$ of angle α satisfies Keane condition if and only if α is irrational. Moreover, the recurrence rate of R_α is

$$r(R_\alpha) = \liminf_{n \to \infty} n \cdot \varepsilon_n(R_\alpha)$$

where $\varepsilon_n(R_\alpha)$ is the length of the smallest domain of continuity of R_α^n.

Partly motivated by this scenario, Ferenczi [Ferenczi (2012)] introduced in 2012 the following notion of Lagrange spectrum[2] for interval exchange transformations of d intervals:

$$L_{d-1} := \left\{ L(T) := \limsup_{n \to \infty} \frac{|I|}{n \cdot \varepsilon_n(T)} \right.$$

$$\left. < \infty : T : I \to I \ d\text{-i.e.t. verifying Keane condition} \right\},$$

where $|I|$ is the length of I and $\varepsilon_n(T)$ is the length of the smallest connected component of the domain of T^n. In this setting, the classical Lagrange spectrum is L_1 and Hurwitz theorem essentially implies that

$$L_1 \cap [0, 2\sqrt{2}) = \{\sqrt{5}\} = \left\{ \ell\left(\frac{1 + \sqrt{5}}{2}\right) \right\}.$$

Furthermore, Ferenczi proved in [Ferenczi (2012)] that

$$L_2 \cap \left[0, \frac{12 + 29\sqrt{3}}{13}\right) = \{2\sqrt{5}\}.$$

In 2017, Boshernitzan–Delecroix [BoDe (2017)] showed that:

Theorem 4.2. *There exists $\varepsilon_0 > 0$ such that $L_k \cap \left[0, k\sqrt{5} + \frac{\varepsilon_0}{k}\right) = \{k\sqrt{5}\}$ for all $k \geq 1$. Moreover, any d-i.e.t. T associated to the length data $\lambda =$*

---

[2]It was originally called *upper Boshernitzan–Lagrange spectrum* in [Ferenczi (2012)].

$(\frac{1+\sqrt{5}}{2}, 1, 1, \ldots, 1) \in \mathbb{R}_+^d$ *and a combinatorial data* $\pi \in Sym_d$ *with* $\pi(1) = d$
satisfies $L(T) = (d-1)\sqrt{5}$.

The idea of the proof of this result can be roughly described as follows.

The first step relies on the connection between interval exchange transformations and *translation surfaces* (see e.g. [MaTa (2002)]). Recall that a *translation surface* is a surface obtained from a collection of $2n$ Euclidean canonically oriented triangles after gluing by translations $3n$ pairs of parallel sides with the same lengths and opposite orientations. These surfaces have a finite number of *conical singularities* with total angles $2\pi k_1, \ldots, 2\pi k_\sigma$, $k_1 + \cdots + k_\sigma = n$, $k_i \geq 2$ for all i, coming from some of the vertices of the $2n$ triangles. Also, these surfaces come equipped with a canonical *vertical translation flow* whose first return maps to appropriate horizontal segments are often interval exchange transformations. Conversely, it is often possible to "suspend" (via the so-called Masur's suspension and/or Veech's zippered rectangles constructions) an interval exchange transformation T to produce a translation surface X such that the first return map of its vertical flow to an appropriate horizontal segment is exactly T. In this setting, it is possible to prove the following generalization of Proposition 1.9 in Chapter 1:

$$L(T) = \limsup_{\substack{|\mathrm{Im}(v)| \to \infty \\ v \in \mathrm{Hol}(X)}} \frac{\mathrm{Area}(X)}{\mathrm{Area}(v)}, \tag{4.1}$$

where $\mathrm{Area}(v) = \mathrm{Re}(v) \cdot \mathrm{Im}(v)$ and $\mathrm{Hol}(X)$ is the set of *holonomy vectors* of X, i.e., the collection of vectors in $\mathbb{R}^2$ representing straight-line segments in X starting and ending at conical singularities.

The second step consists in plugging the information $L(T) \leq n\sqrt{5} + \frac{\varepsilon_0}{n}$ into (4.1) to see that we can apply the so-called *Teichmüller flow* $g_t = \begin{pmatrix} e^t & 0 \\ 0 & e^{-t} \end{pmatrix}$ to X in order to get that certain holonomy vectors $v \in \mathrm{Hol}(X)$ (corresponding to "best approximations") have the property that $g_t(v)$ is an edge of a L^∞-*Delaunay triangulation*[3] of $g_t(X)$ which is close to a *golden triangulation*, i.e., a triangulation derived from n parallelograms isometric to a fundamental domain of the lattice $\sqrt{\frac{\mathrm{Area}(X)}{5}}(\mathbb{Z}(\phi - 1, \phi) \oplus \mathbb{Z}(-1, 1))$ (by dividing these parallelograms into two triangles by cutting along their diagonals).

---

[3]A triangulation T with set of vertices V is L^∞-Delaunay whenever the vertices $p, q, r \in V$ of each triangle of T are contained in an Euclidean square Q with sides parallel to the axes such that $Q \cap V = \partial Q \cap V = \{p, q, r\}$.

The final step is based on a rigidity result (based on a sort of fixed point argument): one shows that if we are close to a golden triangulation, then X is actually produced by a golden triangulation. In particular, this completes the proof of Boshernitzan–Delecroix theorem because it is not hard to use (4.1) to check that $L(T) = n\sqrt{5}$ when X comes from a golden triangulation.

Remark 4.2. The Lagrange spectra L_k have a dynamical interpretation: indeed, the identity (4.1) can be extended (in the spirit of Proposition 1.9 in Chapter 1) to write $L(T) = \limsup\limits_{t\to\infty} \frac{2\cdot\mathrm{Area}(X)}{\mathrm{sys}(g_t(X))^2}$ where $\mathrm{sys}(Y) = \min\{|w| : w \in \mathrm{Hol}(Y)\}$ is the *systole* of the translation surface Y.

4.3 Bottom of the dynamical spectra of geodesic flows on negatively curved manifolds

4.3.1 *Bottom of dynamical spectra of Hecke triangle groups*

We saw in Chapter 1, §1.7 that the classical Lagrange spectrum can be viewed in terms of cusp excursions of geodesics of the modular surface $\mathbb{H}/SL(2,\mathbb{Z})$.

The modular group $SL(2,\mathbb{Z}) = H_3$ is the first of the *Hecke triangle groups* H_q, $q \geq 3$, generated by the matrices

$$\begin{pmatrix} 1 & 2\cos(\pi/q) \\ 0 & 1 \end{pmatrix} \quad \text{and} \quad \begin{pmatrix} 0 & -1 \\ 1 & 0 \end{pmatrix},$$

and the corresponding dynamical Lagrange spectra are

$$L(H_q) := \{\ell_{H_q}(x) < \infty : x \in \mathbb{R}\}$$

where $\ell_{H_q}(x)$ is the best constant for the inequality

$$\left| x - \frac{a(g)}{c(g)} \right| < \frac{1}{\ell_{H_q}(x) \cdot c(g)^2}$$

for $g = \begin{pmatrix} a(g) & b(g) \\ c(g) & d(g) \end{pmatrix} \in H_q$. In this setting, Hurwitz theorem asserts that $\min L(H_3) = \sqrt{5}$ and, more generally, it was proved by Lehner [Lehner (1985)] and Haas–Series [HaSe (1986)] that:

Theorem 4.3 (Lehner and Haas–Series). *If q is even, resp. odd, then* $\min L(H_q) = 2$, *resp.* $\min L(H_q) = 2\sqrt{1 + (1 - \cos(\pi/q))^2}$.

In fact, Lehner used the so-called Rosen continued fractions [R (1954)] to show that $L(H_q) = 2$ when $q \geq 4$ is even and $2 \leq L(H_q) \leq 2\sqrt{1 + \left(\frac{1}{2\cos(\pi/q)-1} - \cos(\pi/q)\right)^2}$ for all q odd, and Haas–Series related $L(H_q)$ to cusp excursions of geodesics of $\mathbb{H}/H_q$ to establish that $L(H_q) = 2\sqrt{1 + (1 - \cos(\pi/q))^2}$ for all $q \geq 3$ odd.

Remark 4.3. The beginning of $L(H_5)$ was studied in more details by Series [S (1988)]: in particular, she showed that there is an explicit constant $\mu = 2.411977\ldots$ such that

$$L(H_5) \cap [2\sqrt{1 + (1 - \cos(\pi/5))^2}, \mu) = \{\delta_0 < \delta_1 < \cdots < \delta_n < \ldots\},$$

where δ_n is an explicit increasing sequence converging to μ.

4.3.2 *Hurwitz constants of negatively curved manifolds with cusps*

Let M be a smooth, complete, non-compact, Riemannian manifold with finite volume and pinched negative sectional curvatures, say $-\infty < -b^2 \leq K \leq -a^2 < 0$.

In the sequel, we fix a cusp e of M and we take a minimizing ray r on M converging to e. The Busemann function $\beta_r : M \to \mathbb{R}$ is the 1-Lipschitz map given by $\beta_r(x) := \lim_{t\to\infty} (t - \text{dist}_M(x, r(t)))$. By a lemma of Margulis, there exists $\eta_0 = \eta_0(r) > 0$ such that if $\widetilde{r}$ is a lift of r to the universal cover $\widetilde{M}$ of $M = \widetilde{M}/\Gamma$, then the quotient of $\beta_{\widetilde{r}}^{-1}([t, \infty))$ by the stabilizer of $\widetilde{r}(\infty) \in \partial\widetilde{M}$ in Γ embeds in M if and only if $t > \eta_0$.

In this context, the *maximal Margulis neighborhood* of e is $\beta_r^{-1}((\eta_0, \infty)) = \beta_e^{-1}((0, \infty))$, where $\beta_e := \beta_r - \eta_0$. Also, we say that a geodesic line starting at e is *rational*, resp. *irrational*, if it converges to e, resp., it is positively recurrent to a compact subset of M. The *depth* $D(r)$ of a rational line is the length of the subpath of r connecting the first and last intersection points of r with the boundary of the maximal Margulis neighborhood of e.

As it is discussed in the article [HePa (2002)] by Hersonsky–Paulin, there is a constant K such that any irrational line ξ starting from e can be approximated by infinitely rational lines r starting from e in the sense that

$$\text{dist}_M(\xi, r) \leq K \exp(-D(r)).$$

The smallest constant $K_{M,e}$ with the property above was called *Hurwitz constant* of M with respect to e. This nomenclature is partly justified by the fact that $K_{M,e} = 1/\sqrt{5}$ when $M = \mathbb{H}/SL(2,\mathbb{Z})$ and $e = i\infty$. Also, from the geometrical point of view, Hersonsky–Paulin showed that the smallest height

$$\mathrm{ht}_e(\gamma) = \max_{s \in \mathbb{R}} \beta_e(\gamma(s))$$

among all closed geodesics γ of M is

$$h_{M,e} = \min_{\gamma \text{ closed geodesic}} \mathrm{ht}_e(\gamma) = \log(1/2K_{M,e}).$$

Remark 4.4. The Hurwitz constant defines a (proper, real-analytic) function $K^{(1,1)} : \mathcal{M}_{1,1} \to \mathbb{R}$ on the moduli space of hyperbolic metrics on once-punctered torii such that $\min K^{(1,1)} = 1/\sqrt{5}$ is attained exactly at the modular torus $\mathbb{H}/SL(2,\mathbb{Z})'$, where $SL(2,\mathbb{Z})'$ is the commutator subgroup of $SL(2,\mathbb{Z})$.

4.3.3   Bottom of the Lagrange spectrum of a certain square-tiled surface

In general, a *square-tiled surface* is a translation surface obtained from a finite cover of the square torus $\mathbb{R}^2/\mathbb{Z}^2$ ramified only at the origin 0: geometrically, this amounts to take a finite collection of unit squares of $\mathbb{R}^2$ and build a surface by gluing sides by translations.

The action of $g_t = \begin{pmatrix} e^t & 0 \\ 0 & e^{-t} \end{pmatrix}$ on the $SL(2,\mathbb{R})$-orbit on a square-tiled surface X is the geodesic flow on the hyperbolic surface $\mathbb{H}/\Gamma(X)$ where $\Gamma(X)$ is the *Veech group* of X, i.e., the finite-index subgroup of $SL(2,\mathbb{Z})$ stabilizing X. Similarly to Remark 4.2 above, we can define the Lagrange spectrum

$$L_X := \left\{ \limsup_{t \to \infty} \frac{2 \cdot \mathrm{Area}(Y)}{\mathrm{sys}(g_t(Y))^2} < \infty : Y \in SL(2,\mathbb{R}) \cdot X \right\},$$

where $\mathrm{sys}(Y)$ is the systole of the translation surface Y.

In the article [HLMU (2015)], Hubert–Lelièvre–Marchese–Ulcigrai studied in the details the Lagrange spectrum L_{B7} associated to the square-tiled surface (of genus 2) obtained from a collection of seven unit squares $S_1, \ldots, S_7$ by gluing by translations the right vertical side of S_i to the left vertical side of $S_{h(i)}$ and the top horizontal side of S_i to the bottom horizontal side of $S_{v(i)}$ where h and v are the permutations $h = (1,2,3,4,5,6,7)$ and $v = (1)(2,4,6)(3,5,7)$. In particular, they proved that the bottom part of L_{B7} has the following structure:

- $L_{B7} \cap [0, \phi_2) = \{\phi_1\}$, where $\phi_1 := 7 + 14 \cdot [0; \overline{3, 1}]$ and $\phi_2 := 14 \cdot [0; 1, 4, \overline{1, 3}]$ is an accumulation point of L_{B7};
- $L_{B7} \cap [\phi_2, \eta_1) = L^\sigma(\Theta_0)$, where $\eta_1 := 7^{\frac{(5 + [0; 1, 4, 2, \overline{1,5}] + [0; 1, 5, 1, \overline{1,5}])}{4}}$ and

$$L^\sigma(\xi) = 7 \cdot \limsup_{n \to \infty} ([\sigma^n(\xi)]_+ + [\sigma^n(\xi)]_-)$$

with $\sigma : \Theta_0 \to \Theta_0$ standing for the first return map to

$$\Theta_0 = \{(\rho_n)_{n \in \mathbb{Z}} \in \{a, b\}^{\mathbb{Z}} : \rho_0 = a = \rho_m \text{ for infinitely many } m \geq 0\}$$

of the shift on $\{a, b\}^{\mathbb{Z}}$, $a = 1, 4, 2, 4$ and $b = 2, 4$, and $[\rho]_+ = [0; 1, 4, \rho_1, \rho_2, \dots]$ and $[\rho]_- = [0; 1, 4, \rho_{-1}, \rho_{-2}, \dots]$.

4.3.4 *Bottom of the Markov spectra of modular billiards*

The classical Markov spectrum can be described in terms of heights of modular billiard trajectories. More precisely, we consider the extended modular group $PGL(2, \mathbb{Z})$ acting on the hyperbolic plane $\mathbb{H}$. The modular billiard trajectory B associated to an oriented hyperbolic geodesic β on $\mathbb{H}$ is the collection of oriented segments of $PGL(2, \mathbb{Z})\beta$ contained in the fundamental domain

$$T := \{z \in \mathbb{H} : 0 \leq \text{Re}(z) \leq 1/2, \text{Im}(z) \geq 1\}$$

of the $PGL(2, \mathbb{Z})$-action on $\mathbb{H}$. In this setting, the classical Markov spectrum is

$$M = \{\lambda_\infty(B) < \infty : B \text{ modular billiard trajectory}\},$$

where $\lambda_\infty(B) = 2 \sup\{\text{Im}(z) : z \in B\}$ is twice of the maximal height of B.

In a recent article, Andersen and Duke [AD (2009)] generalized the classical Markov spectrum by introducing

$$M_z := \{\lambda_z(B) < \infty : B \text{ modular billiard trajectory}\},$$

where $z \in T$ and $\lambda_z(B) := 1/\sinh(\text{dist}_{\mathbb{H}}(z, B))$. In particular, they proved that

$$\min M_{e^{i\pi/3}} = \sqrt{3}$$

is an accumulation point of $M_{e^{i\pi/3}}$, and

$$M_i \cap [0, (3 + \sqrt{21})/3] = \left\{ \frac{\sqrt{21}}{2} < \frac{2\sqrt{14}}{3} < \frac{3 + \sqrt{21}}{3} \right\},$$

where $(3 + \sqrt{21})/3$ is an accumulation point of M_i.

Chapter 5

Intervals and Hall rays in dynamical Lagrange and Markov spectra

This chapter discusses some extensions of the Hall's theorem on the existence of half-infinite rays contained in the classical Lagrange and Markov spectra. In particular, we will see a result of Moreira–Romaña [MR (2016)] about the presence of non-trivial open intervals in the dynamical Lagrange and Markov spectra associated to typical thick horseshoes of surface diffeomorphisms. After that, we will state some theorems by Hubert–Marchese–Ulcigrai [HMU (2015)], Parkkonen–Paulin [PP (2010)] and Artigiani–Marchese–Ulcigrai [AMU (2016)] on the presence of Hall's ray in the dynamical Lagrange and Markov spectra of geodesic flows on negatively curved manifolds.

5.1 Intervals in dynamical spectra of certain thick horseshoes

The general setting of this section is similar to Chapter 3, §3.2 and Chapter 4, §4.1, namely, we have a horseshoe Λ of a C^2-diffeomorphism $\varphi : M^2 \to M^2$ of a surface whose dynamics is coded by a homeomorphism $\Pi : \Sigma_B \to \Lambda$ conjugating $\varphi|_\Lambda$ and the shift map σ acting on a subshift of finite type $\Sigma_B \subset \mathbb{A}^{\mathbb{Z}}$ associated to the transition matrix B of a geometrical Markov partition $\{R_a\}_{a \in \mathbb{A}}$ of small diameter.

The main result of this section (originally proved by Moreira–Romaña in [MR (2016)]) says that the dynamical Lagrange and Markov spectra

$$L(f, \Lambda) = \left\{ \limsup_{n \to \infty} f(\varphi^n(x)) : x \in \Lambda \right\}$$

and

$$M(f, \Lambda) = \left\{ \sup_{n \in \mathbb{Z}} f(\varphi^n(x)) : x \in \Lambda \right\}$$

143

associated to a thick horseshoes Λ of typical surface diffeomorphisms φ and height functions f contain intervals:

Theorem 5.1. *Let Λ be a horseshoe associated to a C^2 surface diffeomorphism φ such that $HD(\Lambda) > 1$. Then, there are C^2-open subsets W arbitrarily close to φ such that, for any $\psi \in W$, one has*

$$int\, L(f, \Lambda_\psi) \neq \emptyset \quad and \quad int\, M(f, \Lambda_\psi) \neq \emptyset,$$

for every height function f in an open and dense set $H_\psi \subset C^1(M, \mathbb{R})$. (Here, Λ_ψ is the continuation of Λ and int A stands for the interior of A.)

For later reference, we recall some facts about the combinatorics and geometry of horseshoes. If $x, y \in \mathbb{A}$, the number $N_n(x, y, B)$ of admissible strings for B of length $n + 1$ is the (x, y) entry b_{xy}^n of the nth power of the transition matrix B. In particular, the transitivity of Σ_B implies that, for all $x, y \in \mathbb{A}$, the minimal number $n(x, y) \in \mathbb{N}^*$ such that $N_{n(x,y)}(x, y, B) > 0$ is well-defined. Thus, we can set $N_0 := \max\{n(x, y) : x, y \in \mathbb{A}\}$, so that for all $x, y \in \mathbb{A}$ there is a word beginning at x and ending with y of length less or equal to $N_0 + 1$. Also, the subshift Σ_B has a local product structure because the local stable and unstable sets

$$W_{loc}^s(\underline{a}) = \{\underline{c} \in \Sigma_B : \forall n \geq 0, \ a_n = c_n\}$$

and

$$W_{loc}^u(\underline{b}) = \{\underline{b} \in \Sigma_B : \forall n \leq 0, \ b_n = c_n\}$$

meet at an unique point $[\underline{a}, \underline{b}] = (\cdots, b_{-n}, \cdots, b_{-1}, b_0, a_1, \cdots, a_n, \cdots)$ whenever $a_0 = b_0$. Moreover, this local product structure is respected by the topological conjugacy Π between $\varphi|_\Lambda$ and $\sigma|_{\Sigma_B}$ in the sense that $\Pi[\underline{a}, \underline{b}] = [\Pi(\underline{a}), \Pi(\underline{b})]$.

Geometrically, this local product structure is captured by the stable and unstable Cantor sets of Λ which live in the boundaries of the rectangles of a geometric Markov partition $\{R_a\}_{a \in \mathbb{A}}$ of small diameter. More precisely, the local stable and unstable sets of Λ given by

$$W^s(\Lambda, R) = \bigcap_{n \geq 0} \varphi^{-n} \left(\bigcup_{a \in \mathbb{A}} R_a \right) \quad and \quad W^u(\Lambda, R) = \bigcap_{n \leq 0} \varphi^{-n} \left(\bigcup_{a \in \mathbb{A}} R_a \right).$$

These local invariant sets of Λ support laminations allowing to define C^r-projections $(\pi_a^u : R_a \to I^u(a))_{a \in \mathbb{A}}$ and $(\pi_a^s : R_a \to I^s(a))_{a \in \mathbb{A}}$ from the rectangles R_a to appropriate arcs $I^u(a)$ and $I^s(a)$ on their boundaries in such a way that if $z, z' \in R_{a_0} \cap \varphi^{-1}(R_{a_1})$ and $\pi_{a_0}^s(z) = \pi_{a_0}^s(z')$, then we have

$$\pi_{a_1}^s(\varphi(z)) = \pi_{a_1}^s(\varphi(z'))$$

(and a similar property holds for π_a^u). In particular, the connected components of $W^s(\Lambda, R) \cap R_a$ are the level lines of π_a^s.

In this context, the unstable Cantor set K^u is the dynamically defined Cantor set associated to the expanding C^r-map (of type Σ_B) g^u given by

$$g^u(\pi_{a_0}(z)) = \pi_{a_1}(\varphi(z))$$

for $z \in R_{a_0} \cap \varphi^{-1}(R_{a_1})$. Note that there is an unique homeomorphism $h^u \colon \Sigma_B^+ \to K^u$ from $\Sigma_B^+ := \{(a_0, a_1, a_2, \dots) \in \mathbb{A}^{\mathbb{N}} : B_{a_i, a_{i+1}} = 1 \; \forall i \geq 0\}$ to K^u such that $h^u(\underline{a}) \in I(a_0)$ for $\underline{a} = (a_0, a_1, \dots) \in \Sigma_B^+$ and $h^u \circ \sigma^+ = g^u \circ h^u$, where $\sigma^+ \colon \Sigma_B^+ \to \Sigma_B^+$ is the left-shift $\sigma^+((a_n)_{n \geq 0}) = (a_{n+1})_{n \geq 0}$. Also, we have similar objects h^s, σ^-, etc. related to K^s. For later use, given a finite word $\underline{a} = (a_0, \cdots, a_n)$, we denote $f_{\underline{a}}^u$ the map given by

$$f_{\underline{a}}^u(z) = h^u(\underline{a}(h^u)^{-1}(z))$$

(with analogous formulas for $f_{\underline{a}}^s$).

In this setting, we have that the horseshoe Λ is locally bi-Lipschitz equivalent to the product of the two dynamically defined Cantor sets K^s and K^u, so that its Hausdorff dimension is

$$HD(\Lambda) = HD(K^s \times K^u) = HD(K^s) + HD(K^u)$$

(cf. [PT (1993), Chap. 4]).

An interesting by-product of the local product structure is the notion of stable and unstable boundaries of Λ:

Definition 5.1. We say that x is a *boundary point* of Λ in the *unstable direction*, if x is a boundary point of $W_\epsilon^u(x) \cap \Lambda$, *i.e.*, if x is an accumulation point only from one side by points in $W_\epsilon^u(x) \cap \Lambda$. If x is a boundary point of Λ in the unstable direction, then, due to the local product structure, the same holds for all points in $W^s(x) \cap \Lambda$. So the boundary points in the *unstable* direction are local intersections of local *stable* manifolds with Λ. For this reason we denote the set of boundary points in the unstable direction by $\partial_s \Lambda$. The *boundary points in the stable direction* are defined similarly. The set of these boundary points is denoted by $\partial_u \Lambda$.

After these preliminaries, let us now get back to the proof of Theorem 5.1. The first step is to build the class H_ψ of adequate height functions.

5.2   Large subsets H_ψ of height functions

Theorem 5.2. *Let Λ be a horseshoe of a C^1-diffeomorphism φ of a surface M. The set*

$$H_\varphi = \{f \in C^1(M, \mathbb{R}) : \#M_f(\Lambda) = 1 \ and, \ for \ z \in M_f(\Lambda), Df_z(e_z^{s,u}) \neq 0\}$$
$$(5.1)$$

is open and dense, where $M_f(\Lambda) = \{z \in \Lambda : f(z) \geq f(y) \ \forall \ y \in \Lambda\}$ and $e_z^{s,u}$ are unit vectors in the subbundles $E_z^{s,u}$ in the definition of hyperbolicity.[1]

Before proving this theorem, we will to show two auxiliary lemmas.

Lemma 5.1. *The set*

$$\mathcal{A}' = \{f \in C^2(M, \mathbb{R}) : there \ is \ z \in M_f(\Lambda) \ with \ Df_z(e_z^{s,u}) \neq 0\}$$

is dense in $C^2(M, \mathbb{R})$, where $e_z^{s,u}$ are unit vectors in $E_z^{s,u}$ respectively.

The proof of Lemma 5.1 relies on the notion of Morse functions. Recall that a C^r function $f \colon M \to \mathbb{R}$, $r \geq 2$, is a *Morse function* if we have that

$$D^2 f(0) \colon T_x M \times T_x M \to \mathbb{R}$$

is nondegenerate[2] for all $x \in M$ with $Df_x = 0$. The set of Morse functions is denoted $\mathcal{M}$. A known result says that the set of Morse functions is open and dense in $C^r(M, \mathbb{R})$, $r \geq 2$.

Note that if $f \in \mathcal{M}$, then the set $\mathrm{Crit}(f) = \{x \in M : Df_x = 0\}$ of critical points of f is discrete. In particular, since Λ is compact, we have that $\#(\mathrm{Crit}(f) \cap \Lambda) < \infty$.

Proof of Lemma 5.1. Note that it is enough to show that $\mathcal{A}'$ is dense in $\mathcal{M}$. Given $f_1 \in \mathcal{M}$, we have $\#\mathrm{Crit}(f_1) < \infty$. Since $\mathrm{int}(\Lambda) = \emptyset$, we can find $f \in \mathcal{M}$ C^2-close to f_1 such that $M_f(\Lambda) \cap \mathrm{Crit}(f) = \emptyset$. Therefore, if $z \in M_f(\Lambda)$, we have $Df_z(e_z^s) \neq 0$ or $Df_z(e_z^u) \neq 0$.

If $Df_z(e_z^s)$ and $Df_z(e_z^u)$ are nonzero at some $z \in M_f(\Lambda)$, then $f \in \mathcal{A}'$. Otherwise, we can suppose that $Df_z(e_z^s) = 0$ and $Df_z(e_z^u) \neq 0$ for any $z \in M_f(\Lambda)$ (because the other case $Df_z(e_z^u) = 0$ and $Df_z(e_z^s) \neq 0$ is entirely analogous). In this setting, we can take a C^2-neighborhood $\mathcal{V}$ of f and a neighborhood U of z such that $Dg_x(e_x^u) \neq 0$ whenever $x \in U \cap \Lambda$ and $g \in \mathcal{V}$. Let R_z be the element of a Markov partition $\mathcal{R} = \{R_a\}_{a \in \mathbb{A}}$ of small diameter containing z. Without loss of generality, we can assume that R_z

---

[1]Cf. Chapter 3, §3.2.2.
[2]I.e., if $D^2 f(0)(v, w) = 0$ for all $w \in T_x M$, then $v = 0$.

is contained in U and, moreover, U is contained in the domain of a C^2-local chart $\phi : \tilde{U} \subset M \to V \subset \mathbb{R}^2$ with $\tilde{U} \cap R' = \emptyset$ for all $R' \in \mathcal{R} \setminus \{R_z\}$.

Observe that, since $Df_z(e_z^u) \neq 0$, we have that $z \in \partial_u \Lambda$. Hence, the possible maximum points of f in $\Lambda \cap R_z$ lie on the stable Cantor set K^s, which has zero Lebesgue measure. Consider the function $\psi^s : K^s \times \mathbb{R} \to \mathbb{R}^2$ defined by

$$\psi^s(x, \alpha) = \nabla(f \circ \phi^{-1})(\phi(x)) - \alpha \begin{pmatrix} 0 & -1 \\ 1 & 0 \end{pmatrix} D\phi_x(e_x^s),$$

where the above matrix is the orthogonal rotation. Since ψ^s extends to a C^1-function, then the Lebesgue measure of $\psi^s(K^s \times \mathbb{R})$ is zero. Therefore, there is $v \in \mathbb{R}^2$ with very small norm such that $v \notin \psi^s(K^s \times \mathbb{R})$. Put $h(y) = f \circ \phi^{-1}(y) - \langle v, y \rangle$ for $y \in V$, so that $D(h \circ \phi)_x e_x^s = Dh_{\phi(x)} D\phi_x e_x^s \neq 0$ for all $x \in K^s$. Since v can be chosen with norm arbitrarily small, then $h \circ \phi$ is C^2-close to f. Because the function increases in the direction of its gradient, the maximum points of h in $\Lambda \cap R_z$ still can only appear in K^s. Thus, $h \in \mathcal{A}'$ and the proof of Lemma is complete $\qquad \square$

Remark 5.1. Since $C^s(M, \mathbb{R})$, $1 \leq s \leq \infty$, is dense in $C^r(M, \mathbb{R})$ for all $0 \leq r < s$, we see that Lemma 5.1 implies that $\mathcal{A}'$ is also dense in $C^1(M, \mathbb{R})$.

Lemma 5.2. *The set*

$$H_1 = \left\{ f \in C^2(M, \mathbb{R}) : \#M_f(\Lambda) = 1 \quad and \quad Df_z(e_z^{s,u}) \neq 0 \text{ at } z \in M_f(\Lambda) \right\}$$

is dense in $C^2(M, \mathbb{R})$, *therefore dense in* $C^1(M, \mathbb{R})$.

Proof. By Lemma 5.1, it is enough to show that H_1 is dense in $\mathcal{A}'$. By definition, given $f \in \mathcal{A}'$, we can find $z \in M_f(\Lambda)$ such that $Df_z(e_z^{s,u}) \neq 0$. Take U a small neighborhood of z and consider a family of non-negative functions $\varphi_\epsilon \in C^2(M, \mathbb{R})$, $\epsilon > 0$, such that

- φ_ϵ is C^2-close to the constant function 0,
- $\varphi_\epsilon = 0$ in $M \setminus U$,
- $\varphi_\epsilon(z) = \epsilon$,
- z is the single maximum of φ_ϵ, and
- $\varphi_\epsilon \xrightarrow{C^2} 0$ as $\epsilon \to 0$.

Define $g_\epsilon = f + \varphi_\epsilon$, so that $g_\epsilon \xrightarrow{C^2} f$ as $\epsilon \to 0$. Since $z \in M_f(\Lambda)$, we have $g_\epsilon(z) = f(z) + \varphi_\epsilon(z) > f(x) + \varphi_\epsilon(x) = g_\epsilon(x)$ for all $x \in \Lambda$, that is, $z \in M_{g_\epsilon}(\Lambda)$ and $\#M_{g_\epsilon}(\Lambda) = 1$. Also, $D(g_\epsilon)_z(e_z^{s,u}) = Df_z(e_z^{s,u}) \neq 0$, that is, $g_\epsilon \in H_1$. $\qquad \square$

Exercise 5.1

Show that the set H_φ defined in (5.1) is open.

At this point, it is easy to complete the proof of Theorem 5.2

Proof of Theorem 5.2. Since $H_1 \subset H_\varphi$, the set H_1 is dense in $C^1(M, \mathbb{R})$ by Lemma 5.2, and $H_\varphi \subset C^1(M, \mathbb{R})$ is open in $C^1(M, \mathbb{R})$ by Exercise 5.1, the proof is complete. $\square$

5.2.1   *Images of thick subhorseshoes inside the dynamical spectra*

In the sequel, we will prove that the dynamical spectra associated to a height function $f \in H_\psi$ contain the image of a subhorseshoe by a real function. The first step in this direction is the following lemma.

Lemma 5.3. *Let $f \in C^1(M, \mathbb{R})$. If $z \in M_f(\Lambda)$ and $Df_z(e_z^{s,u}) \neq 0$, then $z \in \partial_s \Lambda \cap \partial_u \Lambda$.*

Proof. Using local coordinates centered at z, we can work on an open set $U \subset \mathbb{R}^2$ containing 0. By hypothesis, $Df_z \neq 0$, so that $f(z)$ is a regular value of f. Thus, $\alpha := f^{-1}(f(z))$ is a C^1-curve transverse to $W_\epsilon^s(z)$ and $W_\epsilon^u(z)$ in z and, moreover, $\nabla f(z)$ is orthogonal to α in the point z.

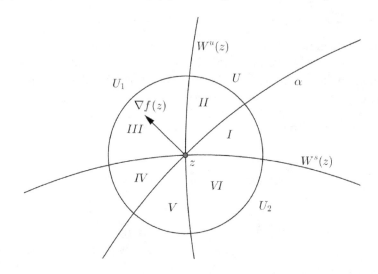

Fig. 5.1   Localization of $z \in M_f(\Lambda)$.

The set $U \setminus \alpha$ has two connected components U_1, U_2 which are distinguished by the fact that $\nabla f(z)$ is pointing in the direction of U_1 (see Figure 5.1). This gives five regions I, II, III, IV and V containing no points of Λ (see Figure 5.1). In fact, since a function increases in the direction of its gradient, there are no points of Λ in II, III and IV (because $z \in M_f(\Lambda)$). Also, if there were points of Λ in I or V, then, by the local product structure, there would exist points of Λ in II or IV, an absurd. Hence, the only region containing points of Λ is VI and, *a fortiori*, $z \in \partial_s \Lambda \cap \partial_u \Lambda$. $\qquad\square$

The following theorem (proved in Palis–Takens book [PT (1993)]) explains why the information in the previous lemma is important: in a nutshell, it relates boundary points and periodic orbits.

Theorem 5.3. *For a horseshoe Λ as above, there is a finite number of (periodic) saddle points $p_1^s, ..., p_{n_s}^s$ such that*

$$\Lambda \cap \left(\bigcup_i W^s(p_i^s) \right) = \partial_s \Lambda.$$

Similarly, there is a finite number of (periodic) saddle points $p_1^u, ..., p_{n_u}^u$ such that

$$\Lambda \cap \left(\bigcup_i W^u(p_i^u) \right) = \partial_u \Lambda.$$

Moreover, both $\partial_s \Lambda$ and $\partial_u \Lambda$ are dense in Λ.

Given $f \in H_\varphi$, let $x_M \in M_f(\Lambda)$. By Lemma 5.3, we have that $x_M \in \partial_s \Lambda \cap \partial_u \Lambda$, so that Theorem 5.3 ensures that

$$x_M \in W^s(p) \cap W^u(q)$$

for some periodic points $p, q \in \Lambda$.

Denote by $(\cdots, a_1, \cdots, a_r, a_1, \cdots, a_r, \cdots)$ and $(\cdots, b_1, \cdots, b_s, b_1, \cdots, b_s, \cdots)$ the symbolic representations of p and q respectively. By definition, there are l symbols $c_1, \cdots, c_l$ such that x_M is symbolically of the form

$\Pi^{-1}(x_M)$
$= (\cdots, b_1, \cdots, b_s, b_1, \cdots, b_s, c_1, \cdots, c_t, \cdots, c_l, a_1, \cdots, a_r, a_1, \cdots, a_r, \cdots)$

where c_t is the zero position of $\Pi^{-1}(x_M)$.

Let $\underline{q}_{\tilde{s}} = (q_{-\tilde{s}}, \cdots, q_0, \cdots, q_{\tilde{s}})$ an admissible word such that $x_M \in R_{\underline{q}_{\tilde{s}}} = \bigcap_{i=-\tilde{s}}^{\tilde{s}} \varphi^{-i}(R_{q_i})$, as in the Figure 5.2, and put a sub-horseshoe

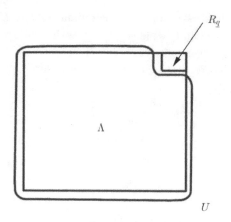

Fig. 5.2   Removing the point of maximum of f.

$\tilde{\Lambda} := \bigcap_{n \in \mathbb{Z}} \varphi^n(\Lambda \setminus R_{q_{\tilde{s}}})$, thus there exists U an open set such that $U \cap \Lambda = \Lambda \setminus R_{q_{\tilde{s}}}$ and

$$\tilde{\Lambda} := \bigcap_{n \in \mathbb{Z}} \varphi^n(U).$$

The next lemma guarantees that if Λ is a thick horseshoe, then $\tilde{\Lambda}$ is also a thick horseshoe:

Lemma 5.4. *If Λ is a horseshoe associated to a C^2-diffeomorphism φ of a surface and $HD(\Lambda) > 1$, then $HD(\tilde{\Lambda}) > 1$ provided $\tilde{s}$ is large enough.*

Proof. The fact that a horseshoe Λ is locally bi-Lipschitz equivalent to the product of its stable and unstable Cantor sets K^s and K^u allows us to reduce our task to prove the following statement.

Let $K = \bigcap_{n \geq 0} \psi^{-n}(\bigcup_{i=1}^{k} I_i)$ be a dynamically defined Cantor set associated to an expanding map ψ acting on a Markov partition $\mathcal{R} = \{I_1, \cdots, I_k\}$. Then, for each $\varepsilon > 0$, there exists a positive integer m_0 such that, for every $m \geq m_0$, and for every finite word $\underline{b} = (b_1, \cdots, b_m)$ of length m, one has

$$HD(K_{\underline{b}}) \geq HD(K) - \varepsilon$$

where $K_{\underline{b}} = \bigcap_{n \geq 0} \psi^{-n}(\bigcup_{i=1}^{k} I_i \setminus I_{\underline{b}})$ and $I_{\underline{b}} = I_{b_1} \cap \psi^{-1}(I_{b_2}) \cap \psi^{-2}(I_{b_3}) \cdots \cap \psi^{-(m-1)}(I_{b_m})$.

In order to show the statement in the previous paragraph, let $\mathcal{R}^n$ be the set of connected components of $\psi^{-(n-1)}(I_i)$, $I_i \in \mathcal{R}$. Let B^n be the set of admissible[3] words of length n, so that $\mathcal{R}^n = \{I_{\underline{b}}, \underline{b} \in B^n\}$. Fix $\tilde{i}, \tilde{j} \leq k$ such that $I_{\tilde{i}} \subset \psi(I_{\tilde{j}})$. Let $X^n = \{\underline{b} = (b_1, \cdots, b_n) \in B^n : b_1 = \tilde{i}, b_n = \tilde{j}\}$. For any positive integer r, and $\underline{b_1}, \underline{b_2}, \ldots, \underline{b_r} \in X^n$, we have $\underline{b_1}\,\underline{b_2}\ldots\underline{b_r} \in X^{nr} \subset B^{nr}$. Let $\tilde{\mathcal{R}}^n = \{I_{\underline{b}}, \underline{b} \in X^n\}$.

For $R \in \mathcal{R}^n$ take $\Lambda_{n,R} = \sup \left|(\psi^n)'_{|R}\right|$. The mixing condition on the Markov partition (cf. Remark 2.7) implies that there is $c_1 > 0$ such that

$$\sum_{R \in \tilde{\mathcal{R}}^n} (\Lambda_{n,R})^{-d} \geq c_1 \sum_{R \in \mathcal{R}^n} (\Lambda_{n,R})^{-d}, \forall d \geq 0, n \geq 1.$$

On the other hand, the Hausdorff dimension bound from Proposition E.2 in Appendix E says that the sequence d_n given by

$$\sum_{R \in \mathcal{R}^n} (\Lambda_{n,R})^{-d_n} = 1$$

converges to $HD(K)$. Notice also that there is $\lambda_1 > 1$ such that $\Lambda_{n,R} \geq \lambda_1^n, \forall n \geq 1$.

Take n be large so that $d_n > \max\{HD(K) - \frac{\varepsilon}{2}, \frac{HD(K)}{2}\}$ and $\lambda_1^{n\varepsilon/2} > 2/c_1$, and let $m_0 = 2n - 1$. Given $m \geq m_0$ and an admissible finite word $\underline{b} = (b_1, \cdots, b_m)$ of length m, define the words $\underline{c_j} = (b_j, b_{j+1}, \ldots, b_{j+n-1}) \in B^n, 1 \leq j \leq n$, and let $L^n = \{\underline{c_j} : 1 \leq j \leq n\}$ and $\hat{\mathcal{R}}^n = \{I_{\underline{c}} : \underline{c} \in X^n \setminus L^n\}$. We have

$$\sum_{R \in \hat{\mathcal{R}}^n} (\Lambda_{n,R})^{-d_n} \geq \sum_{R \in \tilde{\mathcal{R}}^n} (\Lambda_{n,R})^{-d_n} - n\lambda_1^{-nd_n}$$

$$\geq \sum_{R \in \tilde{\mathcal{R}}^n} (\Lambda_{n,R})^{-d_n} - n\lambda_1^{-nHD(K)/2}$$

$$\geq c_1 \sum_{R \in \mathcal{R}^n} (\Lambda_{n,R})^{-d_n} - n\lambda_1^{-nHD(K)/2}$$

$$= c_1 - n\lambda_1^{-nHD(K)/2} > c_1/2,$$

and so

$$\sum_{R \in \hat{\mathcal{R}}^n} (\Lambda_{n,R})^{-(HD(K)-\varepsilon)} > \sum_{R \in \hat{\mathcal{R}}^n} (\Lambda_{n,R})^{-(d_n-\varepsilon/2)} > 1.$$

The previous estimate implies that the dynamically defined Cantor set $\tilde{K} := \bigcap_{r \geq 0} \psi^{-nr}(\cup_{\hat{I} \in \hat{\mathcal{R}}^n} \hat{I})$ associated to ψ^n satisfies

$$\sum_{R \in \overline{\mathcal{R}}^{nr}} (\Lambda_{nr,R})^{-(HD(K)-\varepsilon)} \geq \left(\sum_{R \in \hat{\mathcal{R}}^n} (\Lambda_{n,R})^{-(HD(K)-\varepsilon)}\right)^r > 1,$$

where $\overline{\mathcal{R}}^{nr} = \{I_{\underline{c_1}\underline{c_2}\ldots\underline{c_r}}, \underline{c_j} \in X^n \setminus L^n, \forall j \leq r\}$. Thus, $HD(\tilde{K}) \geq HD(K) - \epsilon$.

---

[3] A word $\underline{b}$ is admissible whenever $I_{\underline{b}} \neq \emptyset$.

This completes the argument because for any positive integer r and $\underline{b_1}, \underline{b_2}, \ldots, \underline{b_r} \in X^n \setminus L^n$, the sequence $(a_1, a_2, \ldots, a_{nr}) = \underline{b_1} \underline{b_2} \ldots \underline{b_r}$ satisfies that $\forall j, 1 \leq j \leq nr - m + 1, (a_j, a_{j+1} \cdots, a_{j+m-1}) \neq \underline{b}$, and so $\tilde{K} \subset K_{\underline{b}}$. In particular, $HD(K_{\underline{b}}) \geq HD(\tilde{K}) \geq HD(K) - \epsilon$, as desired. $\qquad\square$

In the sequel, we will use the thick horseshoe $\tilde{\Lambda}$ to produce images of thick subhorseshoes in the dynamical spectra.

We begin with the case of the dynamical Markov spectrum. Let $d \in \tilde{\Lambda}$ with symbolic representation $\underline{d} = (\cdots, d_{-n}, \cdots, d_0, \cdots, d_n, \cdots)$. Given $\epsilon > 0$ small, take $n_0 \in \mathbb{N}$ such that $\sum_{|n| \geq n_0} 2^{-(2|n|+1)} < \epsilon$ and put $\underline{d}_{n_0} = (d_{-n_0}, \cdots, d_{n_0})$ an admissible finite word. Denote the cylinder $C_{\underline{d}_{n_0}} = \{\underline{w} \in \mathbb{A}^{\mathbb{Z}} : w_i = d_i \text{ for } i = -n_0, \cdots, n_0\}$. Then, the set

$$C_{\underline{d}_{n_0}, B} := \Sigma_B \cap C_{\underline{d}_{n_0}} = \{\underline{w} \in \Sigma_B : w_i = d_i \text{ for } i = -n_0, \cdots, n_0\}$$

is not empty and contains a periodic point.

The transitivity of Σ_B tells us that there are admissible strings $\underline{e} = (e_1, \cdots, e_{k_0-1})$ and $\underline{f} = (f_1, \cdots, f_{j_0-1})$ with $k_0, j_0 < N_0$ joining d_0 with b_1 and a_r with d_1. Since x_M is a unique maximum point of f in Λ, if $\epsilon > 0$ is small enough, we can take $\tilde{\tilde{s}} > \tilde{s}$ and an admissible word $\underline{q}_{\tilde{\tilde{s}}} = (q_{-\tilde{\tilde{s}}}, \cdots, q_0, \cdots, q_{\tilde{\tilde{s}}})$ such that $x_M \in R_{\underline{q}_{\tilde{\tilde{s}}}} = \bigcap_{i=-\tilde{\tilde{s}}}^{\tilde{\tilde{s}}} \varphi^{-i}(R_{q_i}) \subset R_{q_{\tilde{s}}}$ and

$$\sup \tilde{f}|_{\Pi^{-1}(\tilde{\Lambda})_\epsilon} < \inf \tilde{f}|_{\Pi^{-1}(R_{\underline{q}_{\tilde{\tilde{s}}}} \cap \Lambda)}, \tag{5.2}$$

where $\tilde{f} = f \circ \Pi$ and $\Pi^{-1}(\tilde{\Lambda})_\epsilon = \{\underline{x} \in \Sigma_B : \text{dist}(\underline{x}, \Pi^{-1}(\tilde{\Lambda})) < \epsilon\}$.

Let us now fix $k \in \mathbb{N}$, $k > N_0$, and $k(s + r) + l > \tilde{\tilde{s}}$. Going back to the symbolic representation of x_M, we define

$$(a_1, \cdots, a_r)^k = \underbrace{(a_1, \cdots, a_r, \cdots\cdots, a_1, \cdots, a_r)}_{k \text{ times}}$$

and

$$(b_1, \cdots, b_s)^k = \underbrace{(b_1, \cdots, b_s, \cdots\cdots, b_1, \cdots, b_s)}_{k \text{ times}}$$

and we introduce the word

$$\alpha = ((b_1, \cdots, b_s)^k, c_1, \cdots, c_t, \cdots, c_l, (a_1, \cdots, a_r)^k),$$

where c_t is the zero position of the word α. In this context, we can define the following map:

$$C_{\underline{d}_{n_0}, B} \ni \underline{x} \mapsto A(\underline{x}) := (\cdots, x_{-1}, x_0, e_1, \cdots, e_{k_0-1}, (b_1, \cdots, b_s)^k, c_1, \cdots,$$
$$c_t, \cdots, c_l, (a_1, \cdots, a_r)^k, f_1, \cdots, f_{j_0-1}, x_1, x_2, \cdots),$$

where c_t is the zero position of the word $A(\underline{x})$. Here, it is worth to recall that if $|\underline{a}| = n$ stands for the length of a word $\underline{a} = (a_1, \cdots, a_n)$, then

$$|\underline{e}|, |\underline{f}|, \tilde{s} < |\alpha| = k(o + r) + l$$

since $k > N_0 \geq \max\{k_0, j_0\}$.

Let us now study $\sup_{n \in \mathbb{Z}} \tilde{f}(\sigma^n(A(\underline{x})))$ for $\underline{x} \in C_{\underline{d}_{n_0}, B} \cap \Pi^{-1}(\tilde{\Lambda})$. If $\tau := l - t + kr + j_0 - 1$, then our choice of n_0 implies $\text{dist}(\sigma^{\tau + n_0 + n}(A(\underline{x})), \sigma^{n_0 + n}(\underline{x})) < \epsilon$ for all $n \geq 0$. Analogously, call $\eta = -(t + sk + k_0 - 1)$, then $\text{dist}(\sigma^{\eta - n_0 - n}(A(\underline{x})), \sigma^{-n_0 - n}(\underline{x})) < \epsilon$ for all $n \geq 0$. Since $\Pi^{-1}(\tilde{\Lambda})$ is a σ-invariant set, it follows from (5.2) that if $\underline{x} \in \Pi^{-1}(\tilde{\Lambda})$, then

$$\tilde{f}(\sigma^{\tau + n_0 + n}(A(\underline{x}))), \tilde{f}(\sigma^{\eta - n_0 - n}(A(\underline{x}))) < \inf \tilde{f}|_{\Pi^{-1}(R_{q_{\tilde{s}}} \cap \Lambda)} \text{ for all } n \geq 0.$$

Hence, for all $\underline{x} \in C_{\underline{d}_{n_0}, B} \cap \Pi^{-1}(\tilde{\Lambda})$, there is $j \in \{\eta - n_0, \dots, \tau + n_0\}$ such that $\sup_{n \in \mathbb{Z}} \tilde{f}(\sigma^n(A(\underline{x}))) = \tilde{f}(\sigma^j(A(\underline{x})))$. In other words, if we set $\Pi^{-1}(x) = \underline{x}$ and

$$\tilde{\Lambda}_j := \{x \in \tilde{\Lambda} \cap \Pi(C_{\underline{d}_{n_0}, B}) : \sup_{n \in \mathbb{Z}} \tilde{f}(\sigma^n(A(\underline{x}))) = f(\sigma^j(A(\underline{x})))\},$$

then

$$\tilde{\Lambda} \cap \Pi(C_{\underline{d}_{n_0}, B}) = \bigcup_{j = \eta - n_0}^{\eta + n_0} \tilde{\Lambda}_j. \tag{5.3}$$

Note that there is some $i_0 \in \{\eta - n_0, \dots, \tau + n_0\}$ such that $\tilde{\Lambda}_{i_0}$ has non empty interior in $\tilde{\Lambda} \cap \Pi(C_{\underline{d}_{n_0}, B})$ and, *a fortiori*,

$$HD(\tilde{\Lambda}) = HD(\tilde{\Lambda} \cap \Pi(C_{\underline{d}_{n_0}, B})) = HD(\tilde{\Lambda}_{i_0}). \tag{5.4}$$

Therefore, for $\underline{x} \in \Pi^{-1}(\tilde{\Lambda}_{i_0})$ we have

$$\sup_{n \in \mathbb{Z}} \tilde{f}(\sigma^n(A(\underline{x}))) = \tilde{f}(\sigma^{i_0}(A(\underline{x}))). \tag{5.5}$$

This information allows to get that $\{f(\varphi^{i_0}(\tilde{A}(x))) : x \in \tilde{\Lambda}_{i_0}\} \subset M(f, \Lambda)$, where $\tilde{A} = \Pi \circ A \circ \Pi^{-1}$. Unfortunately, this is not quite enough for our purposes: we need to extend $\tilde{A} = \Pi \circ A \circ \Pi^{-1}$ into a *local diffeomorphism*.

Lemma 5.5. *If φ is a C^2-diffeomorphism, then $\tilde{A}$ extends to a local C^1-diffeomorphism defined in neighborhood U_d of d. We may assume without loss of generality (increasing n_0, if necessary) that $U_d \supset \tilde{\Lambda} \cap \Pi(C_{\underline{d}_{n_0}, B})$.*

Proof. Consider the finite word $\beta = \underline{e}\alpha\underline{f}$ and recall that Λ is symbolically the product $\Sigma_B^- \times \Sigma_B^+$. In particular, if $x^u \in W^u_{loc}(d) \cap \Lambda$, resp. $x^s \in W^s_{loc}(d) \cap \Lambda$, then

$$f^u_\beta(x^u) \in W^u(d) \cap \Lambda \quad \text{and} \quad (\Pi^{-1}(f^u_\beta(x^u)))^+ = \beta(\Pi^{-1}(x^u))^+,$$

resp.

$$f^s_\beta(x^s) \in W^s(d) \cap \Lambda \quad \text{and} \quad (\Pi^{-1}(f^s_\beta(x^s)))^- = (\Pi^{-1}(x^s))^-\beta,$$

where f^u_β and f^s_β are adequate inverse branches of the expanding maps g^u and g^s defining the unstable and stable Cantor sets K^u and K^s of Λ. Since the symbol at the zero position of $\Pi^{-1}\left(\varphi^{-|\beta|+1}\left(f^s_\beta(x^s)\right)\right)$ is $(\beta)_0 = e_1$, we have

$$\left(\Pi^{-1}\left(\varphi^{-|\beta|+1}\left(f^s_\beta(x^s)\right)\right)\right)_0 = (\beta)_0 = \left(\Pi^{-1}\left(f^u_\beta(x^u)\right)\right)_0,$$

so that the bracket

$$\left[\Pi^{-1}\left(f^u_\beta(x^u)\right), \Pi^{-1}\left(\varphi^{-|\beta|+1}\left(f^s_\beta(x^s)\right)\right)\right] = (\Pi^{-1}(x^s))^- \beta (\Pi^{-1}(x^u))^+$$

which is equals to $A\left[\Pi^{-1}(x^u), \Pi^{-1}(x^s)\right]$ is well-defined. Because Π respects the local product structure, we have that

$$\left[f^u_\beta(x^u), \varphi^{-|\beta|+1}\left(f^s_\beta(x^s)\right)\right] = \Pi\left(\left[\Pi^{-1}\left(f^u_\beta(x^u)\right), \Pi^{-1}\left(\varphi^{-|\beta|+1}\left(f^s_\beta(x^s)\right)\right)\right]\right)$$

$$= \Pi(A\left[\Pi^{-1}(x^u), \Pi^{-1}(x^s)\right]) = \tilde{A}\left[x^u, x^s\right]. \quad (5.6)$$

In other terms, if $\tilde{A}_1(x^u) := f^u_\beta(x^u)$ and $\tilde{A}_2(x^s) := \varphi^{-|\beta|+1}(f^s_\beta(x^s))$, then $\tilde{A}\left[x^u, x^s\right] = [\tilde{A}_1(x^u), \tilde{A}_2(x^s)]$.

Since the stable and unstable foliations of the horseshoe Λ can be extended to C^1 invariant foliations defined on a full neighborhood of Λ, we can use (5.6) to extend $\tilde{A}$. $\qquad\square$

An immediate consequence of Lemma 5.5 and the equality (5.5) is:

Corollary 5.1. *We can extend $\tilde{A}$ into a local diffeomorphism respecting local stable and unstable manifolds near d such that* $\sup_{n \in \mathbb{Z}} f(\varphi^n(\tilde{A}(x))) = f(\varphi^{i_0}(\tilde{A}(x)))$ *for all* $x \in \tilde{\Lambda}_{i_0}$. *In particular,* $\{f(\varphi^{i_0}(\tilde{A}(x))) : x \in \tilde{\Lambda}_{i_0}\} \subset M(f, \Lambda)$.

Remark 5.2. We have $Df_{x_M}(e^{s,u}_{x_M}) \neq 0$, so this property is true in neighborhood of x_M. Since, for every $x \in \tilde{\Lambda}_{i_0}$, $\varphi^{i_0}(\tilde{A}(x))$ belongs to a small neighborhood of x_M, we see that $Df_{\varphi^{i_0}(\tilde{A}(x))}(e^{s,u}_{\varphi^{i_0}(\tilde{A}(x))}) \neq 0$ for every $x \in \tilde{\Lambda}_{i_0}$. Because $D\varphi^{i_0}_{\tilde{A}(x)}(e^{s,u}_{\tilde{A}(x)}) \in E^{s,u}_{\varphi^{i_0}(\tilde{A}(x))}$ and, by construction of $\tilde{A}$, we also have that $\frac{\partial \tilde{A}}{\partial e^{s,u}_x}$ is parallel to $e^{s,u}_{\tilde{A}(x)}$, it follows that $D(f \circ \varphi^{i_0} \circ \tilde{A})_x(e^{s,u}_x) \neq 0$ for every $x \in \tilde{\Lambda}_{i_0}$.

Next, we establish the analogue of Corollary 5.1 for the dynamical Lagrange spectrum. For this sake, take ϵ, n_0, and $\tilde{\tilde{s}}$ as above, and recall that for each $x \in \tilde{\Lambda} \cap \Pi(C_{\underline{d}_{n_0}, B})$, $\Pi^{-1}(x) = (\cdots, x_{-n}, \cdots, x_0, \cdots, x_n, \cdots)$, there is an admissible string $E_i = (e_1^i, \cdots, e_{s_i}^i)$ joining x_i with x_{-i} the length $|E_i| = m_i - 1 < N_0$ for each i. In this setting, we can define the map

$$C_{\underline{d}_{n_0}, B} \ni \underline{x} \mapsto A_1(\underline{x})$$
$$= (\cdots, x_3, E_3, x_{-3}, x_{-2}, x_{-1}, x_0, \beta, x_1, x_2, E_2, x_{-2}, x_{-1}, x_0, \beta, x_1, E_1, x_{-1}, x_0,$$
$$\beta, x_1, E_1, x_{-1}, x_0, \beta, x_1, x_2, E_2, x_{-2}, x_{-1}, x_0, \beta, x_1, x_2, x_3, E_3, x_{-3}, \cdots)$$

where $\beta = \underline{e}\alpha\underline{f}$ as above. Since $|E_i| < N_0$ for all i, the set of words $\{E_i : i \in \mathbb{N}^*\}$ is finite, say $\{E_i : i \in \mathbb{N}^*\} = \{D_1, \ldots D_m\}$ for some admissible words D_i with $|D_i| < N_0$. By increasing $\tilde{\tilde{s}}$ if necessary, since $|D_i| < N_0$, we can assume that, for each i, there exists a neighborhood $\mathcal{U}_i$ of the periodic sequence $\overline{D_i}$ for which

$$\sup \tilde{f}|_{\sigma^r(\mathcal{U}_i)} < \inf \tilde{f}|_{\Pi^{-1}(R_{q_{\tilde{s}}} \cap \Lambda)} \quad \text{for} \quad |r| \le n_0 + |D_i| < n_0 + N_0. \quad (5.7)$$

Let us now study $\limsup_{n \to \infty} \tilde{f}(\sigma^n(A_1(\underline{x})))$ for $\underline{x} \in C_{\underline{d}_{n_0}, B} \cap \Pi^{-1}(\tilde{\Lambda})$. In this direction, take $m(n) \in \mathbb{N}$ such that

$$(\sigma^{m(n)}(A_1(\underline{x})))^+ = x_1, x_2, \cdots, x_n E_n x_{-n}, \cdots, x_0, \cdots \quad \text{and} \quad n \ge 2n_0.$$

Let us now set $n - k^* = n_0$, so that

$$\text{dist}(\sigma^{m(n)+n_0+j}(A_1(\underline{x})), \sigma^{n_0+j}(\underline{x})) < \epsilon \text{ for all } j = 0, \ldots, k^* - n_0,$$

and

$$\text{dist}(\sigma^{m(n)+n+|E_n|+n_0+j}(A_1(\underline{x})), \sigma^{-k^*+j}(\underline{x})) < \epsilon \text{ for all } j = 0, \ldots, k^* - n_0$$

thanks to our choice of n_0.

By σ-invariance of $\Pi^{-1}(\tilde{\Lambda})$, it follows from (5.2) that if $\underline{x} \in \Pi^{-1}(\tilde{\Lambda})$, then

$$\tilde{f}(\sigma^{m(n)+n_0+j}(A_1(\underline{x}))) < \inf \tilde{f}|_{\Pi^{-1}(R_{q_{\tilde{s}}} \cap \Lambda)} \text{ for all } j = 0, \ldots, k^* - n_0,$$

and

$$\tilde{f}(\sigma^{m(n)+n+|E_n|+n_0+j}(A_1(\underline{x}))) < \inf \tilde{f}|_{\Pi^{-1}(R_{q_{\tilde{s}}} \cap \Lambda)} \text{ for all } j = 0, \ldots, k^* - n_0.$$

Also, we have

$$\sigma^{m(n)+k^*+s}(A_1(\underline{x})) \in \sigma^{|E_n^-|+n_0-s}(\mathcal{U}_{i(n)}) \text{ for all } s = 0, \ldots, n_0 + |E_n^-|,$$

and

$$\sigma^{m(n)+n+|E_n^-|+s}(A_1(\underline{x})) \in \sigma^{-s}(\mathcal{U}_{i(n)}) \text{ for all } s = 0, \ldots, n_0 + |E_n^+|,$$

where $E_n = E_n^- E_n^+$ and $i(n) \in \{1, \ldots, m\}$. Therefore, the inequality (5.7) implies that

$$\tilde{f}(\sigma^{m(n)+k^*+s}(A_1(\underline{x}))) < \inf \tilde{f}|_{\Pi^{-1}(R_{q_{\overline{s}}} \cap \Lambda)} \text{ for all } s = 0, \ldots, n_0 + |E_n^-|,$$

and

$$\tilde{f}(\sigma^{m(n)+n+|E_n^-|+s}(A_1(\underline{x}))) < \inf \tilde{f}|_{\Pi^{-1}(R_{q_{\overline{s}}} \cap \Lambda)} \text{ for all } s = 0, \ldots, n_0 + |E_n^+|.$$

Note that, if $n_0 \le n < 2n_0$, then $k^* < n_0$. Hence, the four last inequalities above implies that, for each $\underline{x} \in C_{\underline{d}_{n_0}, B} \cap \Pi^{-1}(\tilde{\Lambda})$, there exists $j \in \{\eta - n_0, \ldots, \tau + n_0\}$ and a sequence $n_k(j)$ with

$$\limsup_{n \to \infty} \tilde{f}(\sigma^n(A_1(\underline{x}))) = \sup_k \tilde{f}(\sigma^{n_k(j)}(A_1(\underline{x}))) \text{ and } \left(\sigma^{n_k(j)}(A_1(\underline{x}))\right)_0 = (A_1(\underline{x}))_j$$

for all k, where $\eta = -(t + ks + k_0 - 1)$ and $\tau = (l - t + kr + j_0 - 1)$ are the length of negative and positive part of the finite word $\beta = \underline{e}\alpha\underline{f}$, respectively.

In particular, if we set $\Pi^{-1}(x) = \underline{x}$ and we define

$$\Lambda'_j := \{x \in \tilde{\Lambda} \cap \Pi(C_{\underline{d}_{n_0}, B}) : \limsup_{n \to \infty} \tilde{f}(\sigma^n(A_1(\underline{x}))) = \sup_k \tilde{f}(\sigma^{n_k(j)}(A_1(\underline{x})))\},$$

then,

$$\tilde{\Lambda} \cap \Pi(C_{\underline{d}_{n_0}, B}) = \bigcup_{j = \eta - n_0}^{\tau + n_0} \Lambda'_j.$$

Thus, there is $j_0 \in \{\eta - n_0, \ldots, \tau + n_0\}$ such that Λ'_{j_0} has non empty interior in $\tilde{\Lambda} \cap \Pi(C_{\underline{d}_{n_0}, B})$ and, *a fortiori*,

$$HD(\tilde{\Lambda}) = HD(\tilde{\Lambda} \cap \Pi(C_{\underline{d}_{n_0}, B})) = HD(\Lambda'_{j_0}). \tag{5.8}$$

Also, it follows from the definitions that if $\underline{x} \in \Pi^{-1}(\Lambda'_{j_0})$, then

$$\limsup_{n \to \infty} \tilde{f}(\sigma^n(A(\underline{x}))) = \sup_k \tilde{f}(\sigma^{n_k(j_0)}(A_1(\underline{x})))$$

and, hence, there is a subsequence $n_{k_m}(j_0)$ with $n_{k_m}(j_0) \to \infty$, as $m \to \infty$ such that

$$\sup_k \tilde{f}(\sigma^{n_k(j_0)}(A_1(\underline{x}))) = \lim_{m \to \infty} \tilde{f}(\sigma^{n_{k_m}(j_0)}(A_1(\underline{x}))).$$

By construction of A_1, we know that $\lim_{m \to \infty} \sigma^{n_{k_m}(j_0)}(A_1(\underline{x})) = \sigma^{j_0}(A(\underline{x}))$. Therefore,

$$\limsup_{n \to \infty} \tilde{f}(\sigma^n(A_1(\underline{x}))) = \tilde{f}(\sigma^{j_0}(A(\underline{x}))).$$

In summary, we proved that:

Corollary 5.2. *If $x \in \Lambda'_{j_0}$, then*

$$\limsup_{n \to \infty} f(\varphi^n(\tilde{A}_1(x))) = f(\varphi^{j_0}(\tilde{A}(x))), \quad \text{where } \tilde{A}_1 = \Pi \circ A_1 \circ \Pi^{-1}.$$

In particular, $\{f(\varphi^{j_0}(\tilde{A}(x))) : x \in \Lambda'_{j_0}\} \subset L(f, \Lambda)$.

At this point, we have that Corollaries 5.1 and 5.2 reduce the problem of finding intervals on the dynamical spectra to the question of exhibiting intervals in the projections of thick horseshoes. As it turns out, this type of question was studied in the works of Moreira–Yoccoz [MY (2001)] and [MY (2010)], and this is our next topic of discussion.

5.2.2 *Typical projections of thick subhorseshoes*

Following Moreira–Yoccoz [MY (2001)], let us introduce the notion of *limit geometries* of dynamically defined Cantor sets. Let $\mathbb{A}$ be a finite alphabet, $\mathbb{B}$ a subset of $\mathbb{A}^2$, and $\Sigma_{\mathbb{B}}$ the subshift of finite type of $\mathbb{A}^{\mathbb{Z}}$ with allowed transitions $\mathbb{B}$. We will always assume that $\Sigma_{\mathbb{B}}$ is topologically mixing, and that every letter in $\mathbb{A}$ occurs in $\Sigma_{\mathbb{B}}$. Recall that an expansive map g of type $\Sigma_{\mathbb{B}}$ is a map with the following properties:

(i) the domain of g is a disjoint union $\bigcup_{\mathbb{B}} I(a, b)$. Where for each (a, b), $I(a, b)$ is a compact subinterval of $I(a) := [0, 1] \times \{a\}$;

(ii) for each $(a, b) \in \mathbb{B}$, the restriction of g to $I(a, b)$ is a smooth diffeomorphism onto $I(b)$ satisfying $|Dg(t)| > 1$ for all t.

In this context, the dynamically defined Cantor set associated to g is the maximal invariant set

$$K = \bigcap_{n \geq 0} g^{-n} \left(\bigcup_{\mathbb{B}} I(a, b) \right).$$

Recall also that there exists an unique homeomorphism $h \colon \Sigma_{\mathbb{B}}^+ \to K$ between the unilateral subshift $\Sigma_{\mathbb{B}}^+$ and K such that $h(\underline{a}) \in I(a_0)$ for $\underline{a} = (a_0, a_1, \dots) \in \Sigma_{\mathbb{B}}^+$ and $h \circ \sigma^+ = g \circ h$, where $\sigma^+ \colon \Sigma_{\mathbb{B}}^+ \to \Sigma_{\mathbb{B}}^+$ is the left-shift.

For $(a, b) \in \mathbb{B}$, the associated inverse branch of g is denoted by $f_{a,b} := \left[g|_{I(a,b)} \right]^{-1}$. This is a contracting diffeomorphism from $I(b)$ onto $I(a, b)$. In general, if $\underline{a} = (a_0, \cdots, a_n)$ is a word of $\Sigma_{\mathbb{B}}$, then

$$f_{\underline{a}} := f_{a_0, a_1} \circ \cdots \circ f_{a_{n-1}, a_n}$$

is a contracting diffeomorphism from $I(a_n)$ onto a subinterval $I(\underline{a})$ of $I(a_0)$. By definition, if z in the domain of $f_{\underline{a}}$, then $f_{\underline{a}}(z) = h(\underline{a}h^{-1}(z))$.

In the sequel, given $r \in (1, +\infty)$, the space of C^r expansive maps of type Σ endowed with the C^r topology is denoted Ω_Σ^r. Also, we endow $\Omega_\Sigma = \bigcup_{r>1} \Omega_\Sigma^r$ with the inductive limit topology.

In this setting, the key notion of *limit geometries* of K is defined as follows. Consider the symbolic space $\Sigma^- = \{(\theta_n)_{n \leq 0}, (\theta_i, \theta_{i+1}) \in \mathbb{B}$ for $i < 0\}$ equipped with the following ultrametric distance: if $\underline{\theta} \neq \widetilde{\underline{\theta}} \in \Sigma^-$, then

$$d(\underline{\theta}, \widetilde{\underline{\theta}}) = \begin{cases} 1 & \text{if } \theta_0 \neq \widetilde{\theta}_0; \\ |I(\underline{\theta} \wedge \widetilde{\underline{\theta}})| & \text{otherwise} \end{cases}$$

where $\underline{\theta} \wedge \widetilde{\underline{\theta}} = (\theta_{-n}, \dots, \theta_0)$ if and only if $\widetilde{\theta}_{-j} = \theta_{-j}$ for $0 \leq j \leq n$ and $\widetilde{\theta}_{-n-1} \neq \theta_{-n-1}$.

Next, given $\underline{\theta} \in \Sigma^-$ and $n > 0$, let $\underline{\theta}^n = (\theta_{-n}, \dots, \theta_0)$ and $B(\underline{\theta}^n)$ be the affine map from $I(\underline{\theta}^n)$ onto $I(\theta_0)$ such that the diffeomorphism $k_n^{\underline{\theta}} = B(\underline{\theta}^n) \circ f_{\underline{\theta}^n}$ is orientation preserving. It is not hard to show (cf. [MY (2001)]) that, for any $\underline{\theta} \in \Sigma^-$, there is $k^{\underline{\theta}} \in \text{Diff}_+^r(I(\theta_0))$ such that the sequence $k_n^{\underline{\theta}}$ converges in $\text{Diff}_+^{r'}(I(\theta_0))$ for all $r' < r$. Moreover, this convergence is uniform in $\underline{\theta}$ and in a neighborhood of g in Ω_Σ^r. Furthermore, if r is an integer, or $r = +\infty$, $k_n^{\underline{\theta}}$ converge to $k^{\underline{\theta}}$ in $\text{Diff}_+^r(I(\theta_0))$ in the sense that, for each $0 \leq j \leq r - 1$, there is a constant C_j (independent on $\underline{\theta}$) such that

$$\left| D^j \log D \left[k_n^{\underline{\theta}} \circ (k^{\underline{\theta}})^{-1} \right] (x) \right| \leq C_j |I(\underline{\theta}^n)|.$$

In particular, $\underline{\theta} \to k^{\underline{\theta}}$ is Lipschitz, i.e., if $\theta_0 = \widetilde{\theta}_0$, then

$$\left| D^j \log D \left[k^{\widetilde{\underline{\theta}}} \circ (k^{\underline{\theta}})^{-1} \right] (x) \right| \leq C_j \, d(\underline{\theta}, \widetilde{\underline{\theta}}).$$

In this context, the *limit geometries* of K are $k^{\underline{\theta}}(K)$, $\theta \in \Sigma^-$.

The limit geometries of dynamically defined Cantor sets can be used to find stably intersecting configurations among them. More concretely, let K and K' be two dynamically defined Cantor sets associated to two sets of data $(\mathbb{A}, \mathbb{B}, \Sigma, g)$, $(\mathbb{A}', \mathbb{B}', \Sigma', g')$. Given $r \in [1, +\infty]$ and $a \in \mathbb{A}$, let $\mathcal{P}^r(a)$ be the space of C^r-embeddings of interval $I(a)$ into $\mathbb{R}$ (endowed with the C^r topology). The affine group $\text{Aff}(\mathbb{R})$ acts by composition on the left on $\mathcal{P}^r(a)$ and the corresponding quotient space is denoted by $\overline{\mathcal{P}}^r(a)$. We also consider $\mathcal{P}(a) = \bigcup_{r>1} \mathcal{P}^r(a)$ and $\overline{\mathcal{P}}(a) = \bigcup_{r>1} \overline{\mathcal{P}}^r(a)$ endowed with the inductive limit topologies.

The space

$$\mathcal{A} \times \mathcal{A}' = \{(\underline{\theta}, A), (\underline{\theta}', A') : \underline{\theta}, \underline{\theta}' \in \Sigma^- \times \Sigma'^- \text{ and }$$
$$A, A' \text{ are affine embeddings of } I(\theta_0), I(\theta_0') \text{ into } \mathbb{R}\}$$

admits a canonical map

$$\mathcal{A} \times \mathcal{A}' \to \mathcal{P}^r \times \mathcal{P}'^r = \bigcup_\mathbb{A} \mathcal{P}^r(a) \times \bigcup_\mathbb{A} \mathcal{P}'^r(a')$$

$$((\underline{\theta}, A), (\underline{\theta}', A')) \mapsto (A \circ k^{\underline{\theta}}, A' \circ k^{\underline{\theta}'}) \quad (\in \mathcal{P}^r(\theta_0) \times \mathcal{P}'^r(\theta_0')).$$

A pair $(h, h') \in \mathcal{P}(a) \times \mathcal{P}'(a')$ is called a *smooth configuration* for $K(a) = K \cap I(a)$, $K'(a') = K' \cap I(a')$. We say that a smooth configuration $(h, h') \in \mathcal{P}(a) \times \mathcal{P}(a')$ is

- *linked* if $h(I(a)) \cap h'(I(a')) \neq \emptyset$;
- *intersecting* if $h(K(\underline{a})) \cap h'(K(\underline{a}')) \neq \emptyset$, where $K(\underline{a}) = K \cap I(\underline{a})$ and $K(\underline{a}') = K \cap I(\underline{a}')$;
- *stably intersecting* if it is still intersecting after small perturbations of it in $\mathcal{P} \times \mathcal{P}' = \bigcup_A \mathcal{P}(a) \times \bigcup_{A'} \mathcal{P}(a')$ and of (g, g') in $\Omega_\Sigma \times \Omega_{\Sigma'}$.

All these definitions are invariant under the action of the affine group, and therefore make sense for *smooth relative configurations*, i.e., the elements of the quotient Q of the product $\mathcal{P} \times \mathcal{P}'$ by the diagonal action of the affine group $\mathrm{Aff}(\mathbb{R})$. Also, the canonical map $\mathcal{A} \times \mathcal{A}' \to \mathcal{P} \times \mathcal{P}'$ allows to define linked, intersecting, and stably intersecting configurations at the level of $\mathcal{A} \times \mathcal{A}'$ or $\mathcal{C} := (\mathcal{A} \times \mathcal{A}')/\mathrm{Aff}(\mathbb{R})$.

In their seminal work [MY (2001)], Moreira–Yoccoz introduced the class V of pairs $(g, g') \in \Omega_\Sigma \times \Omega_{\Sigma'}$ such that, for any $[(\underline{\theta}, A), (\underline{\theta}', A')] \in \mathcal{A} \times \mathcal{A}'$, the configuration $(R_t \circ A \circ k^{\underline{\theta}}, A' \circ k'^{\underline{\theta}'})$ is a stably intersecting for some translation $R_t(x) = x + t$ in $\mathbb{R}$, and they proved that:

Theorem 5.4 (Moreira–Yoccoz). *The set V is open in $\Omega_\Sigma \times \Omega_{\Sigma'}$, and $V \cap (\Omega_\Sigma^\infty \times \Omega_{\Sigma'}^\infty)$ is dense (for the C^∞-topology) in the set $\{(g, g'), HD(K) + HD(K') > 1\}$. Moreover, if $(g, g') \in V$, then there exists $d^* < 1$ such that, for any $(h, h') \in \mathcal{P} \times \mathcal{P}'$, the set*

$$\mathcal{I}_s = \{t \in \mathbb{R}, (R_t \circ h, h')$$

is a stably intersecting smooth configuration for $(g, g')\}$

is (open and) dense in

$$\mathcal{I} = \{t \in \mathbb{R}, (R_t \circ h, h') \text{ is an intersecting smooth configuration for } (g, g')\}$$

with $HD(\mathcal{I} - \mathcal{I}_s) \leq d^*$. Also, d^* is uniform in a neighborhood of (g, g') in $\Omega_\Sigma \times \Omega_{\Sigma'}$.

This theorem is very important for our purposes because it permits to find intervals in typical projections of thick horseshoes.

Theorem 5.5. *Let K, K' be two Cantor sets defined by expanding map g, g'. Suppose that $HD(K) + HD(K') > 1$ and $(g, g') \in V$. Let $f: U \to \mathbb{R}$ be a C^1-function defined in a neighborhood U of $K \times K'$. If the gradient of f at some point of $K \times K'$ is not parallel to the coordinate axes, then $\mathrm{int}\, f(K \times K') \neq \emptyset$.*

Proof. By hypothesis, we can find a pair of periodic points p_1, p_2 of K and K', respectively, with addresses $\bar{a}_1 = \underline{a}_1\underline{a}_1\underline{a}_1...$ and $\bar{a}_2 = \underline{a}_2\underline{a}_2\underline{a}_2...$, where $\underline{a}_1$ and $\underline{a}_2$ are finite sequences, such that $df(p_1, p_2)$ is not a real multiple of dx nor dy. There are increasing sequences of natural number (m_k), (n_k) such that the intervals $I_{\underline{a}_1^{m_k}}$ and $I'_{\underline{a}_2^{n_k}}$ defined by the finite words $\underline{a}_1^{m_k}$ and $\underline{a}_2^{n_k}$, satisfy

$$\frac{|I_{\underline{a}_1^{m_k}}|}{|I'_{\underline{a}_2^{n_k}}|} \in (C^{-1}, C) \text{ for some } C > 1.$$

Thus, we can assume that $\frac{|I_{\underline{a}_1^{m_k}}|}{|I'_{\underline{a}_2^{n_k}}|} \to \lambda \in [C^{-1}, C]$ as $k \to \infty$, define $\tilde{\lambda} :=$ $-\frac{\frac{\partial f}{\partial x}(p_1, p_2)}{\frac{\partial f}{\partial y}(p_1, p_2)}\lambda$.

Since $(K, K') \in V$, there is $t \in \mathbb{R}$ such that $(\tilde{\lambda}k^{\bar{a}_1} + t, k'^{\bar{a}_2})$ is a stably intersecting configuration. So, there are $\tilde{x} \in I((\underline{a}_1)_0)$ and $\tilde{y} \in I((\underline{a}_2)_0)$ such that $x_0 = k^{\bar{a}_1}(\tilde{x})$ and $y_0 = k^{\bar{a}_2}(\tilde{y})$ with $\tilde{\lambda}x_0 + t = y_0$, where $(\underline{a}_i)_0$ is the zero position of the finite word $\underline{a}_i$, for $i = 1, 2$. Moreover, $\tilde{x} = g^{m_k|\underline{a}_1|-1}(\bar{x})$ and $\tilde{y} = (g')^{n_k|\underline{a}_2|-1}(\bar{y})$, for some $\bar{x} \in I_{\underline{a}_1^{m_k}}$ and $\bar{y} \in I'_{\underline{a}_2^{n_k}}$. By taking k large enough, we can assume that $df(\bar{x}, \bar{y})$ is not a real multiple of dx nor of dy. In particular, $\frac{\partial f}{\partial y}(\bar{x}, \bar{y}) \neq 0$ and the implicit function theorem says that $f(H(x, y)) = y$ for some C^1-diffeomorphism $H(x, y) = (x, g(x, y))$ defined in neighborhood of $(\bar{x}, \bar{y})$. Of course, there is no loss of generality in assuming that H is defined in $I_{\underline{a}_1^{m_k}} \times I'_{\underline{a}_2^{n_k}}$.

Let us now set $g_s(x) := g(x, s)$, so that $g_{s_0}(\bar{x}) = \bar{y}$, where $f(\bar{x}, \bar{y}) := s_0$. By definition, $s \in f((K \cap I_{\underline{a}_1^{m_k}}) \times (K' \cap I'_{\underline{a}_2^{n_k}}))$ is equivalent to $g_s(K \cap I_{\underline{a}_1^{m_k}}) \cap (K' \cap I'_{\underline{a}_2^{n_k}}) \neq \emptyset$, our task reduces to prove that $g_s(K \cap I_{\underline{a}_1^{m_k}})$ and $K' \cap I'_{\underline{a}_2^{n_k}}$ have non-empty intersection for s close to $s_0 = f(\bar{x}, \bar{y})$.

Denote by $B_k : I'_{\underline{a}_2^{n_k}} \to [0, 1]$ and $T_k : I_{\underline{a}_1^{m_k}} \to [0, 1]$ the orientation-preserving affine maps given by $B_k(x) = \frac{1}{b'_k - a'_k}(x - a'_k) = \frac{1}{|I'_{\underline{a}_2^{n_k}}|}(x - a'_k)$ and $T_k(x) = \frac{1}{b_k - a_k}(x - a_k) = \frac{1}{|I_{\underline{a}_1^{m_k}}|}(x - a_k)$, where $I_{\underline{a}_1^{m_k}} = [a_k, b_k]$ and $I'_{\underline{a}_2^{n_k}} = [a'_k, b'_k]$. By the definition of limit geometries, we have that $B_k(K' \cap I'_{\underline{a}_2^{n_k}})$ converges to $k^{\bar{a}_2}(K')$ and $T_k(K \cap I_{\underline{a}_1^{m_k}})$ converges to $k^{\bar{a}_1}(K)$ as regular Cantor sets. Moreover, $B_k(g_{s_0}(K \cap I_{\underline{a}_1^{m_k}})) = B_k \circ g_{s_0} \circ T_k^{-1}(T_k(K \cap I_{\underline{a}_1^{m_k}}))$.

We *claim* that the map $B_k \circ g_{s_0} \circ T_k^{-1}$ converges to $\tilde{\lambda}x + t$ in the C^1

topology. In fact, if $\epsilon_k := b_k - a_k = |I_{\underline{a}_1^{m_k}}|$ and $\epsilon'_k = b'_k - a'_k = |I'_{\underline{a}_2^{n_k}}|$, then

$$B_k \circ g_{s_0} \circ T_k^{-1}(x) = \frac{1}{\epsilon'_k}(g_{s_0}(\epsilon_k x + a_k) - a'_k)$$

$$= \frac{1}{\epsilon'_k}\left(g_{s_0}(a_k) + g'_{s_0}(a_k)\epsilon_k x + r(\epsilon_k x) - a'_k\right)$$

$$= B_k(g_{s_0}(a_k)) + g'_{s_0}(a_k)\frac{\epsilon_k}{\epsilon'_k}x + \frac{\epsilon_k}{\epsilon'_k}\frac{r(\epsilon_k x)}{\epsilon_k}. \quad (5.9)$$

Since $g_{s_0}(\bar{x}) = \bar{y}$, then $B_k(g_{s_0}(\bar{x})) = B_k(\bar{y}) = B_k \circ (g')^{-(n_k|\underline{a}_2|-1)}(\tilde{y})$ and the definition of limit geometries implies that $B_k(g_{s_0}(\bar{x}))$ converges to $k^{\bar{a}_2}(\tilde{y}) = y_0 = \tilde{\lambda}x_0 + t$. Also, $T_k(\bar{x}) = T_k \circ g^{-(m_k|\underline{a}_1|-1)}(\tilde{x})$ converges to $k^{\bar{a}_1}(\tilde{x}) = x_0$. Therefore, (5.9) implies that

$$B_k \circ g_{s_0}(\bar{x}) = B_k \circ g_{s_0} \circ T_k^{-1}(T_k(\bar{x}))$$

$$= B_k(g_{s_0}(a_k)) + g'_{s_0}(a_k)\frac{\epsilon_k}{\epsilon'_k}T_k(\bar{x}) + \frac{\epsilon_k}{\epsilon'_k}\frac{r(\epsilon_k T_k(\bar{x}))}{\epsilon_k}.$$

So, if $k \to +\infty$, then the left side of the equality above converges to $\tilde{\lambda}x_0 + t$. Since $g'_{s_0}(a_k) \to -\frac{\frac{\partial f}{\partial x}(p_1,p_2)}{\frac{\partial f}{\partial y}(p_1,p_2)}$, $\frac{\epsilon_k}{\epsilon'_k} \to \lambda$, $T_k(\bar{x}) \to x_0$ and $\frac{r(\epsilon_k T_k(\bar{x}))}{\epsilon_k} \to 0$, it follows from the definition of $\tilde{\lambda}$ and the equality above that

$$B_k \circ g_{s_0}(a_k) \to \tilde{\lambda}x_0 + t - \tilde{\lambda}x_0 = t. \quad (5.10)$$

Thus, by the equalities (5.9) and (5.10), we have

$$\lim_{k \to +\infty} B_k \circ g_{s_0} \circ T_k^{-1}(x) = \tilde{\lambda}x + t.$$

Moreover, since g_{s_0} is a C^1-function, we get

$$\left(B_k \circ g_{s_0} \circ T_k^{-1}\right)'(x) = \frac{1}{\epsilon'_k}g'_{s_0}(T_k^{-1}(x)).\epsilon_k \to -\frac{\frac{\partial f}{\partial x}(p_1,p_2)}{\frac{\partial f}{\partial y}(p_1,p_2)}\lambda = \tilde{\lambda}.$$

This proves the claim.

Therefore,

$$B_k(g_{s_0}(K \cap I_{\underline{a}_1^{m_k}})) = B_k \circ g_{s_0} \circ T_k^{-1}(T_k(K \cap I_{\underline{a}_1^{m_k}})) \to \tilde{\lambda}k^{\bar{a}_1}(K) + t,$$

and

$$B_k(K' \cap I'_{\underline{a}_2^{n_k}}) \to k^{\bar{a}_2}(K').$$

Since $(\tilde{\lambda}k^{\bar{a}_1} + t, k^{\bar{a}_2})$ is a stably intersecting configuration and this property is open, we can use the fact that $g_s(\cdot)$ is C^1-close to $g_{s_0}(\cdot)$ for s close to s_0 to deduce that, for k large enough, the Cantor sets $B_k(g_s(K \cap I_{\underline{a}_1^{m_k}}))$ and $B_k(K' \cap I'_{\underline{a}_2^{n_k}})$ have non-empty intersection. Hence, $g_s(K \cap I_{\underline{a}_1^{m_k}})$ and $K' \cap I'_{\underline{a}_2^{n_k}}$ have non-empty intersection, so that the argument is complete. $\square$

The following example shows that the property V in the Theorem 5.5 is fundamental.

Example 5.1. Consider the Cantor set $K_\alpha := \bigcap_{n \geq 0} \psi^{-n}(I_1 \cup I_2)$, where

$$\psi(x) = \begin{cases} \frac{2}{1-\alpha}x & \text{if } x \in I_1 := [0, \frac{1-\alpha}{2}]; \\ \\ -\frac{2}{1-\alpha}x + \frac{2}{1-\alpha} & \text{if } x \in I_2 := [\frac{1+\alpha}{2}, 1]. \end{cases}$$

Since $HD(K_\alpha) = -\frac{\log 2}{\log(\frac{1-\alpha}{2})}$ (cf. [PT (1993)]), $HD(K_\alpha) > 1/2$ whenever $\alpha < 1/2$. On the other hand, it is known that $K_\alpha - K_\alpha$ has measure zero for $1/3 < \alpha < 1/2$.

In particular, $HD(K_\alpha \times K_\alpha) > 1$ and $f(x, y) = x - y$ (so that most of the hypothesis of Theorem 5.5 are verified), but int $f(K_\alpha \times K_\alpha) = \emptyset$ for $1/3 < \alpha < 1/2$.

For later reference, let us record here the following immediate corollary of Theorem 5.5:

Corollary 5.3. *Let φ be a C^2-diffeomorphism, and Λ a horseshoe associated to φ. Suppose that K^s, K^u satisfy the hypotheses of Theorem 5.5, and consider the open and dense subset*

$$\mathcal{A}_\Lambda := \{f \in C^1(M, \mathbb{R}) : \exists z = (z^s, z^u) \in \Lambda \quad \text{such that} \quad Df(z).e_z^{s,u} \neq 0 \}$$

of $C^1(M, \mathbb{R})$. Then, for all $f \in \mathcal{A}_\Lambda$, we have int $f(\Lambda) \neq \emptyset$.

5.2.3   *End of proof of Theorem 5.1*

A fundamental result due to Moreira-Yoccoz in [MY (2010)] says that the stable and unstable Cantor sets K^s and K^u attached to a typical thick horseshoe are defined by a pair (g^s, g^u) in the class V:

Theorem 5.6 (Moreira–Yoccoz). *Suppose that we have a horseshoe with Hausdorff dimension > 1 such that its stable and unstable Cantor sets K^s, K^u are defined by g^s, g^u. If $\mathcal{U}$ is a sufficiently small neighborhood of φ_0 in $Diff^\infty(M)$, then there is an open and dense $\mathcal{U}^* \subset \mathcal{U}$ such that for any $\varphi \in \mathcal{U}^*$ the corresponding pair of expanding applications $(g^s_\varphi, g^u_\varphi)$ belongs to V.*

At this point, we are ready to complete the proof of Theorem 5.1. Recall from §5.2.1 that, given a horseshoe Λ associated to a diffeomorphism φ and a height function $f \in H_\varphi$, we defined the subhorseshoe $\tilde{\Lambda} := \bigcap_{n \in \mathbb{Z}} \varphi^n(\Lambda \backslash R_{q_{\tilde{s}}})$

such that $HD(\tilde{\Lambda}) > 1$ (cf. Lemma 5.4). Moreover, we constructed two subsets $\tilde{\Lambda}_{i_0}$ and Λ'_{j_0} with $HD(\tilde{\Lambda}_{i_0})$, $HD(\Lambda'_{j_0}) > 1$ (cf. (5.4) and (5.8)) such that

$$\{f(\varphi^{i_0}(\tilde{A}(x))) : x \in \tilde{\Lambda}_{i_0}\} \subset M(f, \Lambda) \text{ and } \{f(\varphi^{j_0}(\tilde{A}(x))) : x \in \Lambda'_{j_0}\} \subset L(f, \Lambda)$$

(cf. Corollaries 5.1 and 5.2) for height functions $f \circ \varphi^{i_0} \circ \tilde{A}$ and $f \circ \varphi^{j_0} \circ \tilde{A}$ in $\mathcal{A}_{\tilde{\Lambda}_{i_0}}$ and $\mathcal{A}_{\Lambda'_{j_0}}$ (cf. Remark 5.2). Therefore, the desired theorem follows from Theorem 5.6 and Corollary 5.3.

5.3 Hall's ray in dynamical spectra of geodesic flows on negatively curved manifolds

5.3.1 *Hall's ray in the dynamical spectra of Veech surfaces*

Recall that a translation surface X is obtained from a collection of $2n$ Euclidean canonically oriented triangles after gluing by translations $3n$ pairs of parallel sides with the same lengths and opposite orientations. Also, the group $SL(2, \mathbb{R})$ acts on translation surfaces, and the stabilizer $SL(X)$ in $SL(2, \mathbb{R})$ of a translation surface X is called its Veech group.

A Veech surface X is a translation surface whose Veech group is a lattice in $SL(2, \mathbb{R})$. For example, a square-tiled surface (i.e., a translation surface built from a finite collection of unit squares of $\mathbb{R}^2$ with sides identified by translations) is a Veech surface because its Veech group is a finite-index subgroup of $SL(2, \mathbb{Z})$.

The Lagrange spectrum L_X associated to a Veech surface is

$$L_X := \left\{ \limsup_{t \to \infty} \frac{2 \operatorname{Area}(Y)}{\operatorname{sys}(Y)^2} : Y \in SL(2, \mathbb{R}) \cdot X \right\},$$

where $\operatorname{Area}(Y)$ is the total area of Y and $\operatorname{sys}(Y)$ is the systole of Y. In particular, it follows from Proposition 1.9 in Chapter 1 says that $L_{\mathbb{T}^2}$ is the classical Lagrange spectrum, where $\mathbb{T}^2 = \mathbb{R}^2/\mathbb{Z}^2$.

In the article [HMU (2015)], Hubert–Marchese–Ulcigrai showed that the Lagrange spectra L_X associated to square-tiled surfaces X always contain a Hall's ray $[r(X), \infty)$ for some explicit constant $r(X) > 0$. Later on, Artigiani–Marchese–Ulcigrai [AMU (2016)] showed that the Lagrange spectrum L_X of any Veech surface contains a Hall's ray $[R(X), \infty)$ for some explicit constant $R(X) > 0$.

5.3.2   Uniform Hall ray in Lagrange spectra of negatively curved manifolds

We consider again the setting of Chapter 4, §4.3.2, i.e., X is a smooth, complete, non-compact, Riemannian manifold with finite volume and pinched negative curvature, e is a cusp of X, β_e is the Busemann function associated to the maximal Margulis neighborhood of e, and the height of a locally geodesic line γ in X is

$$\mathrm{ht}_e(\gamma) = \sup_{t \in \mathbb{R}} \beta_e(\gamma(t)).$$

In this context, we have a Markov spectrum

$$M_{X,e} := \{0 \le \mathrm{ht}_e(\gamma) < \infty : \gamma \text{ locally geodesic line in } X\}.$$

It was proved by Parkkonen–Paulin in [PP (2010)] that $[4.2, \infty) \subset M_{X,e}$ when X has dimension > 2.

5.3.3   Hall's ray in Lagrange spectra of hyperbolic surfaces

Let X be a hyperbolic surface of finite area, i.e., $X = \mathbb{H}/\Gamma$ where Γ is a lattice of $SL(2, \mathbb{R})$. Given a cusp of X, we can always perform a change of coordinates to assume that its maximal Margulis neighborhood is the projection of $\{z \in \mathbb{H} : 0 \le \mathrm{Re}(z) \le \delta, \mathrm{Im}(z) \ge \eta\}$ to X.

It was proved by Artigiani–Marchese–Ulcigrai in [AMU (2016)] that there is a Hall ray in the Lagrange spectrum

$$L_h := \left\{ \limsup_{t \to \infty} h(\gamma(t)) < \infty : \gamma \text{ geodesic line in } X \right\}$$

associated to the geodesic flow on X and any height function $h : X \to \mathbb{R}$ sufficiently Lipschitz close to $\mathrm{Im}(.)$ in the maximal Margulis neighborhood of the cusp.

Appendix A

Proof of Hurwitz theorem

Given $\alpha \notin \mathbb{Q}$, we want to show that the inequality

$$\left| \alpha - \frac{p}{q} \right| \leq \frac{1}{\sqrt{5}q^2}$$

has infinitely many rational solutions.

In this direction, let $\alpha = [a_0; a_1, \ldots]$ be the continued fraction expansion of α and denote by $[a_0; a_1, \ldots, a_n] = p_n/q_n$. We affirm that, for every $\alpha \notin \mathbb{Q}$ and every $n \geq 1$, we have

$$\left| \alpha - \frac{p}{q} \right| < \frac{1}{\sqrt{5}q^2}$$

for some $\frac{p}{q} \in \{ \frac{p_{n-1}}{q_{n-1}}, \frac{p_n}{q_n}, \frac{p_{n+1}}{q_{n+1}} \}$.

Remark A.1. Of course, this last statement provides infinitely many solutions to the inequality $\left| \alpha - \frac{p}{q} \right| \leq \frac{1}{\sqrt{5}q^2}$. So, our task is reduced to prove the affirmation above.

The proof of the claim starts by recalling Perron's Proposition 1.5:

$$\alpha - \frac{p_n}{q_n} = \frac{(-1)^n}{(\alpha_{n+1} + \beta_{n+1})q_n^2}$$

where $\alpha_{n+1} := [a_{n+1}; a_{n+2}, \ldots]$ and $\beta_{n+1} = \frac{q_{n-1}}{q_n} = [0; a_n, \ldots, a_1]$.

For the sake of contradiction, suppose that the claim is false, i.e., there exists $k \geq 1$ such that

$$\max\{(\alpha_k + \beta_k), (\alpha_{k+1} + \beta_{k+1}), (\alpha_{k+2} + \beta_{k+2})\} \leq \sqrt{5}. \tag{A.1}$$

Since $\sqrt{5} < 3$ and $a_m \leq \alpha_m + \beta_m$ for all $m \geq 1$, it follows from (A.1) that

$$\max\{a_k, a_{k+1}, a_{k+2}\} \leq 2. \tag{A.2}$$

If $a_m = 2$ for some $k \leq m \leq k+2$, then (A.2) would imply that $\alpha_m + \beta_m \geq 2 + [0; 2, 1] = 2 + \frac{1}{3} > \sqrt{5}$, a contradiction with our assumption (A.1).

So, our hypothesis (A.1) forces

$$a_k = a_{k+1} = a_{k+2} = 1. \tag{A.3}$$

Denoting by $x = \frac{1}{\alpha_{k+2}}$ and $y = \beta_{k+1} = q_{k-1}/q_k \in \mathbb{Q}$, we have from (A.3) that

$$\alpha_{k+1} = 1 + x, \quad \alpha_k = 1 + \frac{1}{1+x}, \quad \beta_k = \frac{1}{y} - 1, \quad \beta_{k+2} = \frac{1}{1+y}.$$

By plugging this into (A.1), we obtain

$$\max\left\{\frac{1}{1+x} + \frac{1}{y}, 1 + x + y, \frac{1}{x} + \frac{1}{1+y}\right\} \leq \sqrt{5}. \tag{A.4}$$

On one hand, (A.4) implies that

$$\frac{1}{1+x} + \frac{1}{y} \leq \sqrt{5} \quad \text{and} \quad 1 + x \leq \sqrt{5} - y.$$

Thus,

$$\frac{\sqrt{5}}{y(\sqrt{5} - y)} = \frac{1}{\sqrt{5} - y} + \frac{1}{y} \leq \frac{1}{1+x} + \frac{1}{y} \leq \sqrt{5},$$

and, *a fortiori*, $y(\sqrt{5} - y) \geq 1$, i.e.,

$$\frac{\sqrt{5} - 1}{2} \leq y \leq \frac{\sqrt{5} + 1}{2}. \tag{A.5}$$

On the other hand, (A.4) implies that

$$x \leq \sqrt{5} - 1 - y \quad \text{and} \quad \frac{1}{x} + \frac{1}{1+y} \leq \sqrt{5}.$$

Hence,

$$\frac{\sqrt{5}}{(1+y)(\sqrt{5} - 1 - y)} = \frac{1}{\sqrt{5} - 1 - y} + \frac{1}{1+y} \leq \frac{1}{x} + \frac{1}{1+y} \leq \sqrt{5},$$

and, *a fortiori*, $(1+y)(\sqrt{5} - 1 - y) \geq 1$, i.e.,

$$\frac{\sqrt{5} - 1}{2} \leq y \leq \frac{\sqrt{5} + 1}{2}. \tag{A.6}$$

It follows from (A.5) and (A.6) that $y = (\sqrt{5} - 1)/2$, a contradiction because $y = \beta_{k+1} = q_{k-1}/q_k \in \mathbb{Q}$. This completes the argument.

Appendix B

Proof of Euler's remark

Denote by $[0; a_1, a_2, \ldots, a_n] = \frac{p(a_1, \ldots, a_n)}{q(a_1, \ldots, a_n)} = \frac{p_n}{q_n}$. It is not hard to see that

$$q(a_1) = a_1, \quad q(a_1, a_2) = a_1 a_2 + 1$$

and in general

$$q(a_1, \ldots, a_n) = a_n q(a_1, \ldots, a_{n-1}) + q(a_1, \ldots, a_{n-2}) \ \forall \, n \geq 3.$$

From this formula, we see that $q(a_1, \ldots, a_n)$ is a sum of the following products of elements of $\{a_1, \ldots, a_n\}$. First, we take the product $a_1 \ldots a_n$ of all a_i's. Secondly, we take all products obtained by removing any pair $a_i a_{i+1}$ of adjacent elements. Then, we iterate this procedure until no pairs can be omitted (with the convention that if n is even, then the empty product gives 1). This rule to describe $q(a_1, \ldots, a_n)$ was discovered by Euler.

It follows immediately from Euler's rule that

$$q(a_1, \ldots, a_n) = q(a_n, \ldots, a_1).$$

This proves Proposition 3.1.

Appendix C

Continued fractions, binary quadratic forms, and Markov spectrum

We remember that $\mathcal{Q} = \{q(x,y); q$ is an indefinite binary quadratic form with $\Delta(q) > 0\}$, where $\Delta(q) = b^2 - 4ac$. The main result of this appendix is:

Theorem C.1. *The Markov spectrum*

$$\left\{ \frac{\sqrt{b^2 - 4ac}}{\displaystyle\inf_{(x,y)\in\mathbb{Z}^2\setminus\{(0,0)\}} |q(x,y)|} \in \mathbb{R} : q(x,y) \in \mathcal{Q} \right\}$$

coincides with the set

$$\left\{ \sup_{n\in\mathbb{Z}}([g_n; g_{n+1}, \dots] + [0; g_{n-1}, g_{n-2}, \dots]) < \infty : (g_m)_{m\in\mathbb{Z}} \in (\mathbb{N}^*)^{\mathbb{Z}} \right\}.$$

Remark C.1. Recall that the definition of the Markov spectrum concerns only the real, indefinite, binary quadratic forms $q(x,y) = ax^2 + bxy + cy^2$ with $b^2 - 4ac > 0$ such that the quantity

$$\frac{\sqrt{b^2 - 4ac}}{\displaystyle\inf_{(x,y)\in\mathbb{Z}^2\setminus\{(0,0)\}} |ax^2 + bxy + cy^2|}$$

is *finite*. In particular, we are *excluding* real binary quadratic forms q with $0 \in q(\mathbb{Z}^2 \setminus \{(0,0)\})$.

In the sequel, we follow the books of Dickson [D (1957)] and Cusick–Flahive [CF (1989)] in order to give a proof of this result via the classical *reduction theory of binary quadratic forms*.

C.1 Generalities on binary quadratic forms

From the point of view of linear algebra, a binary quadratic form is

$$q(x,y) = ax^2 + bxy + cy^2 = \langle Mv, v \rangle = v^t M v \qquad \text{(C.1)}$$

where $\langle .,. \rangle$ is the usual Euclidean inner product of $\mathbb{R}^2$, v is the column vector $v = \begin{pmatrix} x \\ y \end{pmatrix}$, and M is the matrix $M = \begin{pmatrix} a & b/2 \\ b/2 & c \end{pmatrix}$.

The *discriminant* of $q(x,y) = ax^2 + bxy + cy^2 = \langle Mv, v \rangle$ is $\Delta(q) = d :=$ $b^2 - 4ac = -4 \det(M)$.

Remark C.2. The values taken by $q(x,y) = bxy$ are very easy to describe: in particular, $0 \in q(\mathbb{Z}^2 \setminus \{(0,0)\})$ in this context. Hence, by Remark C.1, we can focus on $q(x,y) = ax^2 + bxy + cy^2$ with $a \neq 0$ or $c \neq 0$. Moreover, by symmetry (i.e., exchanging the roles of the variables x and y), we can assume that $a \neq 0$. So, *from now on*, we shall assume that $a \neq 0$.

Note that $4aq(x,y) = (2ax + by)^2 - dy^2$. Therefore, q is *definite* (i.e., its values are all positive or all negative) whenever the discriminant is $d \leq 0$. Conversely, when $a \neq 0$, q is *indefinite* (i.e., it takes both positive and negative values) whenever $d > 0$. In view of the definition of the Markov spectrum, we will restrict ourselves *from now on* to the indefinite case $d > 0$.

Observe also that $q(x,y) = a \prod_{a\omega^2 + b\omega + c = 0} (x - \omega y) = a(x - fy)(x - sy)$ where

$$f := \frac{\sqrt{d} - b}{2a} \quad \text{and} \quad s := \frac{-\sqrt{d} - b}{2a} \tag{C.2}$$

are the *first* and *second roots* of $a\omega^2 + b\omega + c = 0$. In particular, $0 \in q(\mathbb{Z}^2 \setminus \{(0,0)\})$ when $f \in \mathbb{Q}$ or $s \in \mathbb{Q}$. So, by Remark C.1, we will suppose *from now on* that $f, s \notin \mathbb{Q}$.

In summary, **from now on**, our standing assumptions on $q(x,y) = ax^2 + bxy + cy^2$ are $a \neq 0$, $d := b^2 - 4ac > 0$ and $f := \frac{\sqrt{d}-b}{2a} \notin \mathbb{Q}$, $s := \frac{-\sqrt{d}-b}{2a} \notin \mathbb{Q}$.

Remark C.3. The first and second roots f, s and the discriminant $d > 0$ determine the coefficients a, b, c of the binary quadratic form $q(x,y)$. Indeed, the formula $q(x,y) = a(x - fy)(x - sy)$ says that it suffices to determine a. On one hand, the modulus $|a|$ is given by $4a^2(f - s)^2 = d$. On the other hand, the fact that f is the 1st root and s is the 2nd root determines the sign of a: if we replace a by $-a$ in the formula $q(x,y) = a(x - fy)(x - sy)$, then we obtain the binary quadratic form $-q$ whose 1st root is s and 2nd root is f; hence, an ambiguity on the sign of a could only occur when $f = s$, i.e., $d = 0$, in contradiction with our assumption $d > 0$.

C.2   Action of $GL_2(\mathbb{R})$ on binary quadratic forms

A key idea (going back to Lagrange) to study the values of a *fixed* binary quadratic form $q(x, y) = ax^2 + bxy + cy^2$ is to investigate the *equivalent* problem of describing the values of a *family* of binary quadratic forms on a fixed vector.

More precisely, if a vector $v = \begin{pmatrix} x \\ y \end{pmatrix} = T(V)$ is obtained from a fixed

vector $V = \begin{pmatrix} X \\ Y \end{pmatrix}$ via a matrix $T = \begin{pmatrix} \alpha & \beta \\ \gamma & \delta \end{pmatrix} \in GL_2(\mathbb{R})$, i.e., $x = \alpha X + \beta Y$,
$y = \gamma X + \delta Y$, then the value of q at v equals the value of $q \circ T$ at V. By (C.1), the binary quadratic forms q and $q \circ T$ are related by:

$$q \circ T(V) = (T(V))^t \cdot M \cdot T(V) = V^t(T^t M T)V \qquad \text{(C.3)}$$

where $M = \begin{pmatrix} a & b/2 \\ b/2 & c \end{pmatrix}$. In other terms, $q \circ T(X, Y) = AX^2 + BXY + CY^2$
where

$$\begin{pmatrix} A & B/2 \\ B/2 & C \end{pmatrix} = T^t M T = \begin{pmatrix} \alpha & \gamma \\ \beta & \delta \end{pmatrix} \begin{pmatrix} a & b/2 \\ b/2 & c \end{pmatrix} \begin{pmatrix} \alpha & \beta \\ \gamma & \delta \end{pmatrix}, \qquad \text{(C.4)}$$

i.e., $A = a\alpha^2 + b\alpha\gamma + c\gamma^2$, $B = 2a\alpha\beta + b(\alpha\delta + \beta\gamma) + 2c\gamma\delta$, and $C = a\beta^2 + b\beta\delta + c\delta^2$.

Observe that (C.3) implies that the discriminant of $q \circ T$ is $\det(T)^2 \cdot d$. Thus, $q \circ T$ and q have the *same* discriminant whenever $\det(T) = \pm 1$.

C.3   Action of $SL_2(\mathbb{Z})$ on binary quadratic forms

The quantity $\inf\{|z| : z \in q(\mathbb{Z}^2 \setminus \{(0,0)\})\}$ appearing in the statement of Theorem C.1 leads us to restrict our attention to the action of $SL_2(\mathbb{Z})$ on $q(x, y) = ax^2 + bxy + cy^2$. In fact, since $q(\lambda x, \lambda y) = \lambda^2 q(x, y)$, we have that

$$\inf\{|z| : z \in q(\mathbb{Z}^2 \setminus \{(0,0)\})\} = \inf\{|z| : z \in q(\mathbb{Z}^2_{prim})\} \qquad \text{(C.5)}$$

where $\mathbb{Z}^2_{prim} := \{(p, q) \in \mathbb{Z}^2 : \gcd(p, q) = 1\}$ is the set of *primitive* vectors of $\mathbb{Z}^2$. So, it is natural to concentrate on $SL_2(\mathbb{Z})$ because it acts transitively on $\mathbb{Z}^2_{prim}$.

Definition C.1. Two binary quadratic forms q and Q are *equivalent* whenever $Q = q \circ T$ for some $T \in SL_2(\mathbb{Z})$.

Note that two equivalent binary quadratic forms have the same discriminant. Furthermore, if $(x - \omega y)$ is a factor of $q(x, y) = ax^2 + bxy + cy^2$

and $x = \alpha X + \beta Y$, $y = \gamma X + \delta Y$ with $T = \begin{pmatrix} \alpha & \beta \\ \gamma & \delta \end{pmatrix} \in SL_2(\mathbb{Z})$, then
$(\alpha X + \beta Y - \omega(\gamma X + \delta Y))$ is a factor of $q \circ T(X, Y) = AX^2 + BXY + CY^2$
and, *a fortiori*, $(X - \frac{-\beta + \omega\delta}{\alpha - \omega\gamma} Y)$ is a factor of $q \circ T$. In particular, the roots
of $A\Omega^2 + B\Omega + C = 0$ are related to the roots of $a\omega^2 + b\omega + c = 0$ via

$$\Omega = \frac{-\beta + \omega\delta}{\alpha - \omega\gamma}. \tag{C.6}$$

In other words, ω and Ω are related to each other via the action of $T^{-1} = \begin{pmatrix} \delta & -\beta \\ -\gamma & \alpha \end{pmatrix}$ by Möbius transformations. Actually, a direct calculation with
the formulas for A and B in (C.4) and the fact that $\alpha\delta - \beta\gamma = 1$ show that
(C.6) respect the order of the roots, i.e.,

$$F = \frac{-\beta + f\delta}{\alpha - f\gamma} \quad \text{and} \quad S = \frac{-\beta + s\delta}{\alpha - s\gamma} \tag{C.7}$$

where f and F are the first roots, and s and S are the 2nd roots.

Remark C.4. Note that $\alpha - \omega\gamma \neq 0$ because $(\alpha, \gamma) \in \mathbb{Z}^2_{prim}$ and we are
assuming that the roots of $a\omega^2 + b\omega + c = 0$ are irrational.

Lemma C.1. *Under our standing assumptions, any binary quadratic form
q is equivalent to some $ax^2 + bxy + cy^2$ with $|b| \leq |a| \leq \sqrt{d/3}$.*

Proof. We will proceed in two steps: first, we apply certain elements of
$SL_2(\mathbb{Z})$ to ensure that $|a| \leq \sqrt{d/3}$, and, after that, we use a parabolic
matrix to obtain $|b| \leq |a|$.

By (C.4), the matrix $H_0 = \begin{pmatrix} h_0 & 1 \\ -1 & 0 \end{pmatrix} \in SL_2(\mathbb{Z})$ converts $q(x, y) = a_0 x^2 + b_0 xy + c_0 y^2$ into $q \circ H_0(X, Y) = a_1 X^2 + b_1 XY + a_0 Y^2$ with $b_1 = 2a_0 h_0 - b_0$.

If $|a_0| > \sqrt{d/3}$, the choice of $h_0 \in \mathbb{Z}$ with $|2a_0 h_0 - b_0| \leq |a_0|$ in the
discussion above leads to $q \circ H_0(X, Y) = a_1 X^2 + b_1 XY + a_0 Y^2$ with

$$4a_1 a_0 = b_1^2 - d \leq b_1^2 = (2a_0 h_0 - b_0)^2 \leq a_0^2 \quad \text{and} \quad -4a_1 a_0 = d - b_1^2 \leq d < 3a_0^2.$$

In other terms, if $|a_0| > \sqrt{d/3}$, then q is equivalent to $a_1 X^2 + b_1 XY + a_0 Y^2$
with $|4a_1 a_0| < 3a_0^2$, i.e., $|a_1| < (3/4)|a_0|$.

By iterating this process, we find a sequence $a_n X^2 + b_n XY + a_{n-1} Y^2$ of
binary quadratic forms equivalent to q such that $|a_n| < (3/4)^n |a_0|$ whenever
$|a_{n-1}| > \sqrt{d/3}$. It follows that q is equivalent to some $ax^2 + \tilde{b}xy + \tilde{c}y^2$ with
$|a| \leq \sqrt{d/3}$.

Finally, by (C.4), the matrix $P = \begin{pmatrix} 1 & k \\ 0 & 1 \end{pmatrix} \in SL_2(\mathbb{Z})$ converts $ax^2 +$
$\tilde{b}xy + \tilde{c}y^2$ into $aX^2 + bXY + cY^2$ with $b = \tilde{b} + 2ka$. Hence, the choice of
$k \in \mathbb{Z}$ with $|\tilde{b} + 2ka| \le |a|$ gives that q is equivalent to $ax^2 + bxy + cy^2$ with
$|b| \le |a| \le \sqrt{d/3}$. $\qquad\square$

C.4 Reduction theory (I)

In general, the study of the dynamics of the action of a group G on a certain
space $\mathcal{M}$ is greatly improved in the presence of nice *fundamental domain*,
i.e., a portion $\mathcal{D} \subset \mathcal{M}$ with good geometrical properties capturing all orbits
in the sense that the G-orbit of any $x \in \mathcal{M}$ intersects $\mathcal{D}$.

In the setting of $SL_2(\mathbb{Z})$ acting on binary quadratic forms, the role of
fundamental domain is played by the notion of *reduced* binary quadratic
form:

Definition C.2. We say that $q(x, y) = ax^2 + bxy + cy^2$ is *reduced* whenever

$$\frac{|\sqrt{d} - b|}{2|a|} = |f| < 1, \quad \frac{|\sqrt{d} + b|}{2|a|} = |s| > 1 \quad \text{and} \quad f \cdot s < 0.$$

Remark C.5. Suppose that q is reduced. Then, $d - b^2 = (\sqrt{d} - b)(\sqrt{d} + b) = -4a^2 f s > 0$ thanks to the condition $f \cdot s < 0$. In particular, $|b| < \sqrt{d}$, so
that $\sqrt{d} - b, \sqrt{d} + b > 0$. Hence, $0 < \sqrt{d} - b < 2|a| < \sqrt{d} + b$ thanks to
the condition $|f| < 1 < |s|$. Note that the inequality $\sqrt{d} - b < \sqrt{d} + b$
says that $b > 0$. In other words, q reduced implies $(0 < b < \sqrt{d}$ and)
$0 < \sqrt{d} - b < 2|a| < \sqrt{d} + b$.

Conversely, the inequality $0 < \sqrt{d} - b < 2|a| < \sqrt{d} + b$ implies that
$(0 < b < \sqrt{d}$ and) q is reduced.

Therefore, q is reduced *if and only if*

$$0 < \sqrt{d} - b < 2|a| < \sqrt{d} + b.$$

Furthermore, the identity $|\sqrt{d} - b| \cdot |\sqrt{d} + b| = 4|ac|$ implies that q is
reduced *if and only if*

$$0 < \sqrt{d} - b < 2|c| < \sqrt{d} + b.$$

In particular, q is reduced if and only $q \circ R$ is reduced where $R \in GL_2(\mathbb{Z})$
is the matrix $R = \begin{pmatrix} 0 & 1 \\ 1 & 0 \end{pmatrix}$ inducing the change of variables $x = Y, y = X$.

Moreover, if q is reduced, then $f \cdot a = (\sqrt{d} - b)/2 > 0$ and $c/a = f \cdot s < 0$,
i.e., the sign of c is opposite to the signs of f and a.

The next result says that reduced forms are a fundamental domain for the $SL_2(\mathbb{Z})$ action on binary quadratic forms:

Theorem C.2. *Under our standing assumptions, any binary quadratic form q is equivalent to some reduced binary quadratic form.*

Proof. By Lemma C.1, we can assume that $q(x, y) = ax^2 + bxy + cy^2$ with $|b| \leq |a| \leq \sqrt{d/3} \leq \sqrt{d}$, so that $|4ac| = d - b^2 \leq d$. Thus, $\min\{2|a|, 2|c|\} \leq \sqrt{d}$. By performing the change of variables $x = Y$, $y = -X$ (i.e., applying the matrix $\begin{pmatrix} 0 & 1 \\ -1 & 0 \end{pmatrix} \in SL_2(\mathbb{Z})$) if necessary, we can suppose that $2|a| \leq \sqrt{d}$.

Since we are assuming that $a \neq 0$, it is possible to choose $k \in \mathbb{Z}$ such that $0 \leq \sqrt{d} - 2|a| \leq \widetilde{b} := b + 2ak \leq \sqrt{d}$. So, if we apply the matrix $\begin{pmatrix} 1 & k \\ 0 & 1 \end{pmatrix} \in SL_2(\mathbb{Z})$ (i.e., $x = X + kY$, $y = Y$) to q, then we get the equivalent binary quadratic form $ax^2 + \widetilde{b}xy + \widetilde{c}$ with $0 \leq \sqrt{d} - \widetilde{b} \leq 2|a| \leq \sqrt{d} \leq \sqrt{d} + \widetilde{b}$.

We affirm that $ax^2 + \widetilde{b}xy + \widetilde{c}$ is reduced. In fact, by Remark C.5, our task is to show that we have strict inequalities

$$0 < \sqrt{d} - \widetilde{b} < 2|a| < \sqrt{d} + \widetilde{b}.$$

However, $0 = \sqrt{d} - b$ or $\sqrt{d} - \widetilde{b} = 2|a|$ or $2|a| = \sqrt{d} + \widetilde{b}$ would imply that 0 or ± 1 is a root of $a\omega^2 + \widetilde{b}\omega + \widetilde{c} = 0$, a contradiction with our assumption that none of its roots is rational. Hence, $ax^2 + \widetilde{b}xy + \widetilde{c}$ is reduced. $\square$

C.5   Reduction theory (II)

In this section, we use the matrices $L_\delta := \begin{pmatrix} 0 & 1 \\ -1 & \delta \end{pmatrix} \in SL_2(\mathbb{Z})$ inducing the change of variables $x = Y$, $y = -X + \delta Y$ (similar to an Euclidean division) to navigate through the set of reduced binary quadratic forms (cf. Lemma C.2 below).

Before explaining this fact, let us introduce the following definition: we say that the binary quadratic forms $q \circ L_\delta$, resp. $q \circ L_\delta^{-1}$, $\delta \in \mathbb{Z}$, are the *right*, resp. *left neighbors* of the binary quadratic form q.

The notion of neighbors is adapted to the investigation of reduced binary quadratic forms because of the following result:

Lemma C.2. *Under our standing assumptions, a reduced binary quadratic form has an unique reduced right neighbor and an unique reduced left neighbor.*

Proof. Let us first show the existence and uniqueness of the reduced right neighbor of a reduced binary quadratic form $q(x, y) = ax^2 + bxy + cy^2$.

Consider the 1st and 2nd roots f and s of $aw^2 + bw + c = 0$. Since q is reduced, i.e., $|f| < 1 < |s|$ and $f \cdot s < 0$, the largest integer $|\delta| < 1/|f|$ is not zero. We take $\delta \in \mathbb{Z}$ with the same sign as f and a and, hence, opposite to c (cf. Remark C.5).

By (C.4) and (C.7), the right neighbor $q \circ L_\delta$ is $cX^2 + \widetilde{b}XY + \widetilde{c}Y^2$ with $\widetilde{b} = -b - 2\delta c$ and the roots of $c\Omega^2 + \widetilde{b}\Omega + \widetilde{c} = 0$ satisfy $F = \delta - \frac{1}{f}$ and $S = \delta - \frac{1}{s}$.

Thus, our choice of δ implies that $|F| < 1$ and F has opposite sign to f and δ. Since $f \cdot s < 0$, s has opposite sign to (f and) δ, we have that S has the same sign of δ and $|S| > |\delta| \geq 1$. In summary, the right neighbor $q \circ L_\delta$ of q is reduced.

Moreover, this reduced right neighbor is unique because $|f|, |F| < 1$ and $F = \delta - \frac{1}{f}$ imply that $\delta \in \mathbb{Z}$ must be the unique integer with the same sign of f and $|\delta| = \lfloor 1/|f| \rfloor$.

Finally, let us establish the existence and the uniqueness of the reduced left neighbor of $q(x, y) = ax^2 + bxy + cy^2$. Consider the matrix $R = \begin{pmatrix} 0 & 1 \\ 1 & 0 \end{pmatrix}$ (with determinant -1) inducing the change of variables $x = Y$ and $y = X$. By Remark C.5, the fact that q is reduced implies that $q \circ R(X, Y) = cX^2 + bXY + aY^2$ is also reduced. In particular, $q \circ R$ has an unique reduced right neighbor $q \circ R \circ L_\delta$. In particular, $q \circ R \circ L_\delta \circ R$ is reduced. Since

$$RL_\delta R = \begin{pmatrix} 0 & 1 \\ 1 & 0 \end{pmatrix} \begin{pmatrix} 0 & 1 \\ -1 & \delta \end{pmatrix} \begin{pmatrix} 0 & 1 \\ 1 & 0 \end{pmatrix} = \begin{pmatrix} \delta & -1 \\ 1 & 0 \end{pmatrix} = L_\delta^{-1},$$

we conclude that $q \circ L_\delta^{-1}$ is the unique left neighbor of q. $\square$

This lemma allows us to organize the reduced binary quadratic forms into chains:

Definition C.3. A *chain* $\ldots, \Phi_{-1}, \Phi_0, \Phi_1, \ldots$ is a sequence of reduced binary quadratic forms such that Φ_{i+1} is the right neighbor Φ_i for all $i \in \mathbb{Z}$.

C.6   Reduction theory (III)

The proof of Lemma C.2 hints to a connection between chains and continued fraction expansions. More concretely, let $\{\Phi_i\}_{i \in \mathbb{Z}}$ be a chain, say $\Phi_i(x, y) = a_i x^2 + b_i xy + c_i y^2$ and the change of variables $x = Y$, $y = -X + \delta_i Y$ converts

Φ_i into Φ_{i+1} (i.e., $\Phi_{i+1} = \Phi_i \circ L_{\delta_i}$). Note that $a_{i+1} = c_i$ for all $i \in \mathbb{Z}$. By Remark C.5, since Φ_i is reduced, it follows that the sign of $a_{i+1} = c_i$ is opposite to the sign of a_i. In particular, up to replacing each Φ_i by its right neighbor Φ_{i+1}, we can assume that $a_0 > 0$.

In this context, we write $\Phi_i(x, y) = (-1)^i A_i + B_i xy + (-1)^{i+1} A_{i+1}$ with $A_i > 0$ for all $i \in \mathbb{Z}$. Since $\Phi_{i+1} = \Phi_i \circ L_{\delta_i}$, we have

$$B_{i+1} + B_i = 2g_i A_{i+1} \tag{C.8}$$

where $g_i = (-1)^i \delta_i$. As $B_i > 0$ (because Φ_i is reduced) and $A_i > 0$ for all $i \in \mathbb{Z}$, the previous equation implies that

$$g_i > 0 \tag{C.9}$$

for all $i \in \mathbb{Z}$. Moreover, $\Phi_{i+1} = \Phi_i \circ L_{\delta_i}$ means that the roots of $a_j \omega^2 + b_j \omega + c_j = 0$ satisfy

$$f_{i+1} = \delta_i - \frac{1}{f_i} \quad \text{and} \quad s_{i+1} = \delta_i - \frac{1}{s_i}.$$

In other words,

$$F_i = g_i + \frac{1}{F_{i+1}} \quad \text{and} \quad S_i = \frac{1}{g_{i-1} + S_{i-1}} \tag{C.10}$$

where $F_i := (-1)^i / f_i$ and $S_i = (-1)^{i+1} / s_i$.

In particular, $F_i = [g_i; g_{i+1}, \dots] := g_i + \cfrac{1}{g_{i+1} + \cfrac{1}{\ddots}}$ and $S_i = [0; g_{i-1}, g_{i-2}, \dots]$ where $g_i \in \mathbb{N}^*$ describe the entries of the continued fraction expansions of

$$\frac{1}{f_0} = [g_0; g_1, g_2, \dots] \quad \text{and} \quad -\frac{1}{s_0} = [0; g_{-1}, g_{-2}, \dots].$$

Moreover, the fact that $f_i = \frac{\sqrt{d} - B_i}{2(-1)^i A_i}$ and $s_i = \frac{-\sqrt{d} - B_i}{2(-1)^i A_i}$ (where d is the common discriminant of Φ_i), we have

$$[g_i; g_{i+1}, \dots] + [0; g_{i-1}, g_{i-2}, \dots] = F_i + S_i = 2A_i \left(\frac{1}{\sqrt{d} - B_i} + \frac{1}{\sqrt{d} + B_i} \right)$$

$$= 2A_i \frac{2\sqrt{d}}{d - B_i^2} = \frac{4A_i\sqrt{d}}{-4(-1)^i A_i (-1)^{i+1} A_{i+1}}$$

$$= \frac{\sqrt{d}}{A_{i+1}},$$

that is

$$[g_i; g_{i+1}, \dots] + [0; g_{i-1}, g_{i-2}, \dots] = \frac{\sqrt{d}}{|\Phi_{i+1}(1, 0)|}. \tag{C.11}$$

C.7   Reduction theory (IV)

Coming back to our study of binary quadratic forms, we saw that, under our standing assumption, any binary quadratic form is equivalent to *some* chain of reduced forms.

The next *fundamental* result says that any equivalence class of binary quadratic forms contains an *unique* chain.

Theorem C.3. *Two equivalent reduced binary quadratic forms belong to the same chain.*

As it turns out, one can deduce Theorem C.1 *directly* from (C.11) and this theorem. For this reason, let us *postpone* the proof of Theorem C.3 to the last portion of this section in order to complete the proof of Theorem C.1.

C.8   End of proof of Theorem C.1 (modulo Theorem C.3)

Lagrange used Theorem C.3 to obtain the following result:

Theorem C.4. *Let $\{\Phi_i\}_{i \in \mathbb{Z}}$ be a chain of reduced binary quadratic forms with discriminant d. Then,*

$$\bigcup_{j \in \mathbb{Z}} \{\Phi_j(n,m) : (n,m) \in \mathbb{Z}^2_{prim}, |\Phi_j(n,m)| < \sqrt{d}/2\} \subset \{\Phi_i(1,0) : i \in \mathbb{Z}\}.$$

Proof. Fix $j \in \mathbb{Z}$ and $(n,m) \in \mathbb{Z}^2_{prim}$ such that $a := \Phi_j(n,m)$ verifies

$$|a| < \sqrt{d}/2.$$

Take $T \in SL_2(\mathbb{Z})$ sending $(1,0)$ to (n,m). By definition, $a = \Phi_j \circ T(1,0)$. In other terms, Φ_j is equivalent to $\Phi_j \circ T(x,y) = ax^2 + Bxy + Cy^2$.

Similarly to the proof of Theorem C.2, we can take $k \in \mathbb{Z}$ such that $b := B + 2ak$ satisfies $\sqrt{d} - 2|a| < b < \sqrt{d}$, so that the matrix $\begin{pmatrix} 1 & k \\ 0 & 1 \end{pmatrix}$ converts $\Phi_j \circ T$ into the equivalent binary quadratic form $Q(X,Y) = aX^2 + bXY + cY^2$. Note that Q is reduced because $2|a| < \sqrt{d}$ and $\sqrt{d} - 2|a| < b < \sqrt{d}$.

In summary, Φ_j and Q are equivalent reduced binary quadratic forms. By Theorem C.3, Φ_j and Q belong to the same chain, i.e., there exists $i \in \mathbb{Z}$ such that $Q = \Phi_i$. By definition, this means that $\Phi_j(n,m) := a = \Phi_i(1,0)$. $\qquad\square$

This result yields the following fact about the values of binary quadratic forms:

Proposition C.1. *Let $q(x,y) = ax^2 + bxy + cy^2$ be a binary quadratic form satisfying our standing assumptions. Denote by $\{\Phi_i\}_{i \in \mathbb{Z}}$ the unique chain in the equivalence class of q (coming from Theorems C.2 and C.3). Then,*

$$\inf\{|z| : z \in q(\mathbb{Z}^2_{prim})\} = \inf\{|\Phi_i(1,0)| : i \in \mathbb{Z}\}.$$

Proof. Since $q(\mathbb{Z}^2_{prim}) = q \circ T(\mathbb{Z}^2_{prim})$ for all $T \in SL_2(\mathbb{Z}^2_{prim})$, there is no loss of generality in applying Theorem C.2 in order to assume that $q(x,y) = ax^2 + bxy + cy^2$ is reduced, say $q = \Phi_j$. In this situation, $4|ac| = d - b^2 < d$, so that $\min\{2|a|, 2|c|\} < \sqrt{d}$, i.e., $\min\{|q(1,0)|, |q(0,1)|\} < \sqrt{d}/2$. Hence,

$$\inf\{|z| : z \in q(\mathbb{Z}^2_{prim})\} = \inf\{|z| : j \in \mathbb{Z}, z \in \Phi_j(\mathbb{Z}^2_{prim}), |z| < \sqrt{d}/2\}.$$

The desired conclusion now follows from Theorem C.4. $\square$

The proof of Theorem C.1 is now very easy to complete: indeed, it suffices to put together (C.5), Proposition C.1 and (C.11).

At this stage, it remains only to prove Theorem C.3. This matter is the purpose of the remainder of this section.

C.9   Reduction theory (IV) bis: proof of Theorem C.3

Before starting our discussion of Theorem C.3, we need to recall the following four basic properties of continued fractions proved in Chapter 1, Section 1.5 above:

Proposition C.2. *Given a sequence $g_i \in \mathbb{N}^*$, $i \in \mathbb{N}$, let $p_k/q_k :=$ $[g_0; g_1, \ldots, g_k]$ be the kth convergent of $[g_0; g_1, \ldots]$. Then:*

- *one has the recurrence relation* $\begin{pmatrix} p_{k+1} & p_k \\ q_{k+1} & q_k \end{pmatrix} \begin{pmatrix} g_{k+2} & 1 \\ 1 & 0 \end{pmatrix} =$ $\begin{pmatrix} p_{k+2} & p_{k+1} \\ q_{k+2} & q_{k+1} \end{pmatrix}$ *with the convention that* $\begin{pmatrix} p_{-1} & p_{-2} \\ q_{-1} & q_{-2} \end{pmatrix} = \begin{pmatrix} 1 & 0 \\ 0 & 1 \end{pmatrix}$;
- $[g_0; g_1, \ldots, g_{k-1}, z] = \frac{zp_{k-1} + p_{k-2}}{zq_{k-1} + q_{k-2}}$;
- $\frac{q_k}{q_{k-1}} = [g_k; g_{k-1}, \ldots, g_1]$ *and* $\frac{p_k}{p_{k-1}} = [g_k; g_{k-2}, \ldots, g_0]$;
- $p_{k+1}q_k - p_kq_{k+1} = (-1)^k$.

Now, let $q(x,y) = ax^2 + bxy + cy^2$ and $Q(x,y) = Ax^2 + Bxy + Cy^2$ be two equivalent reduced binary quadratic forms. By replacing q and/or Q by their right neighbors if necessary, we can assume that $a, A > 0$. In this case, the 1st and 2nd roots of $a\omega^2 + b\omega + c = 0 = A\Omega^2 + B\Omega + C$ satisfy $0 < f, F < 1$ and $s, S < -1$.

Denote by $T - \begin{pmatrix} \alpha & \beta \\ \gamma & \delta \end{pmatrix} \in SL_2(\mathbb{Z})$ the matrix realizing the equivalence between q and $Q = q \circ T$, so that $f = \frac{\alpha F + \beta}{\gamma F + \delta}$ and $s = \frac{\alpha F + \beta}{\gamma F + \delta}$ (cf. (C.7)).

Since there is nothing to prove when T is the identity matrix (i.e., $q = Q$), we will assume that $T \neq \mathrm{Id}$. Up to replacing T by $-T$ if necessary, we can further assume that either $\alpha > 0$ or $\alpha = 0$ and $\gamma > 0$.

Proposition C.3. *One has* $\alpha > 0$.

Proof. Indeed, $\alpha = 0$ and $\gamma > 0$ would imply that $\gamma = 1 = -\beta$ (because $T \in SL_2(\mathbb{Z})$). Therefore, $f = -\frac{1}{F+\delta}$ and $s = -\frac{1}{S+\delta}$, i.e.,

$$-\delta = f + \frac{1}{F} > 1 \quad \text{and} \quad \delta = -S - \frac{1}{s} > 1,$$

a contradiction. $\square$

Since $\frac{1}{f} = \frac{\gamma F + \delta}{\alpha F + \beta}$, $\frac{1}{s} = \frac{\gamma S + \delta}{\alpha S + \beta}$ and $T \in SL_2(\mathbb{Z})$, one has

$$\left(\frac{\alpha}{f} - \gamma \right)(\alpha F + \beta) = 1 = \left(\frac{\alpha}{s} - \gamma \right)(\alpha S + \beta). \qquad \text{(C.12)}$$

Proposition C.4. *If T is not the identity matrix, then* $\beta\gamma \neq 0$.

Proof. If $\beta = 0$, then $\alpha = \delta = 1$ (because $T \in SL_2(\mathbb{Z})$ and Proposition C.3 ensures that $\alpha > 0$). This would imply that $(\frac{1}{s} - \gamma)S = 1$, i.e.,

$$\gamma = \frac{1}{s} - \frac{1}{S} \quad \text{and} \quad -\gamma = \frac{1}{S} - \frac{1}{s}.$$

Since $s, S < -1$, one would get $|\gamma| < 1$, i.e., $\gamma = 0$, so that T would be the identity matrix. Similarly, if $\gamma = 0$, then one would have $\alpha = \delta = 1$, so that $|\beta| = |f - F|$. Since $0 < f, F < 1$, this would force $|\beta| < 1$, i.e., $\beta = 0$, so that T would be the identity matrix. $\square$

Proposition C.5. *If T is not the identity matrix, then* $\beta\gamma > 0$.

Proof. By the previous proposition, our task is to prove that $\beta > 0 \iff \gamma > 0$.

If $\beta > 0$, then $\alpha F + \beta > 1$ (as $F > 0$ and Proposition C.3 says that $\alpha > 0$). By (C.12), it follows that $\frac{\alpha}{f} - \gamma < 1$. Since $0 < f < 1$, one gets that

$$\gamma + 1 > \alpha/f > \alpha \geq 1$$

that is, $\gamma > 0$.

Similarly, if $\gamma > 0$, then $\frac{\alpha}{s} - \gamma < -1$. By (C.12), one gets that $0 > \alpha S + \beta > -1$. Thus, $\beta > 0$ (as $-S > 1$ and $\alpha \geq 1$). This completes the argument. $\qquad\square$

The previous propositions allow us to suppose that $\alpha > 0$ and $\beta\gamma > 0$. Since $\alpha\delta = \beta\gamma + 1$ (as $T \in SL_2(\mathbb{Z})$), we also have $\delta > 0$. Moreover, up to exchanging the roles of q and Q, i.e., replacing $T = \begin{pmatrix} \alpha & \beta \\ \gamma & \delta \end{pmatrix}$ by its inverse $T^{-1} = \begin{pmatrix} \delta & -\beta \\ -\gamma & \alpha \end{pmatrix}$ if necessary, we can assume that $\beta, \gamma > 0$. Finally, the proof of Proposition C.5 also tells us that $\beta, \gamma \geq \alpha$. In particular, the fact that $\alpha\delta = \beta\gamma + 1 > \beta\gamma$ says that $\beta, \gamma < \delta$.

In summary, we can (and do) assume from now on that

$$0 < \alpha \leq \beta, \gamma < \delta.$$

Therefore, $x = \alpha$ and $y = \gamma$ is a solution of the equation

$$\delta x - \beta y = 1$$

with $0 < x \leq \beta$ and $0 < y < \delta$.

Proposition C.6. *The equation $\delta x - \beta y = 1$ has an unique solution with $0 < x \leq \beta$ and $0 < y < \delta$.*

Proof. Since any solution of this equation satisfies $\delta(x - \alpha) = \beta(y - \gamma)$ and we have that $gcd(\delta, \beta) = 1$ (as $T \in SL_2(\mathbb{Z})$), there exists $m \in \mathbb{Z}$ with

$$x - \alpha = \beta m \quad \text{and} \quad y - \gamma = \delta m.$$

On the other hand, if $0 < x \leq \beta$ and $0 < y < \delta$, then $|x - \alpha| < \beta$ and $|y - \gamma| < \delta$. It follows that $m = 0$ and, *a fortiori*, $x - \alpha = 0 = y - \gamma$. This completes the proof. $\qquad\square$

We affirm that α and γ are determined by a continued fraction expansion of δ/β.

In fact, let us write $\delta/\beta = [g_0; g_1, \ldots, g_{i-1}]$ with i *even* and $g_k \geq 1$ for all $0 \leq k \leq i-1$: this is always possible because if i is odd, then we replace the term $g_{i-2} + \frac{1}{g_{i-1}}$ in the continued fraction expansion by $g_{i-2} + 1$, resp. $g_{i-2} + \frac{1}{g_{i-1}-1+\frac{1}{1}}$ when $g_{i-1} = 1$, resp. $g_{i-1} > 1$. Denoting by

$$\frac{y}{x} := [g_0; g_1, \ldots, g_{i-2}],$$

we have from Proposition C.2 that $\delta x - \beta y = (-1)^{i-2} = 1$ and $0 < x < \beta$, $0 < y < \delta$. By Proposition C.6, this implies that $x = \alpha$ and $y = \gamma$, i.e.,

$$\frac{\gamma}{\alpha} = [g_0; g_1, \ldots, g_{i-2}].$$

Moreover, Proposition C.2 also tells us that

$$[g_0; g_1, \ldots, g_{i-1}, 1/F] = \frac{\delta/F + \gamma}{\beta/F + \alpha} = \frac{1}{f}.$$

Since $1/F > 1$, we have that g_k, $0 \le k \le i-1$, are the first entries of the continued fraction expansion of $1/f$. Furthermore, (C.10) says that $1/F = (-1)^i/f_i = 1/f_i$ where f_i is the first root of the ith element Φ_i of the chain containing $\Phi_0 := q$. In other words, Q and Φ_i have the same discriminant and the same first root $F = f_i$.

At this point, we are ready to complete the proof of Theorem C.3. For this sake, it suffices to establish the following fact:

Proposition C.7. *One has $Q = \Phi_i$. In particular, Q and q belong to the same chain.*

Proof. We saw that Q and Φ_i share the same discriminant and 1st root. Since a binary quadratic form is determined by its discriminant and its 1st and 2nd roots (cf. Remark C.3), our task is simply to show that Q and Φ_i have the same 2nd root $S = s_i$.

By Proposition C.2, the fact that $\delta/\beta = [g_0; g_1, \ldots, g_{i-1}]$ and $\frac{\gamma}{\alpha} = [g_0; g_1, \ldots, g_{i-2}]$ implies $\frac{\beta}{\alpha} = [g_{i-1}; g_{i-2}, \ldots, g_1]$, $\frac{\delta}{\gamma} = [g_{i-1}; g_{i-2}, \ldots, g_0]$ and, hence,

$$[g_{i-1}; g_{i-2}, \ldots, g_0, \frac{1}{S_0}] := [g_{i-1}; g_{i-2}, \ldots, g_0, -s] = \frac{-s\delta + \beta}{-s\gamma + \alpha} = -S.$$

Since $-s > 1$, it follows from (C.10) that $-S = (-1)^{i+1} s_i = -s_i$, i.e., $S = s_i$. $\qquad \square$

Appendix D

Freiman's rightmost gap on Markov spectrum

In [Fr75 (1975), Section 10, pp. 66–71], Freiman proved the following result:

Theorem D.1. *One has* $M \cap (\nu, \mu) = \emptyset$ *where*

$$\nu = [4; 3, 1, 3, 1, 3, \overline{4, 4, 4, 3, 2, 3}] + [0; 3, 1, 3, 1, 2, 1, 1, 3, 3, \overline{3, 1, 3, 1, 2, 1}]$$

$$= \frac{19033619 - 2\sqrt{462}}{72101381} + \frac{594524 - \sqrt{243542}}{139318} = 4.527829538...$$

and

$$\mu = [4; 4, 3, 2, 2, \overline{3, 1, 3, 1, 2, 1}] + [0; 3, 2, 1, 1, \overline{3, 1, 3, 1, 2, 1}]$$

$$= 4 + \frac{253589820 + 283798\sqrt{462}}{491993569} = 4.527829566...$$

In the sequel, we explain the proof of this theorem. For this sake, we restrict from now on our attention to the sequences $\underline{a} = (a_n)_{n \in \mathbb{Z}} \in (\mathbb{N}^*)^{\mathbb{Z}}$ such that

$$4 < m(\underline{a}) = \lambda_0(\underline{a}) < 5.$$

Note that these inequalities imply that

$$\underline{a} \in \{1, 2, 3, 4\}^{\mathbb{Z}} \quad \text{and} \quad a_0 \in \{3, 4\}.$$

D.1 Preliminaries

Lemma D.1. *If* $m(\underline{a}) < 4.55$, *then* $\underline{a} \in \{1, 2, 3, 4\}^{\mathbb{Z}}$ *can not contain the subwords* 41, 42 *or their transposes.*

Proof. This happens because $[4; 1, 1] + [0; 4, 1] = 4.7$ and $[4; 2, 1] + [0; 4, 2] > 4.55$. $\square$

Lemma D.2. *If* $m(\underline{a}) < 4.52786$, *then* $\underline{a} \in \{1, 2, 3, 4\}^{\mathbb{Z}}$ *can not contain the subwords* 313131, 313132, 313133, 443131344 *or their transposes.*

Proof. This is a consequence of Lemma D.1 and $[3; 1, 3, 3, 1, 3, 1] + [0; 1, 3, 4, 3] > 4.529$ (this forbids the sequence 313133, and, with more reason, the sequences 313132 and 313131), $[3; 1, 3, 4, 4, 4, 3] + [0; 1, 3, 4, 4, 4, 3] > 4.52786$. □

Lemma D.3. *Suppose that* $m(\underline{a}) < 4.55$. *If the neighborhood of* a_0 *coincide with* 3^*2, 13^*12, 4313^*1343, 44^*4 *or their transposes, then* $\lambda_0(\underline{a}) < 4.5278$.

Proof. This follows from Lemma D.1 and $[3; 2] + [0; 1] = 4.5$, $[3; 1, 2, \overline{1, 3}] + [0; 1, \overline{3, 1}] < 4.5276$, $[3; 1, 3, 4, 3, \overline{1, 3}] + [0; 1, 3, 4, \overline{4, 3}] < 4.5278$, $[4; 4] + [0; 4] = 4.5$. □

Corollary D.1. *Suppose that* $4.5278 < m(\underline{a}) = \lambda_0(\underline{a}) < 4.52786$. *Then,* $\underline{a} \in \{1, 2, 3, 4\}^{\mathbb{Z}}$ *has the form* $\ldots a_{-1}a_0a_1 \cdots = \ldots 343 \ldots$ *or* $\ldots 344 \ldots$ *(up to transposition).*

Proof. Recall that $\underline{a} \in \{1, 2, 3, 4\}^{\mathbb{Z}}$ and $a_0 \in \{3, 4\}$ whenever $4 < m(\underline{a}) = \lambda_0(\underline{a}) < 5$.

We affirm that $a_0 = 4$ under our assumption that $4.5278 < m(\underline{a}) = \lambda_0(\underline{a}) < 4.52786$. Indeed, if $a_0 = 3$, then a first application of Lemma D.3 forces $a_{-1}a_0a_1 = 131$. By Lemmas D.1 and D.3, this means that $a_{-2}a_{-1}a_0a_1a_2 = 31313$. By Lemma D.2, we must have $a_{-3}a_{-2}a_{-1}$ $a_0a_1a_2a_3 = 4313134$. By Lemma D.3, $a_{-4}a_{-3}a_{-2}a_{-1}a_0a_1a_2a_3a_4 = 443131344$, a contradiction with Lemma D.2.

In summary, we saw that $4.5278 < m(\underline{a}) = \lambda_0(\underline{a}) < 4.52786$ implies $a_0 = 4$. By Lemmas D.1 and D.3, we deduce that $a_{-1}a_0a_1 = 343$ or 344 (up to transposition). □

D.2   Extensions of the word 343

Lemma D.4. *If* $m(\underline{a}) < 4.52786$, *then* $\underline{a} \in \{1, 2, 3, 4\}^{\mathbb{Z}}$ *can not contain the subwords* 3432, 134312, 31343132, 21313431312 *or their transposes.*

Proof. This is a consequence of Lemma D.1 and the following estimates:

$$[4; 3, 2] + [0; 3, 1] > 4.53, \quad [4; 3, 1, 2, 1] + [0; 3, 1, 3, 1] > 4.529,$$

$$[4; 3, 1, 3, 2] + [0; 3, 1, 3, 1] > 4.52786,$$

$$[4; 3, 1, 3, 1, 2, \overline{1, 3}] + [0; 3, 1, 3, 1, 2, \overline{1, 3}] > 4.52786.$$

□

Corollary D.2. *If* $m(\underline{a})$ < 4.52786 *and* $a_{-1}a_0a_1 = 343$, *then* $a_{-4}\ldots$ $a_0\ldots a_6 = 13134313134$ *(up to transposition).*

Proof. By Lemmas D.1 and D.4, $a_{-4}\ldots a_0\ldots a_5 = 1313431313$ (up to transposition). By Lemma D.2, we must have $a_{-4}\ldots a_0\ldots a_6 = 13134313134$. $\square$

Lemma D.5. *If* $m(\underline{a})$ < 4.5278296, *then* $\underline{a} \in \{1,2,3,4\}^{\mathbb{Z}}$ *can not contain the subwords* 113134313134, 2213134313134, 2311213134313134, 33112131343131343 *or their transposes.*

Proof. This is a direct consequence of Lemma D.1 and the following inequalities:

$$[4;3,1,3,1,3,4] + [0;3,1,3,1,1,1] > 4.5279,$$

$$[4;3,1,3,1,3,4] + [0;3,1,3,1,2,2] > 4.52784$$

$$[4;3,1,3,1,3,4,4,3] + [0;3,1,3,1,2,1,1,3,2,1,3,1] > 4.5278296,$$

$$[4;3,1,3,1,3,4,3,\overline{1,3}] + [0;3,1,3,1,2,1,1,3,3,\overline{1,3}] > 4.5278296.$$

$\square$

Lemma D.6. *Suppose that* $m(\underline{a})$ < 4.5278296. *If the neighborhood of* a_0 *coincide with* 313134*31313, 211213134*313134, 4311213134*313134 *or their transposes, then* $\lambda_0(\underline{a})$ < 4.5278295.

Proof. The first two claims follow from Lemma D.1 and the inequalities

$$[4;3,1,3,1,3,\overline{4,3}] + [0;3,1,3,1,3,\overline{4,3}] < 4.5277,$$

$$[4;3,1,3,1,3,4,3] + [0;3,1,3,1,2,1,1,2,1] < 4.52782904.$$

The last claim follows from Lemma D.1, Corollary D.2, Lemma D.5, and the estimate $[4;3,1,3,1,3,4,3,1,3,1,2] + [0;3,1,3,1,2,1,1,3,4,\overline{4,3}] <$ 4.5278295. $\square$

Corollary D.3. *If* 4.5278295 < $m(\underline{a}) = \lambda_0(\underline{a})$ < 4.5278296 *and* $a_{-1}a_0a_1 =$ 343, *then* $a_{-9}\ldots a_0\ldots a_7 = 33112131343131344$ *(up to transposition).*

Proof. By Corollary D.2, we have $a_{-4}\ldots a_0\ldots a_6 = 13134313134$ (up to transposition). By a first application of Lemmas D.5 and D.6, we get $a_{-5}\ldots a_0\ldots a_6 = 213134313134$. By a second application of Lemmas D.5 and D.6, we deduce that $a_{-8}\ldots a_0\ldots a_6 = 311213134313134$. A third application of Lemmas D.5 and D.6 implies $a_{-9}\ldots a_0\ldots a_6 = 3311213134313134$. Finally, a last application of Lemma D.5 reveals that $a_{-9}\ldots a_0\ldots a_7 = 33112131343131344$. $\square$

Lemma D.7. *If $m(\underline{a}) < 4.528$, then $\underline{a} \in \{1,2,3,4\}^{\mathbb{Z}}$ can not contain the subwords 334, 223444 or their transposes.*

Proof. This follows from Lemma D.1 and the estimates $[4; 3, 3, \overline{4, 3}] + [0; \overline{4, 3}] > 4.53$, $[4; 4, 4, \overline{4, 3}] + [0; 3, 2, 2, \overline{1, 3}] > 4.528$. $\qquad\square$

Corollary D.4. *If $4.5278295 < m(\underline{a}) = \lambda_0(\underline{a}) < 4.5278296$ and $a_{-1}a_0a_1 = 343$, then $m(\underline{a}) \leq \nu$.*

Proof. By Corollary D.3, we have that $a_{-9} \ldots a_0 \ldots a_7 = 33112131343131344$ (up to transposition). We want to *maximize* $4.5278295 < m(\underline{a}) = \lambda_0(\underline{a}) < 4.5278296$. By Lemma D.1, this means that $a_{-9} \ldots a_0 \ldots a_9 = 3311213134313134443$. By Lemma D.7, we have $a_{-9} \ldots a_0 \ldots a_{11} = 331121313434313134444323$. By Lemma D.4, we derive $a_{-9} \ldots a_0 \ldots a_{11} = 33112131343131344432344$. By repeating this argument, we conclude that $a_{-9} \ldots a_0 \ldots a_7 \cdots = 33112131343131344\overline{432344}$. Similarly, we have from Lemma D.7 that $a_{-10} \ldots a_0 \ldots a_7 = 333112131343131344$. By Lemma D.1, we get $a_{-13} \ldots a_0 \ldots a_7 = 131333112131343131344$. By Lemma D.2, $a_{-15} \ldots a_0 \ldots a_7 = 12131333112131343131344$. By repeating this argument, we get $\ldots a_{-9} \ldots a_0 \ldots a_7 = \overline{1213}1333112131343131344$.

In summary, our assumptions imply the maximal value of $m(\underline{a})$ is ν. $\quad\square$

D.3   Extensions of the word 344

Lemma D.8. *Suppose that $m(\underline{a}) < 4.55$. If the neighborhood of a_0 coincide with 134^*4, 2234^*43, 3234^*4 or their transposes, then $\lambda_0(\underline{a}) < 4.5278291$.*

Proof. This follows from Lemma D.1 and the estimates $[4; 4, \overline{4, 1}] + [0; 3, 1, \overline{1, 4}] < 4.52$, $[4; 4, 3, \overline{1, 3}] + [0; 3, 2, 2, \overline{3, 1}] < 4.5278291$, and $[4; 4, \overline{4, 3}] + [0; 3, 2, 3, \overline{4, 3}] < 4.5276$. $\qquad\square$

Corollary D.5. *If $4.5278291 < m(\underline{a}) = \lambda_0(\underline{a}) < 4.528$ and $a_{-1}a_0a_1 = 344$, then $a_{-3}a_{-2}a_{-1}a_0a_1a_2 = 123443$.*

Proof. This is an immediate consequence of Lemmas D.1, D.7 and D.8. $\quad\square$

Lemma D.9. *If $m(\underline{a}) < 4.5279$, then $\underline{a} \in \{1,2,3,4\}^{\mathbb{Z}}$ can not contain the subwords 1234431, 21234432, 211234432 or their transposes.*

Proof. The first two claims follow from $[4; 4, 3, 1, \overline{1, 4}] + [0; 3, 2, 1, \overline{1, 4}] >$ 4.528 and $[4; 4, 3, 2, \overline{1, 4}] + [0; 3, 2, 1, 2, \overline{4, 1}] > 4.529$. The last claim follows from the previous inequality and the fact that $[4; 4, 3, 2, 1, 1, \overline{1, 4}] + [0; 3, 2, 1, 1, 2, \overline{1, 4}] > 4.5279$. $\square$

Corollary D.6. *If* $4.5278291 < m(\underline{a}) = \lambda_0(\underline{a}) < 4.5279$ *and* $a_{-1}a_0a_1 = 344$, *then the neighborhood of* a_0 *coincides with* 311234^*432.

Proof. By Corollary D.5, $a_{-3}a_{-2}a_{-1}a_0a_1a_2 = 123443$. By Lemmas D.1, D.7 and D.9, we have $a_{-5}a_{-4}a_{-3}a_{-2}a_{-1}a_0a_1a_2a_3 = 311234432$. $\square$

Lemma D.10. *If* $m(\underline{a}) < 4.527832$, *then* $\underline{a} \in \{1, 2, 3, 4\}^{\mathbb{Z}}$ *can not contain the subwords* 3112344323, 3311234432113 *or their transposes.*

Proof. This follows from the estimates $[4; 4, 3, 2, 3, \overline{4, 1}] + [0; 3, 2, 1, 1, 3, \overline{1, 4}] > 4.5279$ and $[4; 4, 3, 2, 1, 1, 3, \overline{4, 1}] + [0; 3, 2, 1, 1, 3, 3, \overline{4, 1}] > 4.527832$. $\square$

Lemma D.11. *Suppose that* $m(\underline{a}) < 4.55$. *If the neighborhood of* a_0 *coincide with* $2311234^*432113$ *or its transpose, then* $\lambda_0(\underline{a}) < 4.5278283$.

Proof. This is a consequence of Lemma D.1 and the estimate $[4; 4, 3, 2, 1, 1, 3, \overline{1, 3}] + [0; 3, 2, 1, 1, 3, 2, \overline{1, 3}] < 4.5278283$. $\square$

Corollary D.7. *If* $4.5278291 < m(\underline{a}) = \lambda_0(\underline{a}) < 4.527832$ *and* $a_{-1}a_0a_1 = 344$, *then* $m(\underline{a}) \geq \mu$.

Proof. By Corollary D.6, the neighborhood of a_0 is 311234^*432. By Lemma D.10, the neighborhood of a_0 is 311234^*4322 or 311234^*4321.

If the neighborhood of a_0 is 311234^*4321, we can apply Lemmas D.1 and D.9 to derive that the neighborhood of a_0 must extend as $311234^*432113$, a contradiction with Lemmas D.11 and D.10. Thus, the neighborhood of a_0 is 311234^*4322.

We want to *minimize* $m(\underline{a})$. For this sake, we apply Lemmas D.1 and D.2 to see that the minimal extension of 311234^*4322 is $\overline{12131}311234^*4322\overline{313121}$. $\square$

D.4 End of the proof of Theorem D.1

The desired result follows directly from Corollaries D.1, D.4 and D.7.

Appendix E

Soft bounds on the Hausdorff dimension of dynamical Cantor sets

In this appendix, we give a proof of Proposition 2.1 along the lines of Palis–Takens book [PT (1993)]. For this sake, we recall that a C^1-dynamical Cantor set K associated to an expanding map $\Psi : I_1 \cup \cdots \cup I_r \to I$ acting on a collection of intervals $\mathcal{R}_1 = \{I_1, \ldots, I_r\}$ is inductively described by the collection $\mathcal{R}_n$ of connected components of $\Psi^{-1}(J)$, $J \in \mathcal{R}_{n-1}$.

For later use, for each $R \in \mathcal{R}_n$, we denote by

$$\lambda_{n,R} := \inf |(\Psi^n)'|_R| \quad \text{and} \quad \Lambda_{n,R} := \sup |(\Psi^n)'|_R|.$$

E.1 Upper bound on the dimension of dynamical Cantor sets

Consider the quantities β_n defined by

$$\sum_{R \in \mathcal{R}_n} \left(\frac{1}{\lambda_{n,R}}\right)^{\beta_n} = 1.$$

Proposition E.1. *For all $n \geq 1$, the box-counting dimension of K satisfies $d(K) \leq \beta_n$.*

Proof. Let $n \geq 1$ and fix $\beta > d(K)$. By definition, there exists $\varepsilon_0 > 0$ such that

$$N_\varepsilon(K) \leq 1/\varepsilon^\beta$$

for all $0 < \varepsilon \leq \varepsilon_0$.

In other words, given $0 < \varepsilon \leq \varepsilon_0$, we can cover K using a collection of intervals $\{J_1, \ldots, J_m\}$ with $m \leq 1/\varepsilon^\beta$ such that every J_l has length $\leq \varepsilon$.

It follows from the definitions that, for each $R \in \mathcal{R}_n$, the pre-images of the intervals J_l under $\Psi^n|_R$ form a covering of $K \cap R$ by intervals of length $\leq \varepsilon/\lambda_{n,R}$. Therefore,

$$N_{\varepsilon/\lambda_{n,R}}(K \cap R) \leq 1/\varepsilon^\beta$$

for all $R \in \mathcal{R}_n$ and $0 < \varepsilon \leq \varepsilon_0$, and, *a fortiori*,

$$N_\delta(K \cap R) \leq 1/(\lambda_{n,R}\delta)^\beta$$

for all $R \in \mathcal{R}_n$ and $0 < \delta \leq \varepsilon_0/\lambda_{n,R}$.

Hence, if we define $\lambda_n = \sup_{R \in \mathcal{R}_n} \lambda_{n,R}$, then

$$N_\delta(K) \leq \frac{1}{\delta^\beta} \left(\sum_{R \in \mathcal{R}_n} \frac{1}{\lambda_{n,R}^\beta} \right)$$

for all $0 < \delta \leq \varepsilon_0/\lambda_n$.

By iterating this argument k times, we conclude that

$$N_\delta(K) \leq \frac{1}{\delta^\beta} \left(\sum_{R \in \mathcal{R}_n} \frac{1}{\lambda_{n,R}^\beta} \right)^k$$

for all $0 < \delta \leq \varepsilon_0/\lambda_n^k$.

Thus,

$$d(K) \leq \beta + \lim_{k \to \infty} \frac{\log \left(\sum_{R \in \mathcal{R}_n} \frac{1}{\lambda_{n,R}^\beta} \right)^k}{\log(\lambda_n^k/\varepsilon_0)} = \beta + \frac{\log \left(\sum_{R \in \mathcal{R}_n} \frac{1}{\lambda_{n,R}^\beta} \right)}{\log \lambda_n}.$$

Since $\beta > d(K)$ is arbitrary, we deduce from the previous inequality that

$$0 \leq \frac{\log \left(\sum_{R \in \mathcal{R}_n} \left(\frac{1}{\lambda_{n,R}} \right)^{d(K)} \right)}{\log \lambda_n}.$$

Because $\lambda_n > 1$, we get that $\sum_{R \in \mathcal{R}_n} \left(\frac{1}{\lambda_{n,R}} \right)^{d(K)} \geq 1$, and, *a fortiori*, $d(K) \leq \beta_n$.   □

E.2   Lower bound on the dimension of dynamical Cantor sets

Let K be a C^1-dynamical Cantor set associated to an expanding map $\Psi : I_1 \cup \cdots \cup I_r \to I$. Consider the sequence β_n defined in the previous section and fix $\beta_\infty \in [0,1]$ be a constant such that $\beta_\infty \geq \beta_n$ for all $n \geq 1$ (e.g., $\beta_\infty = 1$ certainly works).

Take $n_0 \in \mathbb{N}$ such that $\Psi^{n_0+1}(K \cap I_j) = K$ for all $1 \leq j \leq r$ and set

$$C := \sup |(\Psi^{n_0})'|^{\beta_\infty} \geq 1.$$

Remark E.1. If Ψ is a full Markov map, i.e., $\Psi(K \cap I_j) = K$ for all $1 \leq j \leq r$, then we can choose $n_0 = 1$ and $C = 1$.

Consider the quantities α_n given by

$$\sum_{R \in \mathcal{R}_n} \left(\frac{1}{\Lambda_{n,R}}\right)^{\alpha_n} = C.$$

Proposition E.2. *For all $n \geq 1$, the Hausdorff dimension of K satisfies $\alpha_n \leq HD(K)$.*

Proof. Suppose by contradiction that $HD(K) < \alpha_n$ and take $HD(K) < \alpha < \alpha_n$.

By definition, $m_\alpha(K) = 0$, so that for every $\varepsilon > 0$ there is a finite cover $(U_a^{(\varepsilon)})_{a=1}^{c(\varepsilon)}$ of K with

$$\sum_{a=1}^{c(\varepsilon)} \mathrm{diam}(U_a)^\alpha \leq \varepsilon.$$

Note that any interval of length strictly smaller than

$$\kappa_0 := \min\{dist(I, J) : I, J \in \mathcal{R}_n, I \neq J\}/2 > 0$$

intersects at most one $R \in \mathcal{R}_n$.

Thus, if we define $\varepsilon_0 := \kappa_0^{1/\alpha}$, then each element of the cover $\mathcal{U} := (U_a^{(\varepsilon_0)})_{a=1}^{c(\varepsilon_0)}$ of K intersects at most one element of $\mathcal{R}_n$. Hence, if we define

$$\mathcal{U}_R := \{U \in \mathcal{U} : U \cap R \neq \emptyset\},$$

then, given any $R \in \mathcal{R}_n$, one has that $\mathcal{U}_R$ has *fewer* elements than $\mathcal{U}$.

Consider $n_0 \in \mathbb{N}$ such that $\Psi^{n_0+1}(K \cap R) = K$ for all $R \in \mathcal{R}_1$. From the definitions, for each $R \in \mathcal{R}_n$, we see that $(\Psi^{n+n_0}|R)(\mathcal{U}_R)$ is a well-defined cover of K such that

$$\sum_{V \in (\Psi^{n+n_0}|R)(\mathcal{U}_R)} \mathrm{diam}(V)^\alpha \leq (\sup |(\Psi^k)'|)^\alpha \cdot \Lambda_{n,R}^\alpha \cdot \sum_{U \in \mathcal{U}_R} \mathrm{diam}(U)^\alpha.$$

Since $\alpha < \alpha_n < \beta_n \leq \beta_\infty$, we get that

$$\sum_{V \in (\Psi^{n+n_0}|R)(\mathcal{U}_R)} \mathrm{diam}(V)^\alpha \leq C \cdot \Lambda_{n,R}^\alpha \cdot \sum_{U \in \mathcal{U}_R} \mathrm{diam}(U)^\alpha.$$

We want to exploit this estimate to prove that

$$\sum_{V \in (\Psi^{n+n_0}|R_0)(\mathcal{U}_{R_0})} \mathrm{diam}(V)^\alpha < \varepsilon_0$$

for some $R_0 \in \mathcal{R}_n$. In this direction, suppose that

$$\sum_{V \in (\Psi^{n+n_0}|R)(\mathcal{U}_R)} \mathrm{diam}(V)^\alpha \geq \varepsilon_0 \quad \forall\, R \in \mathcal{R}_n.$$

In this case, the discussion above would imply

$$\sum_{U \in \mathcal{U}} \operatorname{diam}(U)^\alpha = \sum_{R \in \mathcal{R}_n} \sum_{U \in \mathcal{U}_R} \operatorname{diam}(U)^\alpha$$

$$\geq \sum_{R \in \mathcal{R}_n} (C \cdot \Lambda_{n,R}^\alpha)^{-1} \left(\sum_{V \in (\Psi^{n+n_0}|_R)(\mathcal{U}_R)} \operatorname{diam}(V)^\alpha \right)$$

$$\geq C^{-1} \left(\sum_{R \in \mathcal{R}_n} \Lambda_{n,R}^{-\alpha} \right) \varepsilon_0,$$

a contradiction because on one hand $\sum_{U \in \mathcal{U}} \operatorname{diam}(U)^\alpha \leq \varepsilon_0$ (by definition of $\mathcal{U}$) and on the other hand $\left(\sum_{R \in \mathcal{R}_n} \Lambda_{n,R}^{-\alpha} \right) > C$ (by our choice of $\alpha < \alpha_n$).

In summary, we assumed that $HD(K) < \alpha_n$, we considered an arbitrary cover $\mathcal{U}$ of K with

$$\sum_{U \in \mathcal{U}} \operatorname{diam}(U)^\alpha \leq \varepsilon$$

and we found a cover $\mathcal{V} := (\Psi^{n+n_0}|_{R_0})(\mathcal{U}_{R_0})$ of K with *fewer* elements than $\mathcal{U}$ such that

$$\sum_{V \in (\Psi^{n+n_0}|_{R_0})(\mathcal{U}_{R_0})} \operatorname{diam}(V)^\alpha < \varepsilon_0.$$

By iterating this argument, we would end up with a cover of K containing no elements, a contradiction. This proves that $\alpha_n \leq HD(K)$. $\square$

E.3   End of the proof of Proposition 2.1

The proof of Proposition 2.1 follows directly from Propositions E.1 and E.2 (and Remarks 2.6 and E.1).

E.4   Slow convergence of towards the Hausdorff dimension

Let K be a C^2-dynamical Cantor set associated to an expanding map $\Psi : I_1 \cup \cdots \cup I_r \to I$. In general, the sequences $\alpha_n \leq HD(K) \leq \beta_n$ discussed above converge slowly towards $HD(K)$:

Proposition E.3. *One has $\beta_n - \alpha_n = O(1/n)$.*

Proof. The so-called *bounded distortion property* (see Theorem 1 in Chapter 4 of Palis-Takens book [PT (1993)]) ensures the existence of a constant $a = a(K) \geq 1$ such that

$$\Lambda_{n,R} \leq a\lambda_{n,R}$$

for all $n \in \mathbb{N}$ and $R \in \mathcal{R}_n$.

Let $\lambda := \inf |\Psi'| > 1$ and, for all $n \geq \log a / \log \lambda$, define

$$\delta_n := \frac{\alpha_n \log a + \log C}{-\log a + n \log \lambda}.$$

In this setting, we have that

$$\sum_{R \in \mathcal{R}_n} \left(\frac{1}{\lambda_{n,R}}\right)^{\alpha_n + \delta_n} \leq a^{\alpha_n + \delta_n} \sum_{R \in \mathcal{R}_n} \left(\frac{1}{\Lambda_{n,R}}\right)^{\alpha_n + \delta_n} \leq \frac{a^{\alpha_n + \delta_n}}{\lambda^{n\delta_n}} \sum_{R \in \mathcal{R}_n} \left(\frac{1}{\Lambda_{n,R}}\right)^{\alpha_n}$$

$$= \frac{a^{\alpha_n + \delta_n}}{\lambda^{n\delta_n}} C = 1.$$

Therefore, $\beta_n \leq \alpha_n + \delta_n$, that is,

$$\beta_n - \alpha_n \leq \frac{\alpha_n \log a + \log C}{-\log a + n \log \lambda} \leq \frac{HD(K) \log a + \log C}{-\log a + n \log \lambda} = O(1/n).$$

This proves the proposition. $\square$

Modulus of continuity of the dimension across classical spectra

In this appendix we will prove a slightly more general version of Theorem 3.4 (restated as Theorem F.1 for the sake of convenience of the reader), and we finish with a discussion on the modulus of continuity of $\Delta^+(t)$ and $d(t)$. In what follows, $T := \lfloor t \rfloor$:

Theorem F.1. *Given $\eta > 0$ and $3 \leq t < +\infty$ with $d(t) := HD(L \cap (-\infty, t)) > 0$, we can find $\delta > 0$ and a Gauss-Cantor set $K(B)$ associated to $\Sigma(B) \subset \Sigma$ such that*

$$\Sigma(B) \subset \Sigma_{t-\delta} \quad \text{and} \quad HD(K(B)) \geq (1-\eta)\Delta^+(t).$$

Proof. Let $\tau = \eta/40$. Since $t > 3$, we have $\Delta^+(t) > 0$, and so we may choose $r_0 \in \mathbb{N}$ large such that, for $r \geq r_0$, $\left| \frac{\log N(t,r)}{r} - \Delta^+(t) \right| < \frac{\tau}{2}\Delta^+(t)$. Let $B_0 := C(t, r_0)$ and $N_0 := N(t, r_0) = |B_0|$. Let $k = 8N_0^2 \lceil 2/\tau \rceil$. Take $\tilde{B} = \{\beta = \beta_1\beta_2 \cdots \beta_k \mid \beta_j \in B_0, 1 \leq j \leq k \text{ and } K_t \cap I(\beta) \neq \emptyset\}$.

Given $\beta = \beta_1\beta_2 \cdots \beta_k \in \tilde{B}$ (with $\beta_i \in B_0$, $1 \leq i \leq k$), we say that j, $1 \leq j \leq k$, is a *right-good* position of β if there are elements $\beta^{(s)} = \beta_1\beta_2 \cdots \beta_{j-1}\beta_j^{(s)}\beta_{j+1}^{(s)} \cdots \beta_k^{(s)}$, $s = 1, 2$, of $\tilde{B}$ such that we have the following inequality of continued fractions: $[0; \beta_j^{(1)}] < [0; \beta_j] < [0; \beta_j^{(2)}]$. We say that j is a *left-good* position if there are elements $\beta^{(s)} = \beta_1^{(s)}\beta_2^{(s)} \cdots \beta_{j-1}^{(s)}\beta_j^{(s)}\beta_{j+1}\beta_{j+2} \cdots \beta_k$, $s = 3, 4$, of $\tilde{B}$ such that $[0; (\beta_j^{(3)})^t] < [0; \beta_j^t] < [0; (\beta_j^{(4)})^t]$. Finally, we say that j is a *good* position if it is both right-good and left-good.

We will show that most positions of most words of $\tilde{B}$ are good. Let us first estimate $|\tilde{B}|$. It is not difficult to show that, for $\beta \in \tilde{B}$, $s(\beta) < (2e^{-r_0})^k < e^{-k(r_0-1)}$. Moreover, since $N(t, k(r_0 - 1)) \geq \frac{1}{T^2}e^{k(r_0-1)\Delta^+(t)}$, $\{I(\beta); \beta \in \tilde{B}\}$ is a covering of K_t by intervals of size smaller than $e^{-k(r_0-1)}$ and the function $h : \tilde{B} \to C(t, k(r_0 - 1))$

defined by $h(\beta) = h((\beta_1\beta_2\ldots\beta_k)) = (\beta_1\beta_2\ldots\beta_j)$, where $j = \min\{i \; ; i \leq k$ and $r((\beta_1\beta_2\ldots\beta_i)) \geq k(r_0 - 1)\}$ is onto, we have:

$$|\tilde{B}| \geq \frac{1}{T^2} e^{k(r_0-1)\Delta^+(t)} > 2\, e^{k(r_0-2)\Delta^+(t)}, \quad \text{since } k \text{ is large}$$

$$\geq 2\, e^{(1-\tau/2)r_0 k\Delta^+(t)}, \quad \text{since } r_0 \text{ is large}$$

$$> 2\, e^{(1-\tau)(1+\tau/2)r_0 k\Delta^+(t)}$$

$$> 2\, N_0^{(1-\tau)k}, \quad \text{since } N(t,r_0) < e^{\left(1+\frac{\tau}{2}\right)\Delta^+(t)r_0}.$$

Now, let us estimate the number of words $\beta \in \tilde{B}$ such that at least $k/20$ positions of β are not right good: we have at most 2^k choices for the set of the $m \geq k/20$ positions which are not right-good. Once we choose this set of positions, if j is such a position and $\beta_1, \beta_2, \ldots, \beta_{j-1} \in B_0$ are already chosen, there are at most two (the largest and the smallest) choices for $\beta_j \in B_0$ such that for some $\beta = \beta_1\beta_2\cdots\beta_{j-1}\beta_j\beta_{j+1}\cdots\beta_k \in \tilde{B}$ the position j is not right good. If j is any other position, we have of course at most $N_0 = |B_0|$ possible choices for β_j, so we have at most $2^m \cdot N_0^{k-m} \leq 2^{k/20} N_0^{19k/20}$ words in $\tilde{B}$ with this chosen set of m positions which are not right-good. Therefore, the number of words $\beta \in \tilde{B}$ for which the number of positions which are not right-good is at least $k/20$ is bounded by $2^k \cdot 2^{k/20} \cdot N_0^{19k/20} = 2^{21k/20} \cdot N_0^{19k/20}$. Analogously, the number of words $\beta \in \tilde{B}$ for which there are at least $k/20$ positions which are not left-good is also bounded by $2^{21k/20} \cdot N_0^{19k/20}$.

This implies that for at least $|\tilde{B}| - 2 \cdot 2^{21k/20} \cdot N_0^{19k/20} > 2N_0^{(1-\tau)k} - 2^{1+21k/20} \cdot N_0^{19k/20} > N_0^{(1-\tau)k}$ words of $\tilde{B}$, the number of good positions is at least $9k/10$. Let us call such an element of $\tilde{B}$ an *excellent* word.

If $\beta = \beta_1\beta_2\cdots\beta_k \in \tilde{B}$ (with $\beta_j \in B_0$, $1 \leq j \leq k$) is an excellent word, we may find $\lceil 2k/5 \rceil$ positions $i_1, i_2, \ldots, i_{\lceil 2k/5 \rceil} \leq k$ with $i_{s+1} \geq i_s + 2$, $\forall s < \lceil 2k/5 \rceil$, such that the positions $i_1, i_1 + 1, i_2, i_2 + 1, \ldots, i_{\lceil 2k/5 \rceil}$, $i_{\lceil 2k/5 \rceil} + 1$ are good. Since $k = 8N_0^2 \lceil 2/\tau \rceil$, we may take, for $1 \leq s \leq 3N_0^2$, $j_s := i_{s\lceil 2/\tau \rceil}$ (notice that $3N_0^2 \lceil 2/\tau \rceil < \frac{16}{5} N_0^2 \lceil 2/\tau \rceil = 2k/5$), so we have $j_{s+1} - j_s \geq 2\lceil 2/\tau \rceil$, $\forall s < 3N_0^2$, and the positions $j_s, j_s + 1$ are good for $1 \leq s \leq 3N_0^2$.

Now, the number of possible choices of $(j_1, j_2, \ldots, j_{3N_0^2})$ is bounded by $\binom{k}{3N_0^2} < 2^k$ and, given $(j_1, j_2, \ldots, j_{3N_0^2})$ the number of choices of $(\beta_{j_1}, \beta_{j_1+1}, \ldots, \beta_{j_{3N_0^2}}, \beta_{j_{3N_0^2}+1})$ is bounded by $N_0^{6N_0^2}$. So, we may choose $\hat{j}_1, \hat{j}_2, \ldots, \hat{j}_{3N_0^2}$ with $\hat{j}_{s+1} - \hat{j}_s \geq 2\lceil 2/\tau \rceil$, $\forall s < 3N_0^2$, and words $\hat{\beta}_{\hat{j}_1}, \hat{\beta}_{\hat{j}_1+1}, \hat{\beta}_{\hat{j}_2}, \hat{\beta}_{\hat{j}_2+1}, \ldots, \hat{\beta}_{\hat{j}_{3N_0^2}}, \hat{\beta}_{\hat{j}_{3N_0^2}+1} \in B_0$ such that the set $X := \{\beta =$

$\beta_1\beta_2\cdots\beta_k \in \tilde{B}$ excellent $|\hat{j}_s,\hat{j}_s+1$ are good positions and $\beta_{\hat{j}_s} = \hat{\beta}_{\hat{j}_s}, \beta_{\hat{j}_s+1} = \hat{\beta}_{\hat{j}_s+1}, \forall s \leq 3N_0^2\}$ has at least $\dfrac{N_0^{(1-\tau)k}}{2^k \cdot N_0^{6N_0^2}} > N_0^{(1-2\tau)k}$ elements, as N_0 and k are large.

Since $N_0 = |B_0|$, there are N_0^2 possible choices for the pairs $(\hat{\beta}_{\hat{j}_s}, \hat{\beta}_{\hat{j}_s+1})$. We will consider, for $1 \leq s < t \leq 3N_0^2$, the projections $\pi_{s,t}: X \to B_0^{\hat{j}_t - \hat{j}_s}$ given by $\pi_{s,t}(\beta_1\beta_2\cdots\beta_k) = (\beta_{\hat{j}_s+1}, \beta_{\hat{j}_s+2}, \ldots, \beta_{\hat{j}_t})$. We will show that the images of many of these projections are large.

For each pair (s,t) with $1 \leq s < t \leq 3N_0^2$ such that $|\pi_{s,t}(X)| < N_0^{(1-10\tau)(\hat{j}_t - \hat{j}_s)}$, we will exclude from $\{1, 2, \ldots, 3N_0^2\}$ the indices $s, s+1, \ldots, t-1$. Let us estimate the total number of indices excluded: the set of excluded indices is the union of the intervals $[s,t)$ (intersected with $\mathbb{Z}$) over the pairs (s,t) as above. Now we use the elementary fact that, given a finite family of intervals, there is a subfamily of disjoint intervals whose sum of lengths is at least half of the measure of the union of the intervals of the original family. We apply this fact to the above intervals $[s,t)$. Suppose that the total number of indices excluded is at least $2N_0^2$. By the above fact, we may find a disjoint collection of intervals $[s,t)$ as above whose sum of lengths is at least N_0^2. Let us call $\mathcal{P}$ the set of these pairs (s,t). Since $\hat{j}_t - \hat{j}_s \geq 2(t-s)\lceil 2/\tau \rceil, \forall t > s$, the sum of $(\hat{j}_t - \hat{j}_s)$ for $(s,t) \in \mathcal{P}$ is at least $2N_0^2 \lceil 2/\tau \rceil$. Since for each pair $(s,t) \in \mathcal{P}$ we have $|\pi_{s,t}(X)| < N_0^{(1-10\tau)(\hat{j}_t - \hat{j}_s)}$, we get

$$N_0^{(1-2\tau)k} < |X| < N_0^{(1-10\tau)\sum\limits_{(s,t)\in\mathcal{P}}(\hat{j}_t - \hat{j}_s)} \cdot N_0^{\#\{i;i\notin[\hat{j}_s,\hat{j}_t),\forall(s,t)\in\mathcal{P}\}}$$
$$< N_0^{(1-10\tau)\cdot 2N_0^2\lceil 2/\tau \rceil} \cdot N_0^{k-2N_0^2\lceil 2/\tau \rceil},$$

since we have at most N_0 choices for β_i for each index i which does not belong to the union of the intervals $[\hat{j}_s, \hat{j}_t)$ associated to these pairs (s,t). However, this is a contradiction, since this inequality is equivalent to $N_0^{20\tau N_0^2\lceil 2/\tau \rceil} < N_0^{2\tau k}$, which cannot hold, because $2\tau k = 16\tau N_0^2\lceil 2/\tau \rceil < 20\tau N_0^2\lceil 2/\tau \rceil$. So, the total number of excluded indices is smaller than $2N_0^2$.

Now, there are at least $N_0^2 + 1$ indices which are not excluded. We will have two non-excluded indices $s < t$ such that $\hat{\beta}_{\hat{j}_s} = \hat{\beta}_{\hat{j}_t}$ and $\hat{\beta}_{\hat{j}_s+1} = \hat{\beta}_{\hat{j}_t+1}$. We claim that, for $B := \pi_{s,t}(X)$, the shift $\Sigma(B)$ satisfies the conclusions of the statement.

Indeed, since s and t are not excluded, we have $|B| \geq N_0^{(1-10\tau)(\hat{j}_t - \hat{j}_s)}$. Moreover, for every $\alpha \in B$ we have

$$|I(\alpha)| = s(\alpha) > (2(T+1)^2 e^{r_0})^{-(\hat{j}_t - \hat{j}_s)} > e^{-(\hat{j}_t - \hat{j}_s)(r_0 + \lceil \log(2(T+1)^2)\rceil)}.$$

So, the Hausdorff dimension of $K(B)$ is at least

$$\frac{(1-10\tau)\log N_0}{r_0 + \lceil \log(2(T+1)^2) \rceil} > \frac{(1-10\tau)r_0}{r_0 + \lceil \log(2(T+1)^2) \rceil} \cdot \left(1 - \frac{\tau}{2}\right)\Delta^+(t)$$

which is greater than

$$(1-12\tau)\,\Delta^+(t) > (1-\eta)\Delta^+(t).$$

On the other hand, if $\tilde{k} := \hat{j}_t - \hat{j}_s$, $\gamma_1 := \hat{\beta}_{\hat{j}_s+1} = \hat{\beta}_{\hat{j}_t+1}$ and $\gamma_2 := \hat{\beta}_{\hat{j}_t} = \hat{\beta}_{\hat{j}_s}$, all elements of B are of the form $\gamma_1\beta_2\beta_3 \cdots \beta_{\tilde{k}-1}\gamma_2$, where $\gamma_1, \beta_2, \beta_3, \ldots, \beta_{\tilde{k}-1}$, $\gamma_2 \in B_0$ and there are γ_1', γ_1'', $\gamma_2', \gamma_2'' \in B_0$ with $[0; \gamma_2'] < [0; \gamma_2] < [0; \gamma_2'']$ and $[0; (\gamma_1')^t] < [0; \gamma_1^t] < [0; (\gamma_1'')^t]$ such that

$$I(\gamma_1'\beta_2\beta_3 \cdots \beta_{\tilde{k}-1}\gamma_2\gamma_1) \cap K_t \neq \emptyset, \quad I(\gamma_1''\beta_2\beta_3 \cdots \beta_{\tilde{k}-1}\gamma_2\gamma_1) \cap K_t \neq \emptyset,$$
$$I(\gamma_2\gamma_1\beta_2\beta_3 \cdots \beta_{\tilde{k}-1}\gamma_2') \cap K_t \neq \emptyset, \quad I(\gamma_2\gamma_1\beta_2\beta_3 \cdots \beta_{\tilde{k}-1}\gamma_2'') \cap K_t \neq \emptyset.$$

We will show that this implies the existence of $\delta > 0$ such that $\Sigma(B) \subset \Sigma_{t-\delta}$. Let $\gamma_1^t = (c_1, c_2, \ldots, c_{m_1})$, with $c_j \in \mathbb{N}^*$, $\forall j \leq m_1$, and $\gamma_2 = (d_1, d_2, \ldots, d_{m_2})$ with $d_j \in \mathbb{N}^*$, $\forall j \leq m_2$. Let $\gamma_1\beta_2\beta_3 \cdots \beta_{\tilde{k}-1}\gamma_2 \in B$ where $\beta_2\beta_3 \cdots \beta_{\tilde{k}-1} = a_1 a_2 \cdots a_{\tilde{m}}$ with $a_j \in \mathbb{N}^*$, $\forall j \leq \tilde{m}$. We want to estimate three kinds of sums of continued fractions. The first one are sums of continued fractions beginning by $[a_j; a_{j+1}, \ldots, a_{\tilde{m}}, \gamma_2, \gamma_1, \ldots] + [0; a_{j-1}, \ldots, a_1, \gamma_1^t, \gamma_2^t, \ldots]$. Let us assume, without loss of generality, that $q_{m_2+\tilde{m}-j}(a_{j+1}, \ldots, a_{\tilde{m}}, \gamma_2) \leq q_{m_1+j-1}(a_{j-1}, \ldots, a_1, \gamma_1^t)$ (the other case, when the reverse inequality $q_{m_1+j-1}(a_{j-1}, \ldots, a_1, \gamma_1^t) \leq q_{m_2+\tilde{m}-j}(a_{j+1}, \ldots, a_{\tilde{m}}, \gamma_2)$ holds, is symmetric). Assume also that $[a_j; a_{j+1}, \ldots, a_{\tilde{m}}, \gamma_2] < [a_j; a_{j+1}, \ldots, a_{\tilde{m}}, \gamma_2']$ (otherwise we change γ_2' by γ_2''). This allows us to exhibit $\delta > 0$ such that, for any $\underline{\theta}^{(i)} \in \{1, 2, \ldots, T\}^{\mathbb{N}}$, $1 \leq i \leq 4$,

$$[a_j; a_{j+1}, \ldots, a_{\tilde{m}}, \gamma_2, \underline{\theta}^{(1)}] + [0; a_{j-1}, \ldots, a_1, \gamma_1^t, \gamma_2^t, \underline{\theta}^{(2)}]$$
$$< [a_j; a_{j+1}, \ldots, a_{\tilde{m}}, \gamma_2', \underline{\theta}^{(3)}] + [0; a_{j-1}, \ldots, a_1, \gamma_1^t, \gamma_2^t, \underline{\theta}^{(4)}] - \delta.$$

Indeed,

$$[a_j; a_{j+1}, \ldots, a_{\tilde{m}}, \gamma_2', \underline{\theta}^{(3)}] - [a_j; a_{j+1}, \ldots, a_{\tilde{m}}, \gamma_2, \underline{\theta}^{(1)}] > Q(T)$$

where

$$Q(T) = \frac{1}{(T+1)(T+2)q_{m_2+\tilde{m}-j}(a_{j+1}, \ldots, a_{\tilde{m}}, \gamma_2)^2}$$

and $\mid [0; a_{j-1}, \ldots, a_1, \gamma_1^t, \gamma_2^t, \underline{\theta}^{(4)}] - [0; a_{j-1}, \ldots, a_1, \gamma_1^t, \gamma_2^t, \underline{\theta}^{(2)}] \mid <$

$$\frac{1}{q_{m_1+m_2+j-1}(a_{j-1}, \ldots, a_1, \gamma_1^t, \gamma_2^t)^2} < \frac{1}{(F_{m_2+1}q_{m_1+j-1}(a_{j-1}, \ldots, a_1, \gamma_1^t))^2} \leq$$

$$\frac{1}{(F_{m_2+1}q_{m_2+\tilde{m}-j}(a_{j+1}, \ldots, a_{\tilde{m}}, \gamma_2))^2} \leq Q(T)/2$$

(here we use that m_2 is large;

(F_n) denotes Fibonacci's sequence, given by

$$F_0 = 0, F_1 = 1, F_{n+2} = F_{n+1} + F_n, \forall n \geq 0).$$

So, the inequality holds with

$$\delta := \frac{1}{(T+1)^{2(m_1+m_2+\tilde{m})}} < Q(T)/2.$$

On the other hand, $I(\gamma_2\gamma_1\beta_2\beta_3 \cdots \beta_{\tilde{k}-1}\gamma_2') \cap K_t \neq \emptyset$, so there are

$$\underline{\theta}^{(3)} \text{ and } \underline{\theta}^{(4)}$$

such that

$$(\underline{\theta}^{(4)})^t\gamma_2\gamma_1\beta_2\beta_3 \cdots \beta_{\tilde{k}-1}\gamma_2'\underline{\theta}^{(3)} \in \Sigma_t,$$

and thus

$$[a_j; a_{j+1}, \ldots, a_{\tilde{m}}, \gamma_2', \underline{\theta}^{(3)}] + [0; a_{j-1}, \ldots, a_1, \gamma_1^t, \gamma_2^t, \underline{\theta}^{(4)}] \leq t,$$

which implies that, for any $\underline{\theta}^{(i)} \in \{1, 2, \ldots, T\}^{\mathbb{N}}, i = 1, 2$,

$$[a_j; a_{j+1}, \ldots, a_{\tilde{m}}, \gamma_2, \underline{\theta}^{(1)}] + [0; a_{j-1}, \ldots, a_1, \gamma_1^t, \gamma_2^t, \underline{\theta}^{(2)}] < t - \delta.$$

The other two kinds of sums of continued fractions we want to estimate are sums of continued fractions beginning by $[d_j; d_{j+1}, \ldots, d_{m_2}, \gamma_1, \ldots] + [0; d_{j-1}, \ldots, d_1, a_{\tilde{m}}, \ldots, a_1, \gamma_1^t, \ldots]$ and, symmetrically, sums of continued fractions beginning by

$$[0; c_{j+1}, \ldots, c_{m_1}, \gamma_2^t, \ldots] + [c_j; c_{j-1}, \ldots, c_1, a_1, \ldots, a_{\tilde{m}}, \gamma_2, \ldots].$$

We have:

$$q_{m_2-j+m_1}(d_{j+1}, \ldots, d_{m_2}, \gamma_1) \leq q_{j-1+\tilde{m}+m_1}(d_{j-1}, \ldots, d_1, a_{\tilde{m}}, \ldots, a_1, \gamma_1^t)$$

(indeed, $\tilde{m}/(m_1 + m_2)$ is large when η and τ are small, depending on T). Assume that $[d_j; d_{j+1}, \ldots, d_{m_2}, \gamma_1] < [d_j; d_{j+1}, \ldots, d_{m_2}, \gamma_1']$ (otherwise we change γ_1' by γ_1''). Since $I(\gamma_2\gamma_1\beta_2\beta_3 \cdots \beta_{\tilde{k}-1}\gamma_2\gamma_1') \cap K_t \neq \emptyset$, estimates analogous to the previous ones imply that, for any $\underline{\theta}^{(i)} \in \{1, 2, \ldots, T\}^{\mathbb{N}}, i = 1, 2$, we have $[d_j; d_{j+1}, \ldots, d_{m_2}, \gamma_1, \underline{\theta}^{(1)}] + [0; d_{j-1}, \ldots, d_1, a_{\tilde{m}}, \ldots, a_1, \gamma_1^t, \gamma_2^t, \underline{\theta}^{(2)}] < t - \delta$.

This implies that the complete shift $\Sigma(B)$ satisfies the conditions of the statement, which concludes the proof of the desired result. $\square$

As we said before, this proof doesn't give any estimate on the modulus of continuity of $d(t)$. Indeed, in the beginning of the proof of Theorem F.1, we used the fact that $u(r) = \log(T^2 N(t, r))$ is subadditive in order to guarantee the existence of $r_0 \in \mathbb{N}$ large such that, for $r \geq r_0$, $\left| \frac{\log N(t,r)}{r} - \Delta^+(t) \right| < \frac{\tau}{2}\Delta^+(t)$ (recall that $\Delta^+(t) = \lim_{m \to \infty} \frac{1}{m} \log(T^2 N(t, m)) = \lim_{m \to \infty} \frac{1}{m} \log(N(t, m)))$. However, this gives no estimate on r_0. Consider, for instance, the function $v(n)$ given by $v(n) = 2n$ for $n \leq M_0$ and $v(n) = n + M_0$ for $n > M_0$, where M_0 is a large positive integer. It is subadditive, increasing, $\lim_{n \to \infty} v(n)/n = 1$ but $v(M_0)/M_0 = 2$, and M_0 can be taken arbitrarily large. However it is possible to adapt the proof in order to give an estimate on the modulus of continuity of $d(t)$, using an idea of [FMM (2018)].

Given $\varepsilon > 0$ (which we may assume to be smaller than $\frac{1}{7} < \frac{1}{10 \log 2}$), we want to obtain $\delta \in (0, 1)$ as an explicit function of ε such that $\Delta^+(t - \delta) > \Delta^+(t) - \varepsilon$. Of course there is no loss of generality in assuming $\Delta^+(t) \geq \varepsilon$. We may also assume that $T = \lfloor t \rfloor < 4 + \varepsilon^{-1}/\log 2$ (and thus $t < T + 1 < 3\varepsilon^{-1}$) since, by the proof of Theorem 3.3, if $\lfloor t \rfloor \geq 4 + \varepsilon^{-1}/\log 2 \geq 14$, for $m = \lfloor t \rfloor - 4 \geq \max\{9, \varepsilon^{-1}/\log 2\}$, we have $\Delta^+(t) - 1 \geq HD(C(m)) > 1 - \frac{1}{m \log 2} > 1 - \varepsilon$ (and so $\Delta^+(t - 1) > \Delta^+(t) - \varepsilon$).

Under these hypothesis, we will apply the conclusions of Theorem F.1 for $\eta = \varepsilon$. In its proof, in this case, it is enough to assume $r_0 \geq 1/\tau^2$ and that, for $k = 8N(t, r_0)^2 \lceil 2/\tau \rceil$, $\frac{\log N(t, r_0)}{r_0} < (1 + \tau/2) \frac{\log N(t, k(r_0 - 1))}{k(r_0 - 1)}$ (indeed, assuming the above bounds for t, it is not difficult to check that, except for this inequality relating $N(t, r_0)$ and $N(t, k(r_0 - 1))$, the claims in other parts of the proof of the theorem that use the assumptions that r_0 and k are large are satisfied provided $r_0 \geq 1/\tau^2$).

We define a sequence $(c_n)_{n \geq 0}$ recursively by $c_0 = \lceil \frac{1}{\tau^2} \rceil$ and, for every $n \geq 0$, $c_{n+1} = 8N(t, c_n)^2 \lceil \frac{2}{\tau} \rceil (c_n - 1)$. We claim that, for some integer $s_0 < (1 + \frac{2}{\tau}) \log(4/\varepsilon)$, we will have $\frac{\log N(t, c_{s_0})}{c_{s_0}} < (1 + \frac{\tau}{2}) \frac{\log N(t, c_{s_0+1})}{c_{s_0+1}} = (1 + \frac{\tau}{2}) \frac{\log N(t, k(c_{s_0} - 1))}{k(c_{s_0} - 1)}$, with $k = 8N(t, c_{s_0})^2 \lceil \frac{2}{\tau} \rceil$. Indeed, if it is not the case, then $\frac{\log N(t, c_{n+1})}{c_{n+1}} \leq (1 + \frac{\tau}{2})^{-1} \frac{\log N(t, c_n)}{c_n}$ for $0 \leq n < (1 + \frac{2}{\tau}) \log(4/\varepsilon)$, and so, for $M = \lceil (1 + \frac{2}{\tau}) \log(4/\varepsilon) \rceil$, we would have $\frac{\log N(t, c_M)}{c_M} \leq (1 + \frac{\tau}{2})^{-M} \cdot \frac{\log N(t, c_0)}{c_0} < (\varepsilon/4) \cdot \frac{\log N(t, c_0)}{c_0}$, since $(1 + \frac{\tau}{2})^{-(1+\frac{2}{\tau})} < e^{-1}$. On the other hand, it follows from elementary properties of continued fractions (see Lemma A.3 in [Mor1 (2018)]) that, for every $m \geq c_0$, $N(t, m) < (T + 1)^2 e^m < e^{2m}$ (recall that $c_0 = \lceil \frac{1}{\tau^2} \rceil = \lceil \frac{1600}{\varepsilon^2} \rceil$), and so $\frac{\log N(t, c_M)}{c_M} \leq (\varepsilon/4) \cdot \frac{\log N(t, c_0)}{c_0} < \varepsilon/2$. This leads to a contradiction since, for every positive integer m, $\varepsilon \leq$

$\Delta^+(t) \leq \frac{\log(T^2 N(t,m))}{m}$ and, in particular, $\frac{\log N(t,c_M)}{c_M} > \varepsilon - \frac{2\log T}{c_M} \geq \varepsilon/2$, since $c_M \geq c_0 \geq \frac{1600}{\varepsilon^2}$ and $T < 3/\varepsilon$.

Now let $r_0 = c_{s_0}$. By the previous discussion, the proof of Theorem F.1 works for this r_0 (and $k = 8N(t,r_0)^2\lceil 2/\tau\rceil$), so, for

$$\delta = \frac{1}{(T+1)^{2(m_1+m_2+\tilde m)}} \geq \frac{1}{(T+1)^{2k \cdot \max\{|\beta|, \beta \in C(t,r_0)\}}} \geq \frac{1}{(T+1)^{2k \cdot r_0/\log 2}},$$

we have $\Delta^+(t-\delta) > (1-\varepsilon)\Delta^+(t) > \Delta^+(t) - \varepsilon$. We will now give an explicit positive lower bound for δ in terms of ε. In order to do this we define recursively, for each integer $n \geq 0$ and $x \in \mathbb{R}$, the functions $\mathcal{T}(x,n)$ and $\mathcal{T}(n)$ by $\mathcal{T}(x,0) = x$, $\mathcal{T}(x,n+1) = e^{\mathcal{T}(x,n)}$ and $\mathcal{T}(n) = \mathcal{T}(1,n)$. We have, for every $n \geq 0$,

$$c_{n+1} = 8N(t,c_n)^2 \left\lceil \frac{2}{\tau}\right\rceil (c_n - 1) \leq 8e^{4c_n} \cdot \frac{3}{\tau} \cdot c_n < e^{e^{c_n}},$$

since, for every $n \geq 0$, $c_n \geq c_0 \geq \frac{1}{\tau^2} = \frac{1600}{\varepsilon^2}$ and $N(t,c_n) \leq e^{2c_n}$, therefore $r_0 = c_{s_0} < \mathcal{T}(c_0, 2s_0) = \mathcal{T}(\lceil\frac{1}{\tau^2}\rceil, 2s_0)$ and

$$2\log(T+1) \cdot k \cdot \frac{r_0}{\log 2} = 16\log(T+1) \cdot N(t,r_0)^2 \left\lceil\frac{2}{\tau}\right\rceil \cdot \frac{r_0}{\log 2}$$

$$\leq 16\log(3/\varepsilon) \cdot e^{4r_0} \cdot \frac{3}{\tau} \cdot \frac{r_0}{\log 2} < e^{e^{r_0}},$$

so

$$\delta \geq \frac{1}{(T+1)^{2k \cdot r_0/\log 2}} = e^{-2\log(T+1) \cdot k \cdot r_0/\log 2} > e^{-e^{e^{r_0}}} > \frac{1}{\mathcal{T}(c_0, 2s_0+3)}.$$

Finally, since $2^k \geq k^2$ for every $k \geq 4$, it follows by induction that, for every $n \geq 4$, $\mathcal{T}(n) \geq (n+1)^6$ for every $n \geq 0$. Indeed, $\mathcal{T}(4) > 2^{16} > 5^6$ and, for $n \geq 4$, $\mathcal{T}(n+1) > 2^{\mathcal{T}} \geq \mathcal{T}(n)^2 \geq (n+1)^{12} > (n+2)^6$. This implies that $\mathcal{T}(\lfloor 1/\varepsilon\rfloor) \geq (1/\varepsilon)^6 > 1601/\varepsilon^2 > \lceil\frac{1600}{\varepsilon^2}\rceil = \lceil\frac{1}{\tau^2}\rceil = c_0$ (recall that $0 < \varepsilon < 1/7$), so $\mathcal{T}(c_0, 2s_0+3) < \mathcal{T}(\mathcal{T}(\lfloor 1/\varepsilon\rfloor), 2s_0+3) = \mathcal{T}(\lfloor 1/\varepsilon\rfloor+2s_0+3)$, and, since $s_0 < (1+\frac{2}{\tau})\log(4/\varepsilon)$, we have

$$\lfloor 1/\varepsilon\rfloor + 2s_0 + 3 < 3 + \lfloor 1/\varepsilon\rfloor + 2\left(1+\frac{2}{\tau}\right)\log(4/\varepsilon)$$

$$\leq 3 + 1/\varepsilon + 2\left(1+\frac{2}{\tau}\right)\log(4/\varepsilon)$$

$$= 3 + 1/\varepsilon + 2\left(1+\frac{80}{\varepsilon}\right)\log(4/\varepsilon) < \frac{161}{\varepsilon}\log(4/\varepsilon),$$

and therefore

$$\delta > \frac{1}{\mathcal{T}(c_0, 2s_0+3)} > \frac{1}{\mathcal{T}(\lfloor 1/\varepsilon\rfloor + 2s_0 + 3)} \geq \frac{1}{\mathcal{T}(\lfloor\frac{161}{\varepsilon}\log(4/\varepsilon)\rfloor)}.$$

Appendix G

Closedness of the dynamical spectra associated to horseshoes

Let Λ be a horseshoe of a diffeomorphism $\varphi : M \to M$ (see §3.2.2 for the definitions) and consider a continuous height function $f : \Lambda \to \mathbb{R}$. In this appendix, we will extend the arguments in Chapter 3 of Cusick–Flahive book [CF (1989)] to show that the dynamical Lagrange and Markov spectra

$$L(f, \Lambda) = \left\{ \limsup_{n \to \infty} f(\varphi^n(x)) : x \in \Lambda \right\} \text{ and } M(f, \Lambda) = \left\{ \sup_{n \in \mathbb{Z}} f(\varphi^n(x)) : x \in \Lambda \right\}$$

are closed subsets of the real line related to the values of the height function f along periodic and eventually periodic orbits of $\varphi|_\Lambda$.

G.1 Closedness of the dynamical Markov spectra

We begin by showing that $L(f, \Lambda) \subset M(f, \Lambda) \subset f(\Lambda)$ (i.e., by solving Exercise 3.2). Given $a \in L(f, \Lambda)$, there exists $x_0 \in \Lambda$ such that $a = \limsup_{n \to +\infty} f(\varphi^n(x_0))$. Since Λ is compact, there exist a subsequence $(\varphi^{n_k}(x_0))_k$ of $(\varphi^n(x_0))_n$ such that $\lim_{k \to +\infty} \varphi^{n_k}(x_0) = y_0 \in \Lambda$ and $a = \limsup_{n \to +\infty} f(\varphi^n(x_0)) = \lim_{k \to \infty} f(\varphi^{n_k}(x_0)) = f(y_0)$.

We claim that $f(y_0) = \sup_{n \in \mathbb{Z}} f(\varphi^n(y_0))$. Otherwise, there would exist an integer m, such that $f(y_0) < f(\varphi^m(y_0))$; since f is a continuous function, if we pose $\varepsilon = f(\varphi^m(y_0)) - f(y_0) > 0$, then there exists a neighbourhood U of y_0 such that

$$f(y_0) + \frac{\varepsilon}{2} < f(\varphi^m(z)), \text{ for all } z \in U.$$

Thus, since $\varphi^{n_k}(x_0) \to y_0$, there exists $k_0 \in \mathbb{N}$ such that $\varphi^{n_k}(x_0) \in U$, for all $k \geq k_0$. Therefore,

$$f(y_0) + \frac{\varepsilon}{2} < f(\varphi^{m+n_k}(z)), \text{ for all } k \geq k_0,$$

and this contradicts the definition of $a = f(y_0)$. So, $L(f, \Lambda) \subset M(f, \Lambda)$. This completes the argument because the remaining inclusion $M(f, \Lambda) \subset f(\Lambda)$ follows easily from the compactness of Λ and the continuity of f.

The arguments above can be adapted to show that:

Proposition G.1. $M(f, \Lambda)$ *is a closed set.*

Proof. We claim that if $x = m_{f.\Lambda}(y) = \sup_{n \in \mathbb{Z}} f(\varphi^n(y))$, then there exists y_0 such that $x = f(y_0) = m_{f,\Lambda}(y_0)$. In fact, if the supremum above is attained along the orbit of y, then we are done. Otherwise, an argument similar to the proof of the inclusion $L(f, \Lambda) \subset M(f, \Lambda)$ shows the claim.

Let us now take a sequence $x_k \in M(f, \Lambda)$ converging to some x. We can assume that $x_k = f(y_k) = m_{f,\Lambda}(y_k)$, $y_k \in \Lambda$. Since Λ is compact and f is continuous, there exists a subsequence (y_{k_j}) such that $y_{k_j} \to y_0 \in \Lambda$ and $f(y_{k_j}) \to f(y_0) = x$. We affirm that $x = f(y_0) = m_{f,\Lambda}(y_0) \in M(f, \Lambda)$. Indeed, suppose that there exists $N \in \mathbb{Z}$ such that $f(\varphi^N(y_0)) > h > f(y_0)$, for some $h \in \mathbb{R}$. By continuity, we have $f(\varphi^N(y_{k_j})) \to f(\varphi^N(y_0))$. If j is large enough, we get

$$f(\varphi^N(y_{k_j})) > h > f(y_{k_j}),$$

and this contradicts the definition of y_{k_j}. $\qquad\square$

G.2 Closedness of $L(f, \Lambda)$

In the classical case, the proof in Cusick–Flahive book [CF (1989)] that the Lagrange spectrum is a closed subset of $\mathbb{R}$ is not direct: in fact, the argument requires showing first that L is related to Markov values at periodic points, i.e., $L = \overline{P}$, where $P = \{m(\underline{\theta}) : \underline{\theta} \in \Sigma$ is a periodic point$\}$.

Nevertheless, it was shown in [PP0 (2009)] that this approach can be generalized to dynamical Lagrange and Markov spectra associated to geodesic flows on negatively curved manifolds and, in what follows, we will extend these ideas to the context of horseshoes:

Proposition G.2. *Let* $P(f, \Lambda) = \{m_{f,\Lambda}(x) : x \in \Lambda$ *is a periodic point*$\}$. *Then,* $L(f, \Lambda) = \overline{P(f, \Lambda)}$.

This proposition relies on the next lemma (based on standard facts in hyperbolic dynamics [HK (1995)]):

Lemma G.1. *Consider a sequence* $(y_n)_{n \in \mathbb{N}}$ *in* Λ *such that* $\lim\limits_{n \to \infty} d(\varphi(y_n), y_{n+1}) = 0$. *Then, there exists* $z \in \Lambda$, *so that* $\lim\limits_{n \to \infty} d(\varphi^n(z), y_n) = 0$.

Proof. Let $\gamma > 0$ be given by the Stable Manifold Theorem for Λ. By the Shadowing Lemma, there exists a $\beta > 0$ such that every β-pseudo-orbit in Λ is $(\gamma/2)$-shadowed by a point of Λ. Take $k \in \mathbb{N}$ such that $d(\varphi(y_m), y_{m+1}) < \beta$, for all $m \geq k$. Consider the β-pseudo-orbit in Λ, given by $(y_m)_{m \geq k}$. Thus, there exists $z_0 \in \Lambda$ whose orbit $(\gamma/2)$-shadows the previous pseudo-orbit, that is:

$$d(\varphi^j(z_0), y_{k+j}) < \frac{\gamma}{2}, \text{ for all } j \geq 0. \tag{G.1}$$

We claim that $\lim\limits_{j \to +\infty} d(\varphi^j(z_0), y_{k+j})) = 0$. Indeed, let $0 < \theta < \gamma/2$. Then there exists $\overline{\beta} > 0$ such that every $\overline{\beta}$-pseudo-orbit in Λ is θ-shadowed by a point of Λ. Let $l > k$ be a natural number large enough, so that $d(\varphi(y_h), y_{h+1}) < \overline{\beta}$, for all $h \geq l$. Consider the $\overline{\beta}$-pseudo-orbit in Λ, given by $(y_h)_{h \geq l}$. Thus, there exists $w_\theta \in \Lambda$ whose orbit θ-shadows the previous pseudo-orbit, that is:

$$d(\varphi^i(w_\theta), y_{l+i}) < \theta, \text{ for all } i \geq 0. \tag{G.2}$$

By (G.1) and (G.2) for all $i \geq 0$, we have $d(\varphi^i(w_\theta), \varphi^{l-k+i}(z_0)) < \gamma$. Thus, $\varphi^{l-k}(z_0) \in W_\gamma^s(w_\theta)$ and so $\lim\limits_{i \to +\infty} d(\varphi^i(w_\theta), \varphi^i(\varphi^{l-k}(z_0))) = 0$. Take $i_0 \in \mathbb{N}$, such that $d(\varphi^i(w_\theta), \varphi^i(\varphi^{l-k}(z_0))) < \theta$, for all $i \geq i_0$. By (G.2), we have that for $i \geq i_0$:

$$d(\varphi^{i+(l-k)}(z_0), y_{k+[i+(l-k)]}) = d(\varphi^i(\varphi^{l-k}(z_0)), y_{l+i}) < 2\theta.$$

This finishes the proof of the claim. Therefore, $z = \varphi^{-k}(z_0)$ satisfies the desired conclusion. $\square$

Let us now prove the relation between $L(f, \Lambda)$ and $P(f, \Lambda)$.

Proof of Proposition G.2. In order to prove $L(f, \Lambda) \subset \overline{P(f, \Lambda)}$, let $l = l_{f,\Lambda}(x)$, $x \in \Lambda$. For any $\varepsilon > 0$, we shall find a periodic point p in Λ such that $|l - m_{f,\Lambda}(p)| < \varepsilon$.

Since Λ is a horseshoe for φ, let $\varepsilon_0 > 0$ be an expansivity constant of φ on Λ. By uniform continuity we may take $0 < \delta < \varepsilon_0/2$, such that $d(x, y) < \delta$ implies $|f(x) - f(y)| < \varepsilon/2$. According to the Shadowing Lemma, there exists $\alpha > 0$ such that every α-pseudo-orbit in Λ is δ-shadowed by a point of Λ.

By definition of l and compactness there exists a subsequence $(\varphi^{n_k}(x))_k$ such that $f(\varphi^{n_k}(x)) \to l$ and $\varphi^{n_k}(x) \to y$. Take k big enough so that, for all $n \geq n_k$:

$$f(\varphi^n(x)) < l + \frac{\varepsilon}{2}, \quad d(\varphi^{n_k}(x), \varphi^{n_k+1}(x)) < \alpha \quad \text{and} \quad |f(\varphi^{n_k}(x)) - l| < \frac{\varepsilon}{2}.$$

Consider the following α-pseudo-orbit periodic in Λ:

$$\cdots \underbrace{\varphi^{n_k}(x), \varphi^{n_k+1}(x), \cdots, \varphi^{n_{k+1}-1}(x)}_{\text{period}}, \varphi^{n_k}(x), \varphi^{n_k+1}(x), \cdots, \varphi^{n_{k+1}-1}(x) \cdots.$$

There exists $p \in \Lambda$, whose orbit δ-shadows the above pseudo-orbit. This means that, for all $j \geq 0$:

$$d(\varphi^j(p), \varphi^{n_k+\bar{j}}(x)) < \delta, \text{ where } 0 \leq \bar{j} < d := n_{k+1} - n_k \text{ and } \bar{j} \equiv j (\mathrm{mod}\ d).$$

The case $j < 0$ is similar. Thus, by expansivity, $p = \varphi^d(p)$ is a periodic point and by uniform continuity of f we have $|l - m_{f,\Lambda}(p)| < \varepsilon$.

In order to prove $\overline{P(f,\Lambda)} \subset L(f,\Lambda)$, let $(x_n)_{n\in\mathbb{N}}$ be a sequence of periodic points (each x_n has period p_n) in Λ, such that $r_n = f(x_n) = m_{f,\Lambda}(x_n)$ and $r_n \to s$. We shall to show that $s \in L(f,\Lambda)$. By compactness, we may assume that $x_n \to x$ and so $r_n = f(x_n) \to f(x) = s$. Consider the sequence $(y_n)_{n\in\mathbb{N}}$, given by:

$$x_0, \varphi(x_0), \cdots, \varphi^{p_0-1}(x_0), x_1, \varphi(x_1), \cdots, \varphi^{p_1-1}(x_1), x_2, \cdots.$$

Since $\lim_{n\to\infty} d(x_n, x_{n+1}) = 0$, Lemma G.1 implies that there exists $z \in \Lambda$, such that $\lim_{n\to\infty} d(\varphi^n(z), y_n) = 0$, that is:

$$\lim_{n\to\infty} d(\varphi^n(z), \varphi^{\bar{n}}(x_{r_n})) = 0, \text{ where } 0 \leq \bar{n} < p_{r_n} \text{ and } n = p_0 + \cdots + p_{(r_n - 1)} + \bar{n}.$$

In particular, we get $\lim_{n\to+\infty} d(x_n, \varphi^{p_0+\cdots+p_{n-1}}(z)) = 0$. The uniform continuity of f implies that $\lim_{n\to+\infty} f(\varphi^{p_0+\cdots+p_{n-1}}(z)) = \lim_{n\to+\infty} f(x_n) = f(x) = s$. Now, suppose that $l_{f,\Lambda}(z) = m > s$, so there exists a subsequence $(\varphi^{n_k}(z))$ such that $f(\varphi^{n_k}(z)) \to m$. Taking $\varepsilon = m - s > 0$, by above claim and uniform continuity there exists k sufficiently large such that

$$|f(\varphi^{n_k}(z)) - m| < \frac{\varepsilon}{4}, \quad |f(\varphi^{n_k}(z)) - f(\varphi^{\overline{n_k}}(x_{r_k}))| < \frac{\varepsilon}{4} \text{ and } |f(x_{r_k}) - s| < \varepsilon/4,$$

where $0 < \overline{n_k} < p_{r_k}$ and $n_k = p_k + \cdots + p_{(r_{n_k}-1)} + \overline{n_k}$. Thus, $f(\varphi^{\overline{n_k}}(x_{r_k})) > f(x_{r_k})$ and this contradicts the definition of x_{r_k}. Therefore, $m_{f,\Lambda}(z) = s$. $\square$

Two immediate consequences of Proposition G.2 are:

Corollary G.1. *The set $L(f,\Lambda)$ is closed in $\mathbb{R}$, and any isolated point l of $L(f,\Lambda)$ is associated to a periodic point, i.e., there exists a periodic point $p \in \Lambda$ such that $m_{f,\Lambda}(p) = l_{f,\Lambda}(p) = l$.*

Corollary G.2. *We have* $L(f, \varphi|_\Lambda) = L(f, \varphi^{-1}|_\Lambda)$, *that is:*

$$\left\{ \limsup_{n \to +\infty} f(\varphi^n(x)) : x \in \Lambda \right\} = \left\{ \limsup_{n \to -\infty} f(\varphi^n(x)) : x \in \Lambda \right\}$$

G.0 Alternative proof of the closedness of dynamical Markov spectra

The classical Markov spectrum is also closely related to periodic points (cf. Chapter 3 of Cusick–Flahive book [CF (1989)]): if $B := \{m(\underline{\theta}) : \underline{\theta} \in \Sigma$ is eventually periodic on both sides$\}$, then $M = \overline{B}$.

As it turns out, this description of the classical Markov spectrum extends to the dynamical Markov spectra of horseshoes. More concretely, let us say that a point x in Λ is *eventually periodic* when the ω and α limit sets $\omega(x)$ and $\alpha(x)$ of x coincide with the orbits of periodic points p_1 and p_2 of φ. In this context, we have:

Proposition G.3. *Let* $B(f, \Lambda) = \{m_{f,\Lambda}(x) : x \in \Lambda$ *is eventually periodic*$\}$. *Then,* $M(f, \Lambda) = \overline{B(f, \Lambda)}$.

Proof. Since $B(f, \Lambda) \subset M(f, \Lambda)$ and $M(f, \Lambda)$ is closed, we get the inclusion $\overline{B(f, \Lambda)} \subset M(f, \Lambda)$. To prove the inclusion $M(f, \Lambda) \subset \overline{B(f, \Lambda)}$, we consider $x \in \Lambda$ such that $f(x) = m_{f,\Lambda}(x)$. For any $\varepsilon > 0$, we shall construct an asymptotically periodic point $y \in \Lambda$ for which $|m_{f,\Lambda}(x) - m_{f,\Lambda}(y)| < \varepsilon$.

By uniform continuity there exists $0 < \delta < \min\{\varepsilon_0/2, \gamma/2\}$, where ε_0 is an expansivity constant of φ on Λ and γ is given by the Stable Manifold Theorem, such that $d(x, y) < \delta$ implies $|f(x) - f(y)| < \varepsilon$. By the Shadowing Lemma, there exists $\alpha > 0$ for which every α-pseudo-orbit is δ-shadowed by some point of Λ. By compactness, there are convergent subsequences $(\varphi^{n_k}(x))_{k \in \mathbb{N}}$ of $(\varphi^n(x))_{n \geq 0}$ and $(\varphi^{m_k}(x))_{k \in \mathbb{N}}$ of $(\varphi^m(x))_{m < 0}$. Thus, there are n_k and $-m_k$ big enough, such that:

$$d(\varphi^{n_k}(x), \varphi^{n_{k+1}}(x)) < \alpha \quad \text{and} \quad d(\varphi^{m_k}(x), \varphi^{m_{k+1}}(x)) < \alpha.$$

Take the following eventually periodic on both sides α-pseudo-orbit:

$$\cdots \varphi^{m_k-1}(x), \underbrace{\varphi^{m_{k+1}}(x), \cdots, \varphi^{m_k-1}(x)}_{\text{left period}}, \varphi^{m_k}(x), \varphi^{m_k+1}(x), \cdots, \varphi^{-1}(x), x,$$

$$\varphi(x), \cdots, \varphi^{n_k-1}(x), \underbrace{\varphi^{n_k}(x), \cdots, \varphi^{n_{k+1}-1}(x)}_{\text{right period}}, \varphi^{n_k}(x) \cdots$$

Thus, there exists a $y \in \Lambda$ that δ-shadows the above pseudo-orbit, this means:

$$d(\varphi^j(x), \varphi^j(y)) < \delta \text{ for all } m_k \le j \le n_k - 1,$$

$$d(\varphi^l(y), \varphi^{\bar{l}+n_k}(x)) < \delta \text{ for all } l > n_k - 1, \ \bar{l} \in \{0, \cdots, d_1 - 1\},$$

$$d(\varphi^t(y), \varphi^{m_k - \hat{t}}(x)) < \delta \text{ for all } t < m_k, \ \hat{t} \in \{0, \cdots, d_2 - 1\},$$

where $d_1 = n_{k+1} - n_k$, $d_2 = m_k - m_{k+1}$, $l - n_k \equiv \bar{l}(\mathrm{mod} \ d_1)$ and $t - m_k \equiv -\hat{t}(\mathrm{mod} \ d_2)$.

By the Shadowing Lemma, we can find p_1 and p_2 periodic points in Λ, such that $\varphi^{n_k}(y) \in W_\gamma^s(p_1)$ and $\varphi^{m_k-1}(y) \in W_\gamma^u(p_2)$. Then, y is eventually periodic. Moreover, the uniform continuity gives to us that $\sup_{n \in \mathbb{Z}} f(\varphi^n(y)) < m_{f,\Lambda}(x) + \varepsilon$ and $|f(x) - f(y)| < \epsilon$. Hence, $|m_{f,\Lambda}(x) - m_{f,\Lambda}(y)| < \varepsilon$. $\qquad \square$

Bibliography

Aigner, M. (2013). *Markov's theorem and 100 years of the uniqueness conjecture, A mathematical journey from irrational numbers to perfect matchings,* Springer, Cham.

Andersen, N. and Duke, W. (2009). *Markov spectra for modular billiards* Math. Ann. 373, no. 3-4, 1151–1175.

Arnoux, P. (1994). *Le codage du flot géodésique sur la surface modulaire,* Enseign. Math. (2) 40, no. 1-2, 29–48.

Artigiani, M., Marchese, L., and Ulcigrai, C. (2016). *The Lagrange spectrum of a Veech surface has a Hall ray,* Groups Geom. Dyn. 10 (2016), no. 4, 1287–1337.

Baragar, A. (1998). *The exponent for the Markoff-Hurwitz equations,* Pacific J. Math. 182, no. 1, 1–21.

Berstein, A. (1973). *The connections between the Markov and Lagrange spectra,* Number-theoretic studies in the Markov spectrum and in the structural theory of set addition, pp. 16–49. Kalinin. Gos. Univ., Moscow.

Berstein, A. (1973). *The structure of the Markov spectrum,* Number-theoretic studies in the Markov spectrum and in the structural theory of set addition, pp. 50–78. Kalinin. Gos. Univ., Moscow.

Bombieri, E. (2007). *Continued fractions and the Markoff tree,* Expo. Math. 25, no. 3, 187–213.

Bourgain, J., Gamburd, A. and Sarnak, P. (2016). *Markoff triples and strong approximation,* C. R. Math. Acad. Sci. Paris 354, no. 2, 131–135.

Bourgain, J., Gamburd, A. and Sarnak, P. (2016), *Markoff triples and strong approximation: 1,* preprint available at arXiv:1607.01530.

Boshernitzan, M. and Delecroix, V. (2017). *From a packing problem to quantitative recurrence in [0, 1] and the Lagrange spectrum of interval exchanges,* Discrete Anal., Paper No. 10, 25 pp.

de Bruijn, N. G. (1946). *A combinatorial problem,* Nederl. Akad. Wetensch., Proc. 49, 758–764, Indagationes Math. 8, 461–467.

Bumby, R. (1982). *Hausdorff dimensions of Cantor sets,* J. Reine Angew. Math. 331, 192–206.

Cerqueira, A., Matheus, C. and Moreira, C. G. (2018). *Continuity of Hausdorff dimension across generic dynamical Lagrange and Markov spectra*, J. Mod. Dyn. 12, 151–174.

de Courcy-Ireland M. and Lee, S. (2020). *Experiments with the Markoff Surface*, to appear in Experiment. Math.

Cusick, T. (1975). *The connection between the Lagrange and Markoff spectra*, Duke Math. J. 42, 507–517.

Cusick, T. and Flahive, M. (1989). *The Markoff and Lagrange spectra*, Mathematical Surveys and Monographs, 30. American Mathematical Society, Providence, RI, x+97 pp.

Davenport, H. and Schmidt, W. (1970) *Dirichlet's theorem on diophantine approximation*, 1970 Symposia Mathematica, Vol. IV (INDAM, Rome, 1968/69) pp. 113–132 Academic Press, London.

Delecroix, V., Matheus, C. and Moreira, C. G. (2019). *Approximations of the Lagrange and Markov spectra*, Preprint available at arXiv:1908.03773, to appear in Math. Comp.

Dickson, L. E. (1957). *Introduction to the theory of numbers*, (1957 reprint of 1929 first edition), Dover, New York.

Dietz, B. (1985). *On the gaps of the Lagrange spectrum*, Acta Arith. 45, no. 1, 59–64.

Dirichlet, P. G. (1842). *Verallgemeinerung eines Satzes aus der Lehre von den Kettenbrüchen nebst einigen Anwendungen auf die Theorie der Zahlen*, pp. 633–638 Bericht über die Verhandlungen der Königlich Preussischen Akademie der Wissenschaften. Jahrg., S. 93–95.

Diviš B. and Novák, B. (1971). *A remark on the theory of Diophantine approximations*, Comment. Math. Univ. Carolinae 12, 127–141.

Falconer, K. (1986). *The geometry of fractal sets*, Cambridge Tracts in Mathematics, 85. Cambridge University Press, Cambridge, xiv+162 pp.

Ferenczi, S. (2012). *Dynamical generalizations of the Lagrange spectrum*, J. Anal. Math. 118, 19–53.

Ferenczi, S., Mauduit, C. and Moreira, C. G. (2018). *An algorithm for the word entropy*, Theoret. Comput. Sci. 743, 1–11.

Freiman, G. (1968). *Non-coincidence of the spectra of Markov and of Lagrange*, Mat. Zametki 3, 195–200.

Freiman, G. (1973). *Non-coincidence of the spectra of Markov and of Lagrange*, Number-theoretic studies in the Markov spectrum and in the structural theory of set addition (Russian), pp. 10–15, 121–125. Kalinin. Gos. Univ., Moscow.

Freiman, G. (1975). *Diophantine approximations and the geometry of numbers (Markov's problem)*, Kalinin. Gosudarstv. Univ., Kalinin. 144 pp.

Gamburd, A., Magee, M. and Ronan, R. (2019). *An asymptotic formula for integer points on Markoff-Hurwitz varieties*, Ann. of Math. (2) 190, no. 3, 751–809.

Gayfulin, D. (2017). *Attainable numbers and the Lagrange spectrum*, Acta Arith. 179, no. 2, 185–199.

Gayfulin, D. (2019). *Admissible endpoints of gaps in the Lagrange spectrum*, Mosc. J. Comb. Number Theory 8, no. 1, 47–56.

Gbur, M. (1976). *On the lower Markoff spectrum*, Monatsh. Math. 84, 95–107.

Gbur, M. (1977). *Accumulation points of the Lagrange and Markov spectra*, Monatsh. Math 84, 91–108.

Haas, A. and Series, C. (1986). *The Hurwitz constant and Diophantine approximation on Hecke groups*, J. London Math. Soc. (2) 34, no. 2, 219–234.

Hall, M. (1947). *On the sum and product of continued fractions*, Ann. of Math. (2) 48, 966–993.

Hasselblatt, B. and Katok, A. (1995). *Introduction to the modern theory of dynamical systems*, With a supplementary chapter by Katok and Leonardo Mendoza. Encyclopedia of Mathematics and its Applications, 54. Cambridge University Press, Cambridge. xviii+802 pp.

Hensley, D. (1992). *Continued fraction Cantor sets, Hausdorff dimension, and functional analysis*, J. Number Theory 40, no. 3, 336–358.

Hersonsky, S. and Paulin, F. (2002). *Diophantine approximation for negatively curved manifolds*, Math. Z. 241, no. 1, 181–226.

Hochman, M. and Shmerkin, P. (2012). *Local entropy averages and projections of fractal measures*, Ann. of Math. (2) 175, no. 3, 1001–1059.

Hubert, P., Lelièvre, S., Marchese, L. and Ulcigrai, C. (2015). *The Lagrange spectrum of some square-tiled surfaces*, Israel J. Math. 225, no. 2, 553–607.

Hubert, P., Marchese, L. and Ulcigrai, C. (2015). *Lagrange spectra in Teichmüller dynamics via renormalization*, Geom. Funct. Anal. 25, no. 1, 180–255.

Hurwitz, A. (1891). *Ueber die angenäherte Darstellung der Irrationalzahlen durch rationale Brüche*, Math. Ann. 39, no. 2, 279–284.

Ivanov, V. A. (1978). *Dirichlet's theorem in the theory of diophantine approximations*, MatŻametki, 24:4, 459–474; Math. Notes, 24:4, 747–755.

Ivanov, V. A. (1980). *On the ray origin in the Dirichlet spectrum of a problem of the theory of Diophantine approximations*, Studies in number theory, 6. Zap. Nauchn. Sem. Leningrad. Otdel. Mat. Inst. Steklov. (LOMI) 93, 164–185, 227.

Jarník, V. (1929). *Zur metrischen Theorie der diophantischen Approximationen*, Prace Mat.-Fiz. 36, 91–106.

Jenkinson, O. and Pollicott, M. (2001). *Computing the dimension of dynamically defined sets: E2 and bounded continued fractions*, Ergodic Theory Dynam. Systems 21, no. 5, 1429–1445.

Jenkinson, O. and Pollicott, M. (2018). *Rigorous effective bounds on the Hausdorff dimension of continued fraction Cantor sets: a hundred decimal digits for the dimension of E_2*, Adv. Math. 325, 87–115.

Khinchin, A. (1964). *Continued fractions*, The University of Chicago Press, Chicago, Ill.-London xi+95 pp.

Kopetzky, H. G. (1989). *Some results on Diophantine approximation related to Dirichlet's theorem*, Number theory (Ulm, 1987), 137–149, Lecture Notes in Math., 1380, Springer, New York.

Korkine, A. and Zolotareff, G. (1873). *Sur les formes quadratiques*, Math. Ann. 6, no. 3, 366–389.

Lehner, J. (1985). *Diophantine approximation on Hecke groups*, Glasgow Math. J. 27, 117–127.

Lévy, P. (1936). *Sur le développement en fraction continue d'un nombre choisi au hasard*, Compositio Math. 3, 286–303.

Lima, D., Matheus, C., Moreira, C. G., and Vieira, S. (2019). $M \setminus L$ *near 3*, Preprint available at arXiv:1904.00269.

Lima, D., Matheus, C., Moreira, C. G., and Vieira, S. (2019). $M \setminus L$ *is not closed*, Preprint available at arXiv:1912.02698, to appear in Int. Math. Res. Not.

Mahler, K. (1946). *On lattice points in n-dimensional star bodies. I. Existence theorems*, Proc. Roy. Soc. London Ser. A. 187, 151–187.

Malyshev, A. V. (1977). *Markov and Lagrange spectra (a survey of the literature)*, Studies in number theory (LOMI), 4. Zap. Nauchn. Sem. Leningrad. Otdel. Mat. Inst. Steklov. (LOMI) 67, 5–38, 225.

Markoff, A. (1880). *Sur les formes quadratiques binaires indéfinies*, Math. Ann. 17, no. 3, 379–399.

Marstrand, J. (1954). *Some fundamental geometrical properties of plane sets of fractional dimensions*, Proc. London Math. Soc. (3) 4, 257–302.

Masur, H. and Tabachnikov, S. (2002). *Rational billiards and flat structures*, Handbook of dynamical systems, Vol. 1A, 1015–1089, North-Holland, Amsterdam.

Matheus, C. and Moreira, C. G. (2019). *Markov spectrum near Freiman's isolated points in $M \setminus L$*, J. Number Theory 194, 390–408.

Matheus, C. and Moreira, C. G. (2019). $HD(M \setminus L) > 0.353$, Acta Arith. 188, no. 2, 183–208.

Matheus, C. and Moreira, C. G. (2020). *Fractal geometry of the complement of Lagrange spectrum in Markov spectrum*, preprint available at arXiv:1803.01230, to appear in Comment. Math. Helv.

Maucourant, F. (2003). *Sur les spectres de Lagrange et de Markoff des corps imaginaires quadratiques*, Ergodic Theory Dynam. Systems 23, no. 1, 193–205.

McCluskey, H. and Manning, A. (1983). *Hausdorff dimension for horseshoes*, Ergodic Theory Dynam. Systems 3, no. 2, 251–260.

Moreira, C. G. (2018). *Geometric properties of the Markov and Lagrange spectra*, Ann. of Math. (2) 188, no. 1, 145–170.

Moreira, C. G. (2016). *Geometric properties of images of cartesian products of regular Cantor sets by differentiable real maps*, Preprint available at arXiv:1611.00933.

Moreira, C. G. (2020). *On the minima of Markov and Lagrange dynamical spectra*, Astérisque No. 415, 45–57.

Moreira, C. G. and Romaña, S. (2017). *On the Lagrange and Markov dynamical spectra*, Ergodic Theory and Dynam. Systems 37, 1570–1591.

Moreira, C. G. and Yoccoz, J.-C. *Stable intersections of regular Cantor sets with large Hausdorff dimensions*, Ann. of Math. (2) 154 (2001), no. 1, 45–96.

Moreira, C. G. and Yoccoz, J.-C. (2010). *Tangences homoclines stables pour des ensembles hyperboliques de grande dimension fractale*, Ann. Sci. Éc. Norm. Supér. (4) 43, no. 1, 1–68.

Morrison, T. (2012). *A glimpse of the Markoff spectrum*, Preprint available at https://dca.ue.ucsc.edu/system/files/dca/2012/193/193.pdf.

Newhouse, S. (1979). *The abundance of wild hyperbolic sets and nonsmooth stable sets for diffeomorphisms*, Inst. Hautes Études Sci. Publ. Math. No. 50, 101–151.

Palis, J. and Takens, F. (1993). *Hyperbolicity and sensitive chaotic dynamics at homoclinic bifurcations*, Fractal dimensions and infinitely many attractors. Cambridge Studies in Advanced Mathematics, 35. Cambridge University Press, Cambridge, x+234 pp.

Parkkonen, J. and Paulin, F. (2009). *On the closedness of approximation spectra*, J. Théor. Nombres Bordeaux 21, no. 3, 701–710.

Parkkonen, J. and Paulin, F. (2010). *Prescribing the behaviour of geodesics in negative curvature*, Geom. Topol. 14, no. 1, 277–392.

Rosen, D. (1954). *A class of continued fractions associated with certain properly discontinuous groups*, Duke Math. J. 21, 549–563.

Schecker, H. (1977). *Über die Menge der Zahlen, die als Minima quadratischer Formen auftreten*, J. Number Theory 9, no. 1, 121–141.

Series, C. (1988). *The Markoff spectrum in the Hecke group G_5*, Proc. London Math. Soc. (3) 57, no. 1, 151–181.

Shub, M. (1987). *Global stability of dynamical systems*, With the collaboration of Albert Fathi and Rémi Langevin. Translated from the French by Joseph Christy. Springer-Verlag, New York, xii+150 pp.

Vieira, S. (2020). *Markov and Lagrange spectra*, PhD thesis, IMPA, Brazil.

Witte Morris, D. (2005). *Ratner's theorems on unipotent flows*, Chicago Lectures in Mathematics. University of Chicago Press, Chicago, IL, 2005. xii+203 pp.

You, Y.-C. (1976). *Role of the rationals in the Markov and Lagrange spectra*, Thesis (Ph.D.)–Michigan State University 44 pp.

Zagier, D. (1982). *On the number of Markoff numbers below a given bound*, Math. Comp. 39, no. 160, 709–723.

Zagier, D. (1981). *Eisenstein series and the Riemann zeta function*, Automorphic forms, representation theory and arithmetic (Bombay, 1979), pp. 275–301, Tata Inst. Fund. Res. Studies in Math., 10, Tata Inst. Fundamental Res., Bombay.